Lattice Gauge Theory
A Challenge in
Large-Scale Computing

NATO ASI Series

Advanced Science Institutes Series

A series presenting the results of activities sponsored by the NATO Science Committee, which aims at the dissemination of advanced scientific and technological knowledge, with a view to strengthening links between scientific communities.

The series is published by an international board of publishers in conjunction with the NATO Scientific Affairs Division

A	**Life Sciences**	Plenum Publishing Corporation
B	**Physics**	New York and London
C	**Mathematical and Physical Sciences**	D. Reidel Publishing Company Dordrecht, Boston, and Lancaster
D	**Behavioral and Social Sciences**	Martinus Nijhoff Publishers
E	**Engineering and Materials Sciences**	The Hague, Boston, and Lancaster
F	**Computer and Systems Sciences**	Springer-Verlag
G	**Ecological Sciences**	Berlin, Heidelberg, New York, and Tokyo

Recent Volumes in this Series

Lattice Gauge Theory

A Challenge in Large-Scale Computing

Edited by
B. Bunk
K. H. Mütter
and
K. Schilling
Gesamthochschule
Wuppertal, Federal Republic of Germany

Plenum Press
New York and London
Published in cooperation with NATO Scientific Affairs Division

Proceedings of a NATO Workshop on
Lattice Gauge Theories: A Challenge in Large-Scale Computing,
held November 5–7, 1985,
in Wuppertal, Federal Republic of Germany

———

Library of Congress Cataloging in Publication Data

NATO Workshop on Lattice Gauge Theories—A Challenge in Large-Scale
    Computing (1985: Wuppertal, Germany)
    Lattice gauge theory.

    (NATO ASI series. Series B, Physics; v. 140)
    "Proceedings of a NATO Workshop on Lattice Gauge Theories: A Challenge
in Large-Scale Computing, held November 5–7, 1985, in Wuppertal, Federal
Republic of Germany."
    "Published in cooperation with NATO Scientific Affairs Division."
    Includes bibliographies and index.
    1. Gauge fields (Physics)—Data processing—Congresses. 2. Lattice theory
—Data processing—Congresses. 3. Quantum chromodynamics—Data pro-
cessing—Congresses. 4. Paticles (Nuclear physics)—Data processing—Con-
gresses. I. Bunk, B. II. Mütter, K. H. III. Schilling, K. IV. North Atlantic Treaty
Organization. Scientific Affairs Division. V. Title. VI. Series.
QC793.3.F5N383   1985   530.1′43   86-15155

    ISBN-13: 978-1-4612-9308-8   e-ISBN-13: 978-1-4613-2231-3
    DOI: 10.1007/978-1-4613-2231-3

———

© 1986 Plenum Press, New York
Softcover reprint of the hardcover 1st edition 1986

A Division of Plenum Publishing Corporation
233 Spring Street, New York, N.Y. 10013

PREFACE

This volume presents the contributions to the international workshop entitled "Lattice Gauge Theory - a Challenge in Large Scale Computing" that was held in Wuppertal from November 4 to 7, 1985. This meeting was the third in a series of European workshops in this rapidly developing field.

The meeting intended to bring together both active university researchers in this field and scientists from industry and research centers who pursue large scale computing projects on problems within lattice gauge theory. These problems are extremely demanding from the point of view of both machine hardware and algorithms, for the verification of the continuum fields theories like Quantum Chromodynamics in four-dimensional Euclidean space-time is quite cumbersome due to the tremendously large number of degrees of freedom. Yet the motivation of theoretical physicists to exploit computers as tools for the simulation of complex systems such as gauge field theories has grown considerably during the past years. In fact, quite a few prominent colleagues of ours have even gone into machine building, both in industry and research institutions: more parallelism, and more dedicated computer architecture are their design goals to help them boost the Megaflop rate in their simulation processes. The workshop contained several interesting seminars with status reports on such supercomputer projects like the Italian APE (by E. Marinari), the IBM project GF-11 (by D. Weingarten), and the Danish projects MOSES and PALLAS (by H. Bohr).

The main physics questions of the workshop were centered on dynamical fermions in lattice QCD, the nature of the phase transitions in lattice thermodynamics, the different Higgs phases, and renormalisation group techniques on the lattice. Some of the speakers agreed to review the state of the art in a variety of topics, as the reader will recognize from the table of contents.

Much of the meeting was informal in character, and therefore is missing in this volume, like the many lively discussions both in the plenum and on the corridors. In this context we would like to mention in particular the vivid response to the talk presented by Prof. G. Seegmüller, chairman of the Computer Committee of the Deutsche Forschungsgemeinschaft, on the supercomputer installation politics in the German Federal Republic.

Prof. R. Dalitz, senior participant and chairman of the last session, praised the youth of the audience in his concluding remarks and emphasized the vigour with which the application of large scale computing methods is pursued by a new generation of theoretical physicists. In view of the material support necessary for this novel approach to fundamental problems in theoretical physics, we think it was quite gratifying to have so many representatives of the German science administration in the audience to assess the new developments in this area of research!

We appreciate the sponsorship of NATO within their Advanced Study Institute Programme. Moreover we thank Dr. K. Peters, Chancellor of the University of Wuppertal, for his support.

We finally thank Dr. I. Schmitt (Scientific Secretary) and our Collaborators Dipl.-Phys. K. Hornfeck and K. Meier (Technical Assistance) and Mrs. V. Drees (Conference Secretariat) and last but not least Mrs. Tammen and her staff from the CVJM Bildungsstätte which helped very much to organize a successful meeting.

Wuppertal,   B. Bunk
February 1986   K.H. Mütter
                                                    K. Schilling

CONTENTS

QCD AT FINITE TEMPERATURE AND BARYON NUMBER DENSITY

Frithjof Karsch

University of Illinois at Urbana-Champaign
1110 West Green Street
Urbana, IL 61801

ABSTRACT

We discuss the status of Monte Carlo simulations for the thermodynamics of QCD. Recent results for pure SU(N) gauge theories at finite
temperature and the influence of dynamical fermions on the phase structure
at finite temperature as well as problems related to the introduction of
non-zero chemical potentials are analyzed.

I. INTRODUCTION

QCD is expected to describe the behavior of strongly interacting
matter at arbitrary temperatures and densities. As the theory is
asymptotically free at high temperature and/or density [1] it has been
speculated that hadronic matter undergoes a fundamental change in behavior
once the density is increased far above typical hadronic scales, i.e. for
densities much larger than nuclear matter density. Quarks and gluons,
which are confined at low temperatures and/or densities are then expected
to approach asymptotically an ideal gas of free fermions and bosons.

During recent years Monte Carlo (MC) simulation techniques [2] for
lattice regularized quantum field theories [3] have been developed, which
opened the possibility to study non-perturbative aspects of these theories.
The development of efficient simulation techniques for theories with
dynamical fermions is still in progress. However, existing algorithms
like the pseudofermion algorithm [4] or microcannonical scheme [5], seem
to be promising and lead to consistent results. They allow a thorough non
perturbative analysis of the behavior of QCD at finite temperature. The
simulation of QCD at finite baryon number density, however, still suffers
from the problem that suitable simulation techniques for complex actions
are not well developed. At finite chemical potential the SU(3) fermion
determinant is complex. This leads to a complex action and thus excludes
standard simulation techniques, which rely on a probability interpretation
of the path integral that represents the partition function. First
attempts do deal with this problem have been undertaken [6,7] and may
hopefully lead to results in the near future.

During the last few years MC simulations and analytic approaches led
to a rather detailed picture of the phase structure of the pure gauge

sector of SU(N) lattice gauge theories, i.e. QCD without dynamical fermions. The existence of a deconfinement phase transition has been proven rigorously [8] and the nature of the transition is well understood in terms of the breaking of a global Z(N) symmetry of the pure gauge action. In fact the order of the transition seen in MC experiments for different color groups [9-12] as well as reduced large N models [13] agrees well with expectations based on universality arguments [14,15] and also the limiting behavior in both phases, ideal gase at high temperatures and a glueball gas at low temperatures agrees with our expectations. However, many of the results in the pure gauge sector have been obtained on small lattices at intermediate couplings. Recent MCRG studies [16] for SU(3) and analysis of observables like the string tension on large lattices [17] have shown that couplings larger than $\beta \equiv 6/g^2 = 6.0$ are necessary to make contact with continuum physics. At present detailed studies of the deconfinement transition on large lattices are performed to get a quantitative measure of the critical temperature in the continuum regime [18,19].

In the presence of dynamical fermions the situation becomes more complicated. The global Z(N) symmetry of the pure gauge theory is explicitly broken by the fermions and can no longer be associated with a deconfinement transition. Analysis of effective spin models [20] and MC simulations with heavy fermions [21] suggested that the deconfinement transition disapears for any finite quark mass in the case of SU(2) and below a critical mass for SU(N), N ≥ 3. MC simulations with light fermions [22-24], however, have shown that there is still a very rapid change in the behavior of thermodynamic observables. This may be induced by a chiral symmetry restoring phase transition at zero quark mass. However, there have been also some indications for a non-trivial phase structure at large number of flavors at zero temperature [26,27].

The analysis of the influence of a finite chemical potential on the phase structure of lattice QCD is just coming on the way. Exploratory simulations have been performed in the quenched sector [28] and for the SU(2) color group [29,30]. Simulations in the presence of dynamical fermions are difficult for SU(3) due to the appearance of a complex action. As mentioned above suitable methods to deal with this problem are currently tested [6,7].

This paper is organized as follows. In chapter 2 we discuss the present status of the analysis of the deconfinement transition in the pure gauge sector. The influence of dynamical fermions on this transition and the chiral transition are discussed in chapter 3. Chapter 4 then presents some recent results on the thermodynamics in the presence of a finite chemical potential and chapter 5 contains our conclusions.

II. THERMODYNAMICS OF PURE SU(N) GAUGE THEORIES

The thermodynamics of pure SU(N) gauge theories has been analyzed for different color groups. In the case of SU(2) results indicate the existence of a second order deconfinement transition [9,10]. Moreover, critical exponents of the transition have been determined [31], which are in agreement with universality [14]. Simulations for SU(3) [11,32-35], SU(4) [12] and reduced large N models [15] indicate, that for all N ≥ 3 the deconfinement transition is first order. In the following we will restrict our discussion on the thermodynamics of SU(3), which certainly is most relevant for QCD.

Introducing a lattice of size $N_\sigma^3 \times N_\tau$ and lattice cut-offs $a(a_\tau)$ in space (time) directions such that temperature and volume are given by $1/T = N_\tau a_\tau$ and $V = (N_\sigma a)^3$ the lattice regularized partition function can be written as

$$Z(T,V) = \int \prod_{x,\mu} dU_{x,\mu} \, e^{-S_G(U)} \tag{2.1}$$

with $U_{x,\mu} \in SU(N)$, $dU_{x,\mu}$ the Haar measure of the group and the Euclidean action, S_G, given by

$$S_G(U) = 2N\Big[\xi \, g_\sigma^{-2} \sum_{\{P_\sigma\}} \Big(1 - \frac{1}{N} \, \mathrm{Re \, Tr} \, U_{x,\mu} U_{x+\mu,\nu} U_{x+\nu,\mu}^\dagger U_{x,\nu}^\dagger\Big)$$

$$+ \xi^{-1} g_\tau^{-2} \sum_{\{P_t\}} \Big(1 - \frac{1}{N} \, \mathrm{Re \, Tr} \, U_{x,o} U_{x+o,\mu} U_{x+\mu,o}^\dagger U_{x,\mu}^\dagger\Big)\Big] \tag{2.2}$$

In eq.(2.2) the summation runs over all space-like ($\{P_\sigma\}$) and time-like ($\{P_\tau\}$) plaquettes of a hypercubic lattice and $\xi = a/a_\tau$ is the ratio of lattice spacings in space and time directions. Notice that in this formulation the temperature dependence is hidden in the asymmetry ξ and the thermal extend of the lattice, N_τ [10]. For a lattice with equal lattice spacings in time and space directions the couplings $g_{\sigma(\tau)}$ are identical, $g_\sigma(a,1) = g_\tau(a,1) \equiv g(a)$.

The phase structure of the theory can be analyzed by studying the behavior of suitable order parameters for a phase transition or directly from the behavior of thermodynamic observables. The latter can be obtained from the partition function by taking derivatives with respect to temperature or volume. For instance the energy density is given by

$$\varepsilon = \frac{\xi}{N_\sigma^3 N_\tau a^4} \frac{\partial}{\partial \xi} \ln Z \Big|_{\xi=1}$$

$$= 6 N \Big\{g^{-2} \big(\langle P_\sigma\rangle - \langle P_\tau\rangle\big) - c_\sigma'\langle P_\sigma\rangle - c_\tau'\langle P_\tau\rangle\Big\} \tag{2.3}$$

where $\langle P_{\sigma(\tau)}\rangle$ denotes the average values for space (time) like plaquettes and the coefficients $c_{\sigma(\tau)}'$ are the derivatives of the space (time) like couplings with respect to ξ. They can be evaluated in weak coupling perturbation theory and for $\xi = 1$ they are given by [36]

$$c_\sigma' = 4N\Big\{\frac{N^2-1}{32N^2} 0.586844 + 0.000499\Big\}$$

$$c_\tau' = 4N\Big\{-\frac{N^2-1}{32N^2} 0.586844 + 0.005306\Big\} \tag{2.4}$$

The energy density given by eq. (2.3) should be normalized by subtracting the zero temperature contribution [10], which in practice is taken to be the result of eq. (2.3) evaluated for $N_\sigma = N_\tau$. In fig. 1 we show the energy density for the pure SU(3) theory calculated on a $8^3 \times 3$ lattice [32]. It clearly shows the existence of a first order phase transition from a low temperature phase with small energy density to a

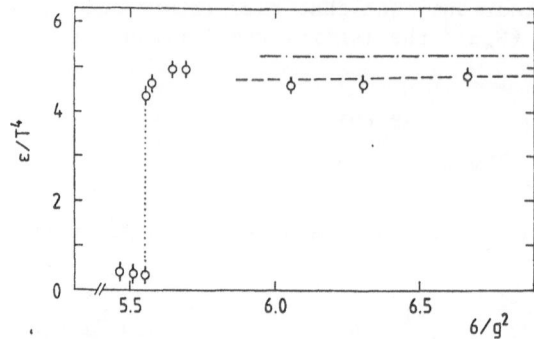

Figure 1: Energy density of the SU(3) gauge theory on a $8^3 \times 3$ lattice versus coupling $\beta = 6/g^2$. Also shown is the lowest order (-·-) and $O(g^2)$ (- -) weak coupling perturbative result for the high temperature phase on a finite lattice [37].

high temperature phase. Soon after the transition the energy density agrees well with the asymptotic ideal gas behavior

$$\varepsilon_{SB} = \frac{N^2-1}{15} \pi^2 T^4 \qquad (2.5)$$

An order parameter for the deconfinement transition is the Polyakov line operator

$$L(\vec{x}) = \prod_{x_o=0}^{N_\tau} U_{(x_o,\vec{x}),o} \qquad (2.6)$$

The pure gauge action, eq. (2.2) is invariant under global Z(N) transformations of all temporal links in a fixed time hyperplane

$$U_{x,o} \to z \, U_{x,o}, \quad z \varepsilon \, Z(N), \quad x \equiv (x_o,\vec{x}) \text{ with } x_o \text{ fixed} \qquad (2.7)$$

However, the Polyakov line, eq. (2.6), transforms non-trivially under this transformation and thus has vanishing expectation value as long as this symmetry is not spontaneously broken. In addition it measures the excess free energy of a static color source in a gluonic heat bath, $\langle TrL \rangle \sim \exp\{-F/T\}$. Thus there is a close relation between the deconfinement transition and the breaking of the global Z(N) centre symmetry:

$$\langle L \rangle \begin{cases} = 0, \text{ confined phase} \leftrightarrow \text{unbroken } Z(N) \text{ symmetry} \\ \neq 0, \text{ deconfined phase} \leftrightarrow \text{broken } Z(N) \text{ symmetry} \end{cases} \qquad (2.8)$$

In fig. 2 we show $\langle L \rangle = \langle N^{-1} \, TrL(\vec{x}) \rangle / N_\sigma^3$, for the SU(3) gauge theory on a $8^3 \times 2$ and $8^3 \times 4$ lattice [34]. Again the existence of a first order transition is clearly visible. Notice, however that the gap in $\langle L \rangle$ at β_c decrease drastically by going from $N_\tau = 2$ to $N_\tau = 4$. The same is true for the jump in the energy density. In units of the lattice spacing a the latent heat of the transition [33,35,38]

$$\Delta \varepsilon a^4 = 6N \frac{dg^{-2}}{d \ell n \, a} \Delta \langle P_\sigma + P_\tau \rangle \qquad (2.9)$$

should decrease as N_τ^{-4} in order that the ratio of latent heat $\Delta \varepsilon$ and critical temperature T_c remains constant in the continuum limit. Thus it becomes increasingly difficult to detect the deconfinement transition on lattices with large extend in time direction. This, however, is necessary in order to push the critical coupling β_c to larger values and to allow an

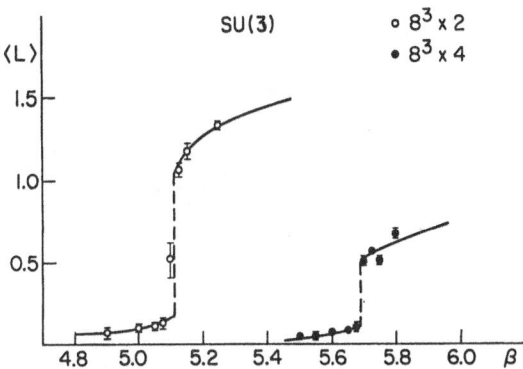

Figure 2: Expectation value of the Polyakov line versus coupling β on lattices of size $8^3 \times 2$ and $8^3 \times 4$. Lines are drawn to guide the eye.

analysis of the thermodynamics deeper in the continuum regime. Recently large scale simulations have been performed to determine the deconfinement temperature on large lattices [18,19]. The results of these calculations are collected in fig. 3 and table I.

Table I: Critical couplings for the SU(3) deconfinement transition on large lattices. Data for N_σ = 16 are from ref. 19, the other data are taken from ref. 18.

N_τ	N_σ	β_c
8	19	6.02 ± .02
10	16	6.065 ± .027
	17	6.15 ± .03
12	16	6.261 ± .020
	19	6.32 ± .03
14	16	6.355 ± .026
	19	6.47 ± .03

One sees that for $N_\tau \gtrsim 10$, i.e. $\beta \gtrsim 6.1$, the critical temperature in units of Λ_L stays constant, which implies that it scales according to the renormalization group equation

$$a\Lambda_L = \exp\left\{ -\frac{24\pi^2}{11Ng^2} - \frac{51}{121} \ln\left(\frac{11Ng^2}{48\pi^2}\right) \right\} \qquad (2.10)$$

There are slight discrepancies between the results obtained in refs. [18] ($T_c/\Lambda_L = 51.1 \pm 1.0$) and [19] ($T_c/\Lambda_L = 46.6 \pm 0.7$) which may partly be due to the different criteria used to define the transition point. This is somewhat ambigious as on large lattices a direct observation of a jump in physical observables is difficult.

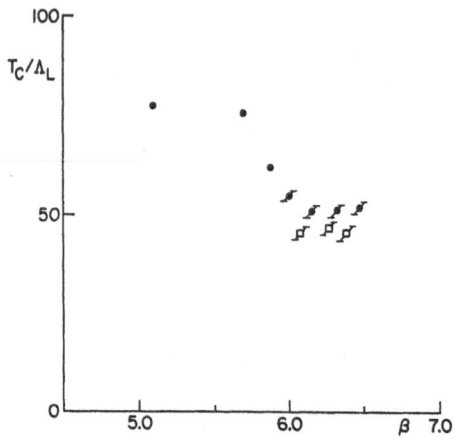

Figure 3: The deconfinement temperatue T_c/Λ_L versus the critical coupling β_c. The data points represent results for the critical couplings for lattices with temporal extend N_τ = 2,4,6,8,10,12 and 14. They are taken from ref. [18] (●) and ref. [19] (□).

These calculations nicely indicate that the deconfinement temperature indeed stays finite in the continuum limit. Using recent data for the string tension [17] , $\sigma^{1/2}/\Lambda_L$, one can remove Λ_L. Using $\sigma^{1/2}$ = 400 Mev one then obtains

$$T_c = (200-230) \text{ MeV} \tag{2.11}$$

for the deconfinement transition temperature.

In the future attempts should be undertaken to analyze other observables on such large lattices. This would be especially interesting for observables like the energy density or latent heat as these quantities had to be corrected for finite size effects on small lattices, which will be unnecessary on large lattices. In addition the energy density involved terms which are known only perturbatively and thus will become more accurate in the continuum regime.

III. THE INFLUENCE OF DYNAMICAL FERMIONS

In the presence of dynamical fermions the partition function becomes

$$Z(T,V) = \int \prod_{x,\mu} dU_{x,\mu} \, d\chi_x \, d\bar{\chi}_x \, e^{-S_G - S_F} \tag{3.1}$$

with the gluonic action, S_G, given by eq. (2.2) and the fermionic action

$$S_F = ma \sum_x \bar{\chi}_x \chi_x + \frac{1}{2} \sum_{x,\mu} \bar{\chi}_x \eta_\mu(x) \left[U_{x,\mu} \chi_{x+\mu} - U_{x-\mu,\mu}^\dagger \chi_{x-\mu} \right] \tag{3.2}$$

Eq. (3.2) is the action for staggered fermions of mass m, the phase factors η_μ are given by $\eta_\mu(x) = (-1)^{x_o + \ldots + x_{\mu-1}}$, $\eta_o(x) = 1$. This action describes four flavors of mass m. However, after integrating out the Grassmann fields χ, $\bar{\chi}$ one obtains a fermion determinant and the number of flavors n_f, can be varied formally by taking an appropriate power of this determinant

$$Z = \int \prod_{x,\mu} dU_{x,\mu} \left[\det(m^2 + D^2) \right]^{n_f/8} e^{-S_G} \qquad (3.3)$$

where $D \equiv \sum_\mu D_\mu$ and

$$D^\mu_{xy} = \frac{1}{2} \eta_\mu(x) \left[U_{x,\mu} \delta_{y,x+\mu} - U^\dagger_{y,\mu} \delta_{y,x-\mu} \right] \qquad (3.4)$$

The action $S = S_G + S_F$ is now no longer Z(N) invariant and thus the Polyakov line is no longer an order parameter for the deconfinement transition. In the strong coupling, large fermion mass limit it can be shown that fermions in lattice gauge models play a similar role like external fields in spin models. It thus has been speculated that the deconfinement transition disappears for light enough quarks in the SU(3) theory [20,21]. However, in the light mass regime the thermodynamics is influenced by a chiral phase transition, which is expected to occur for massless quarks. In the zero mass limit the action has a global U(f) × U(f), $f = n_f/4$, chiral symmetry which is spontaneously broken at low temperatures and is expected to be restored at high temperatures. For the SU(2) lattice theory the existence of a finite temperature chiral transition has been proven rigorously [39]. An order parameter for the restoration of chiral symmetry is the mesonic condensate

$$\langle \bar{\chi}\chi \rangle = \frac{n_f}{4} \langle \mathrm{Tr}(m+D)^{-1} \rangle \qquad (3.5)$$

In the presence of dynamical fermions the energy density ecieves additional contributions from the fermion action

$$\varepsilon = \varepsilon_G + \varepsilon_F \qquad (3.6)$$

where ε_G is given by eq.(2.3) and ε_F is

$$\varepsilon_F a^4 = \frac{n_f}{4} \langle \mathrm{tr}\, D^4 (m+D)^{-1} \rangle - \left\{ \frac{N n_f}{16} - \frac{m}{4} \langle \bar{\chi}\chi \rangle_{T=0} \right\} \qquad (3.7)$$

Notice that the expectation values now have to be computed in the presence of the fermion determinant and thus also the gluonic part, ε_G, feels the presence of dynamical fermions.

The main problem in the simulation of dynamical fermions is the evaluation of the determinant appearing in eq. (3.3), or more precisely the change of the determinant under a change of the gauge configuration. There have been several suggestions how to deal with this problem and in fact the analysis of thermodynamic quantities has been used to test some of these approaches. In the staggered fermion formulation pseudofermion [22-24], microcanonical [25] and a recently developed hybrid scheme [40] have been used to analyze the influence of dynamical fermions on the thermodynamics. It is reasuring that these different approaches lead to qualitatively consistent results. A recent quantitative comparison of the pseudofermion algorithm and the microcanonical scheme indeed shows also good quantitative agreement of the results [41].

In fig. 4 we show the energy density for SU(3) with three dynamical quark flavors of mass ma = 0.1 obtained from a simulation with pseudofermions on a $8^3 \times 4$ lattice [24]. As can be seen the transition seems to be smoother than in the pure gauge case which is shown in fig. 1. The transition appears to be continous. At the same place where a rapid change in the energy density is visible a chiral phase transition seems to

occur in the zero mass limit. This is shown in fig. 5 where also results
for the Polyakov line are given. A similar behavior has been obtained
from microcanonical simulations with 4 flavors [25] and recently also by
applying the hybrid algorithm [40]. In fig. 6 we show a comparison
between microcanonical results [25] and results obtained with a pseudofermion simulation [41] on a $8^3 \times 4$ for 4 flavors. Clearly there is good
agreement between both approaches. The crossover behavior for the 4
flavor theory seems to be sharper than in the 3 flavor case. Indeed for
larger numbers of flavors, i.e. for 8 [27,41] and 12 flavors [43]
respectively, evidence for a first order phase transition has been
found. However, it is up to now not clear whether this transition is

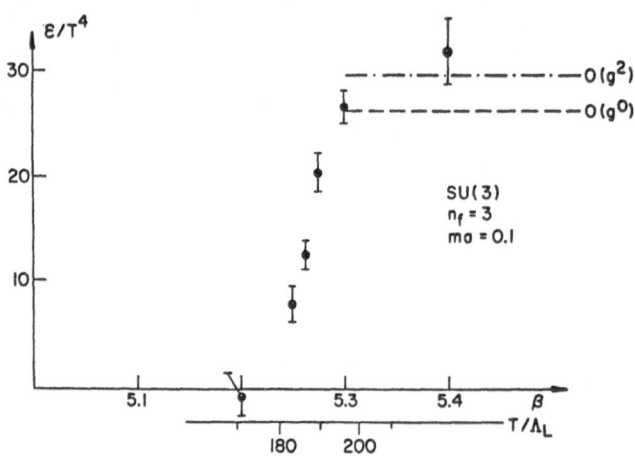

Figure 4: Energy density ε/T^4 versus coupling β for SU(3) with 3 flavors
of mass ma = 0.1 on a 8×4 lattice. Also shown are the lowest order (--)
and $O(g^2)$ (-·-) weak coupling perturbative results [42].

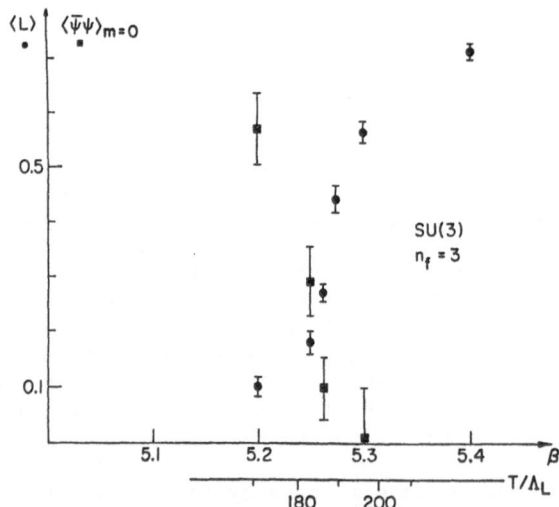

Figure 5: The Polyakov line expectation value (●) and the chiral order
parameter $\langle \bar{\chi} \chi \rangle_{m=0}$ (■) versus β. All other parameters are the same as in
fig. 4. $\langle \bar{\chi} \chi \rangle_{m=0}$ has been obtained from a linear extrapolation of data at
ma = 0.075 and 0.1.

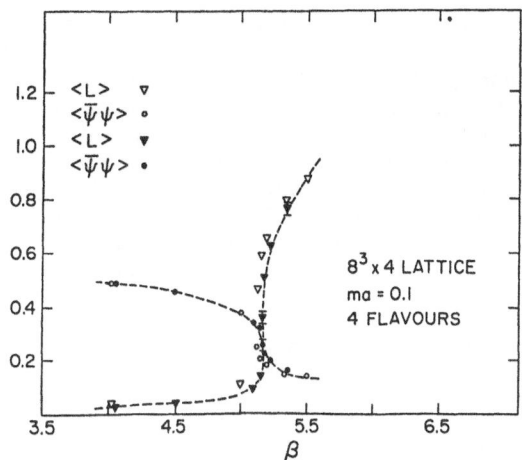

Figure 6: Chiral order parameter $\langle\bar{\chi}\chi\rangle$ and Polyakov loop $\langle L\rangle$ versus coupling β on a $8^3 \times 4$ lattice. The open circle ($\langle\bar{\chi}\chi\rangle$) and open triangle ($\langle L\rangle$) are microcanonical results [25], while those with full dots and triangles are pseudofermion results [41].

a finite temperature effect or related to a non trivial structure of the $T = 0$ theory with many flavors. It has been speculated earlier [26] that SU(N) gauge theories with large numbers of flavors undergo a first order chiral phase transition at zero temperature. The existence of such a bulk transition may lead to spurious first order phase transitions and may also influence the behavior of the three and four flavor theories at inter-mediate couplings. This question can be answered by performing simula-tions on larger lattices, which will show whether the critical couplings for the 8 and 12 flavor theory are independent of the temporal size N_τ of the lattice ($T = 0$ bulk transition) or move to larger values when N_τ is increased (finite temperature transition). Such calculations are certainly very important in order to understand the influence of dynamical fermions on the finite temperature transitions.

IV. LATTICE QCD AT FINITE BARYON NUMBER DENSITY

We have seen that MC simulations at finite temperature gave con-vincing evidence for the existence of a phase transition from a low temperature hadronic phase to a quark-gluon plasma phase at high tempera-tures. A similar transition is expected to occur at finite baryon number density when the density becomes conciderably larger than nuclear matter density.

The possibility of a chiral phase transition at finite density has been studied in quenched MC simulations [28] and also by means of analytic calculations in the strong coupling ($\beta = 0$) limit [44]. MC simulations with dynamical fermions, however, are difficult for the SU(3) theory. The fermion determinant becomes complex and standard simulation techniques are no longer applicable. There have been attempts to deal with this problem by either neglecting the imaginary part of the determinant as a first approximation [6] or to use a simulation based on the Langevin approach [7]. However, at present results for the SU(3) theory are not very conclusive.

For the SU(2) theory these problems do not occur as the action is still real.[*] Exploratory simulations with dynamical fermions have been performed in this case [29]. In the following we will thus concentrate on a discussion of the SU(2) theory. Furthermore we will restrict ourself to a discussion of the strong coupling, $\beta = 0$, limit of the theory, where we can compare MC simulations with mean field calculations. For finite chemical potential μ the fermionic part of the action gets modified. The chemical potential can be introduced in analogy to the continuum theory, where it enters the action like the zeroth component of a constant imaginary abelian field. On the lattice this leads to [28,45]

$$S_F = \sum_x \{ma\bar{\chi}_x\chi_x + \frac{1}{2} \sum_{i=1}^{3} \eta_i(x)\bar{\chi}_x[U_{x,i}\chi_{x+i} - U^{\dagger}_{x-i,i}\chi_{x-i}]$$

$$+ \frac{1}{2} \bar{\chi}_x[e^{\mu a}U_{x,o}\chi_{x+o} - e^{-\mu a}U^{\dagger}_{x-o,o}\chi_{x-o}]\} \qquad (4.1)$$

In the strong coupling limit the partition function is given by

$$Z(T,V) = \int \prod_x d\bar{\chi}_x d\chi_x \prod_{x,\mu} dU_{x,\mu} e^{-S_F} \qquad (4.2)$$

This partition function can then be analyzed either by MC simulations or meanfield techniques similar to those used for the $\mu = 0$ strong coupling action [46]. In the meanfield calculations at $\mu \neq 0$ one has, however, to be careful in dealing with the mesonic and baryonic sector of the theory, which are influenced differently by the chemical potential and thus can take on different mean values. Moreover, as the chemical potential introduces an anisotropy between the space and time directions of the lattice one can use a mean field ansatz which allows for different mean-field values of the mesonic (σ^1) and baryonic ($\sigma^{2,3}$) fields in space and time directions. The phase diagram in the chemical potential (μ)-quark mass (m) plane resulting from such a meanfield analysis for the $T = 0$, $\beta = 0$, SU(2) gauge theory is shown in fig. 7 (for a more detailed discussion see ref. [30]). The phase diagram reflects the basic proper-ties of a fermionic system at $T = 0$ and finite chemical potential. As long as the chemical potental is so small that the fermi energy is smaller than the energy of the lowest lying excited baryonic state above the vaccum no particles can be created and we are just probing the vacuum (region I). All thermodynamic observables will coincide with there corresponding $\mu = 0$ value. In particular the energy density and baryon number density continue to be zero in this regime. The threshold value, μ_o, is determined by the lowest baryonic state which for SU(2) is actually degenerate with the lightest mesonic state. Thus μ_o is given by

$$\mu_o(ma) = \frac{1}{2} m_o(ma) \qquad (4.3)$$

where m_o is the strong coupling value for the pion mass ($\equiv$ nucleon mass for SU(2)) [46]

---

[*] For SU(2) det(m+D) is real also for finite chemical potential due to the following symmetry: $\sigma_1 U \sigma_1 = U^{\dagger}$ for all $U \in$ SU(2) and σ_1 being the standard Pauli matrix.

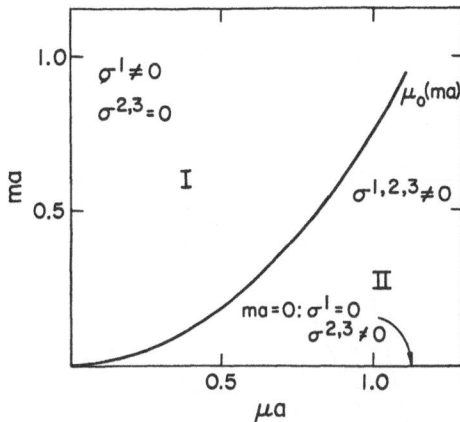

Figure 7: The phase diagram of the T = 0, β = 0 SU(2) gauge theory with staggered fermions. The threshold line μ_o(ma) shown is given by eq. (4.3). It seperates the vacuum regime (I) from the thermodynamic regime (II). Also shown are the meanfield values in the different regions.

$$m_o = \ell n\left[1 + d(\bar{\lambda}^2-1) + (2d(\bar{\lambda}^2-1) + d^2(\bar{\lambda}^2-1)^2)^{1/2}\right]$$

$$\bar{\lambda} = \frac{ma}{\sqrt{2d}} + (1 + \frac{(ma)^2}{2d})^{1/2}, \quad d = 4 \qquad (4.4)$$

Only for chemical potentials μ larger than μ_o we enter the thermodynamic regime of the phase diagram (region II) with a non-zero energy and baryon number density. It is in this regime where we would expect a possible chiral phase transition to occur. However, neither meanfield nor MC simulations gave any evidence for a phase transition in this regime [30]. Chiral symmetry seems to be broken for all finite μ, although the breaking pattern is quite peculiar. Meanfield as well as MC simulations indicate that for any finite chemical potential the mesonic condensate vanishes in the zero mass limit. We show these results for three different values of μ in fig. 8. Notice, that in all cases $\langle\bar{\chi}\chi\rangle$ extrapolates to zero and that for large masses $\langle\bar{\chi}\chi\rangle$ is independent of μ. This indicates that one enters the vacuum region I in the phase diagram of fig. 7. The MC results shown in fig. 8b have been obtained from simulations with dynamical fermions which have been introduced using the pseudofermion algorithm [4]. We would like to stress that the good agreement between meanfield calculations and these Monte Carlo results [30] have been achieved by introducing different meanfields in the mesonic and baryonic sector. This was an essential ingredient and explains the difference to other meanfield calculations [44] where evidence for a chiral symmetry restoring transition in the SU(2) gauge theory with μ ≠ 0 has been reported.

It turns out that at nonzero μa chiral symmetry remains broken due to the appearance of a non vanishing baryonic condensate in the thermodynamic regime II of the phase diagram [30]. As in SU(2) baryons are in fact bosons this condensate is a bose condensate which is expected to disappear at large temperatures. We thus expect that a chiral symmetry restoring transition will occur in the strong coupling SU(2) theory at finite μ as soon as the temperature becomes non zero. This question as well as the question in how far these results can be generalized for the SU(3) theory are at present under investigation.

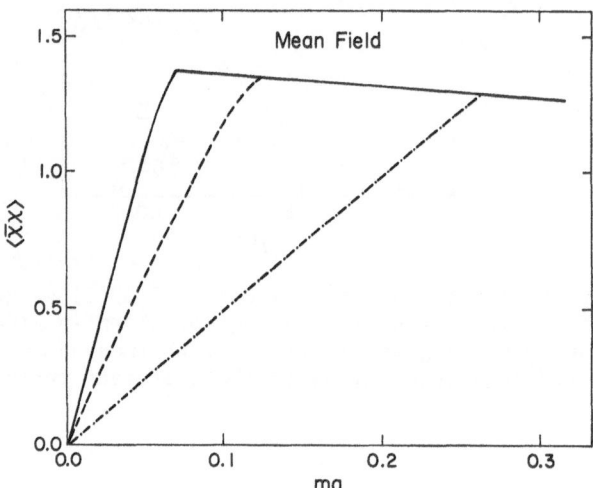

Figure 8a: Meanfield results for the mesonic condensate $\langle \bar{\chi}\chi \rangle$ versus ma for 3 values of the chemical potential, $\mu a = 0.3$ (––), 0.4 (– –) and 0.6 (–•–).

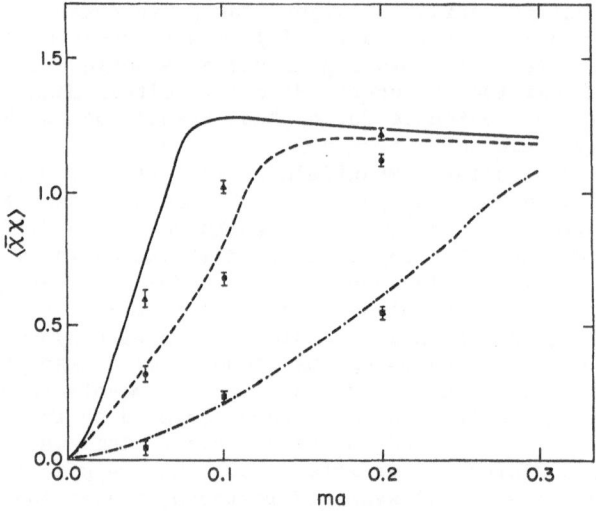

Fig. 8b: Monte Carlo results for $\langle \bar{\chi}\chi \rangle$ versus ma. The values used for μa are the same as in fig. 8a.

V. CONCLUSIONS

We have discussed recent developments in the analysis of the thermodynamics of QCD. Finite temperature phase transitions in the pure gauge sector as well as in the presence of dynamical fermions are at present investigated with large computational effort. Simulations on large lattices gave evidence for asymptotic scaling behavior of the deconfinement temperature. This allows a quite accurate determination of the transition in the pure gauge sector. Thermodynamics in the presence of dynamical fermions has been studied only on comparatively small lattices. These calculations concentrated on a qualitative understanding of the SU(3) phase diagram and its dependence on the quark mass and number of flavors. They gave evidence for a chiral phase transition for massless fermions and a rather abrupt change of thermodynamic observables for light quarks. However, the behavior at intermediate couplings may be influenced by spurious bulk transitions of the T = 0, large flavor theory. Thus larger lattices are needed to judge the order and abruptness of the transition in the SU(3) theory with 3 light flavors.

First attempts have been undertaken to analyze QCD at finite baryon number density. The analysis of the SU(3) theory with dynamical fermions, however, requires new simulation techniques which can deal with complex actions. A satisfactory algorithm to deal with this problem is still missing. The analysis of the SU(2) theory in the strong coupling limit shows that a generalization of $\mu = 0$ results is not straightforward. The breaking pattern of chiral symmetry becomes more complicated. Chiral symmetry remains broken although $\langle \bar{\chi}\chi \rangle$ seems to be zero for any non-zero chemical potential. Indeed, a careful analysis of the quenched SU(3) data [28] indicates that this behavior is not specific for SU(2) but may also be present in SU(3).

ACKNOWLEDGEMENT

This work is partially supported by the National Science Foundation under grant number NSF PHY82-01948.

REFERENCES
1) J. C. Collins and M. J. Perry, Phys. Rev. Lett. 34:1353 (1975).
2) For a collection of early references see, C. Rebbi, Lattice Gauge Theory and Monte Carlo Simulation, World Sientific (1983).
3) K. G. Wilson, Phys. Rev. D10:2445 (1974).
4) F. Fucito, E. Marinari, G. Parisi and C. Rebbi, Nucl. Phys. B180 [FS2]:369 (1981); E. Marinari, G. Parisi and C. Rebbi, Nucl. Phys. B190 [FS3]:734 (1981); H. W. Hamber, E. Marinari, G. Parisi and C. Rebbi, Phys. Lett. 124B:99 (1983).
5) J. Polonyi and H. W. Wyld, Phys. Rev. Lett. 51:2257 (1983).
6) J. Engels and H. Satz, Phys. Lett. 159B:151 (1985).
7) F. Karsch and H. W. Wyld, Phys. Rev. Lett. 55:2242 (1985).
8) C. Borgs and E. Seiler, Commun. Math. Phys. 91:195 (1983).
9) L. McLerran and B. Svetitsky, Phys. Lett. 98B:195 (1981); J .Kuti, J. Polonyi and K. Szlachanyi, Phys. Lett. 98B:199 (1981).
10) J. Engels, F. Karsch, I. Montvay and H. Satz, Phys. Lett. 101B:83 (1981) and Nucl. Phys. B205 [FS5]:545 (1982).
11) K. Kajantie, C. Montonen and E. Pietarinen, Z. Phys. C9:148 (1982); I. Montvay and E. Pietarinen, Phys. Lett. 110B:148 (1982).
12) A. Gocksch and M. Okawa, Phys. Rev. Lett. 52:1751 (1984); G. G. Batrouni and B. Svetitsky, Phys. Rev. Lett. 52:2205 1984).

13) K. Fabricius, O. Haan and F. Klinkhamer, Phys. Rev. D to be published; S. R. Das and J. B. Kogut, Illinois preprint, ILL-(TH)-85-#4.

14) B. Svetitsky and L. Yaffe, Nucl. Phys. B210 [FS6]:423 (1984).

15) B. Svetitsky, Phys. Rep. to appear.

16) For a recent review see F. Karsch, ILL-(TH)-85-#73, Oct. 1985.

17) A. Hasenfratz, P. Hasenfratz, U. Heller and F. Karsch, Z. Phys. C25:191 (1984); D. Barkai, K. J. M. Moriarty and C. Rebbi, Phys. Rev. D30:1293 (1984).

18) A. D. Kennedy, J. Kuti, S. Meyer and B. J. Pendleton, Phys. Rev. Lett. 54:87 (1985); S. A. Gottlieb, J. Kuti, D. Toussaint, A. D. Kennedy, S. Meyer, B. J. Pendleton and R. L. Sugar, Phys. Rev. Lett. 55:1958 (1985).

19) N. H. Christ and A. E. Terrano, Columbia preprint, CU-TP-325 (1985).

20) T. Banks and A. Ukawa, Nucl. Phys. B225 [FS9]:145 (1983); J. Bartholomew, D. Hochberg, P. H. Damgaard and M. Gross, Phys. Lett. 133B:218 (1983); F. Green and F. Karsch, Nucl. Phys. B238:297 (1984).

21) P. Hasenfratz, F. Karsch and I. O. Stamatescu, Phys. Lett. 133B:221 (1983).

22) R. V. Gavai, M. Lev and P. Peterson, Phys. Lett. 140B:397 (1984) and Phys. Lett. 149B:492 (1984).

23) F. Fucito and S. Solomon, Phys. Lett. 140B:387 (1984); F. Fucito, C. Rebbi and S. Solomon, Nucl. Phys. B248:615 (1984) and Phys. Rev. D31:1461 (1985).

24) R. V. Gavai and F. Karsch, Nucl. Phys. B261:273 (1984).

25) J. Polonyi, H. W. Wyld, J. B. Kogut, J. Shigemitsu and D. K. Sinclair, Phys. Rev. Lett. 53:644 (1984).

26) J. Banks and A. Zaks, Nucl. Phys. B196:189 (1982).

27) J. Kogut, J. Polonyi D. K. Sinclair and H. W. Wyld, Phys. Rev. Lett. 54:1475 (1985).

28) J. Kogut, H. Matsuoka, M. Stone, H. W. Wyld, S. Shenker, J. Shigemitsu and D. K. Sinclair, Nucl. Phys. B225 [FS9]:93 (1983).

29) A. Nakamura, Phys. Lett. 149B:391 (1984).

30) E. Dagotto, F. Karsch and A. Moreo, Illinois preprint, ILL-(TH)-86-#4, January 1986.

31) R. V. Gavai and H. Satz, Phys. Lett. 145B:248 (1984); G. Curci, and R. Tripiccione, Phys. Lett. 151B:145 (1985).

32) T. Celik, J. Engels and H. Satz, Phys. Lett. 125B:411 (1983).

33) T. Celik, J. Engels and H. Satz, Phys. Lett. 129B:323 (1983).

34) J. Kogut, M. Stone, H. W. Wyld, W. R. Gibbs, J. Shigemitsu, S. H. Shenker and D. K. Sinclair, Phys. Rev. Lett. 50:393 (1983).

35) J. Kogut, H. Matsuoka, M. Stone, H. W. Wyld, S. H. Shenker, J. Shigemitsu and D. K. Sinclair, Phys. Rev. Lett. 51:869 (1983).

36) F. Karsch, Nucl. Phys. B205 [FS5]:285 (1982).

37) U. Heller and F. Karsch, Nucl. Phys. B251 [FS13]:254 (1985).

38) F. Fucito and B. Svetitsky, Phys. Lett. 131B:165 (1985).

39) E. T. Tomboulis and L. G. Yaffe, Phys. Rev. Lett. 52:2115 (1984).

40) S. Duane and J. B. Kogut, Phys. Rev. Lett. 55:2774 (1985).

41) R. V. Gavai, Brookhaven preprint, BNL-37214, 1985.

42) U. Heller and F. Karsch, Nucl. Phys. B258:29 (1985).

43) J. B. Kogut, Illinois preprint, ILL-(TH)-85-#75, October 1985.

44) P. H. Damgaard, D. Hochberg and N. Kawamoto, Phys. Lett. 158B:239 (1985); E. M. Ilgenfritz and J. Kripfganz, Z. Phys. C29:79 (1985).

45) P. Hasenfratz and F. Karsch, Phys. Lett. 125B:308 (1983).

46) H. Kluberg-Stern, A. Morel and B. Petersson, Nucl. Phys. B215 [FS7]:527 (1983).

DECONFINING PHASE TRANSITION AND THE CONTINUUM LIMIT OF

LATTICE QUANTUM CHROMODYNAMICS*

S.A. Gottlieb, J. Kuti, and D. Toussaint

Department of Physics, University of California
La Jolla, CA 92093

A.D. Kennedy

Institute for Theoretical Physics, University of California
Santa Barbara, CA 93106

S. Meyer

Universität Kaiserslautern, Fachbereich Physik
D 6750 Kaiserslautern, BRD

B.J. Pendleton

Physics Department, University of Edinburgh
Edinburgh EH 9 3JZ, Scotland

R.L. Sugar

Department of Physics, University of California
Santa Barbara, CA 93106

ABSTRACT

We present a large-scale Monte Carlo calculation of the deconfining phase-transition temperature in lattice quantum chromodynamics without fermions. Using the Wilson action, we find that the transition temperature as a function of the lattice coupling g is consistent with scaling behaviour dictated by the perturbative β-function for $6/g^2 > 6.15$.

A detailed treatment appeared in Phys. Rev. Lett. 55 (1985), 1958.

---

*Presented by S. Meyer; work supported in part by DFG.

CRITICAL BEHAVIOUR IN BARYONIC MATTER[*]

Helmut Satz

Fakultät für Physik
Universität Bielefeld
Bielefeld
F.R. Germany

and

Physics Department
Brookhaven National Laboratory
Upton, Long Island, N.Y.
USA

Abstract

First we consider the phenomenology of deconfinement and chiral symmetry restoration for strongly interacting matter at non-vanishing baryon number density. Subsequently, we present numerical results obtained by a Monte Carlo evaluation of statistical QCD on an $8^3 \times 3$ lattice, using Wilson fermions with $N_f = 2$, in fourth order hopping parameter expansion, and suppressing the imaginary part of the fermion action. We consider baryonic chemical potentials up to $\mu a = 0.6$ $(\mu/\Lambda_L \simeq 200)$; in this range, the critical parameters for deconfinement and chiral symmetry restoration are found to coincide.

---

[*] Joint work with B. Berg (Florida State University, Tallahassee), J. Engels, E. Kehl and B. Waltl (Fakultät für Physik, Universität Bielefeld), to be published in Z. Phys. C.

I. INTRODUCTION

The prediction of the phase structure of strongly interacting matter is one of the most challenging problems in statistical QCD. We expect that with increasing density, hadronic matter will be transformed into a plasma of coloured, massless quarks and gluons: it should undergo deconfinement and chiral symmetry restoration. The increase in density can be achieved either by heating or by compression, and hence the phase of the system will depend on both temperature and baryon number density.

In the case of vanishing baryon number density, deconfinement and chiral symmetry restoration have been investigated in a variety of lattice evaluation schemes, and the thermodynamics of such "mesonic" matter is slowly emerging[1]. Quantitative studies of the baryon number dependence, however, have been initiated only quite recently[2,3]; and this topic will form the main subject of our paper.

To clarify the phenomena which we want to investigate, it seems helpful to first consider a simple schematic model, stripped of all but the essential dynamics. This will be the topic of the first section. Following it, we will turn to lattice QCD at non-vanishing baryonic chemical potential μ. Treating the quarks (we will consider two flavours) as Wilson fermions in low order hopping parameter expansion, we will calculate the basic thermodynamic observables and study the pattern of deconfinement and chiral symmetry restoration in baryonic matter.

II. BASIC PHENOMENOLOGY

Let us first look at hadron and quark systems of vanishing baryon number density. Consider an ideal gas of massless pions. Its pressure is

$$P_\pi = \frac{\pi^2}{90} \times 3 \times T^4 = \frac{\pi^2}{30} T^4 \quad , \tag{1}$$

taking into account the three possible charge states. For an ideal plasma of massless quarks, antiquarks and gluons, the pressure becomes

$$P_q = \frac{\pi^2}{90} [\frac{7}{8} \times 2 \times 2 \times 2 \times 3 + 8 \times 2] T^4 = \frac{37\pi^2}{90} T^4 \quad , \tag{2}$$

including two flavours (u and d), two spin orientations and three colours for quarks and antiquarks, eight colours and two spin orientations for gluons. Comparing the two states, we note that the pressure of the plasma - with more degrees of freedom – always exceeds that of the pion gas. Matter in equilibrium would thus always be in the plasma phase. The essential dynamical input to change this is the (non-perturbative) bag pressure B, which reduces P_q:

$$P'_q = P_q - B \quad .$$

(3)

The result is shown in fig. 1; there now is a cross-over, which determines

$$T_c = (\frac{90}{34\pi^2} B)^{1/4} \simeq 0.72 \, B^{1/4}$$

(4)

as transition temperature. - Including the pion mass and/or further resonant states, such as ρ and ω, does not lead to any qualitative change of this picture; neither do perturbative corrections to the quark-gluon plasma.

For baryonic matter, we will now consider the other extreme: T = 0 at non-zero baryonic chemical potential μ. A perfect Fermi gas of massless protons and neutrons has the pressure

$$P_N = \frac{1}{24\pi^2} \times 2 \times 2 \times \mu^4 = \frac{\mu^4}{6\pi^2} \quad .$$

(5)

The ideal quark plasma, with coloured u and d quarks, gives

$$P_q = \frac{1}{24\pi^2} \times 2 \times 2 \times 3 \times \mu_q^4 = \frac{\mu_q^4}{2\pi^2} \quad .$$

(6)

At equilibrium, $\mu_q = \mu/3$ and hence

$$P_q = \frac{1}{27} \frac{\mu^4}{6\pi^2} \quad .$$

(7)

Here we find that the nuclear matter phase dominates at all μ. To change this, we have to take into account the repulsion between nucleons, which puts a bound on the compression of nuclear matter. For nucleons with hard cores of volume V_N, the nuclear pressure becomes

$$P'_N = \frac{P_N}{1 + n \, V_N} \quad ;$$

(8)

here

$$n = \frac{2}{3\pi^2} \mu^3$$

(9)

is the density of a perfect Fermi gas of nucleons. Hence we find

$$P'_N = \frac{\mu^4}{6\pi^2 + 4\mu^3 \, V_N} \quad .$$

(10)

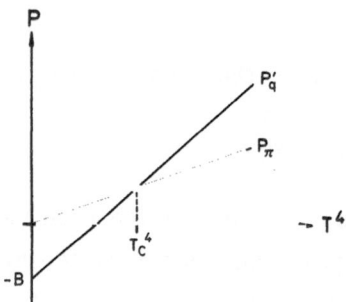

Figure 1 : Ideal pion gas (P_π) vs. ideal quark-gluon plasma with bag pressure (P_q') .

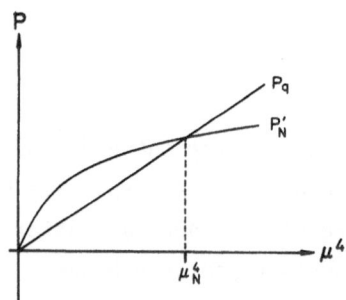

Figure 2 : Ideal Fermi gas of quarks (P_q) vs. ideal Fermi gas of massless nucleons with hard core repulsion (P_N') .

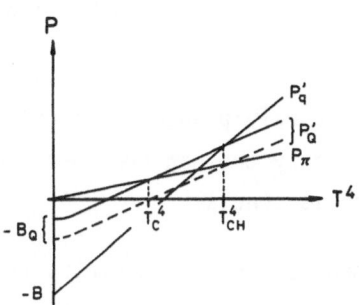

Figure 3 : Constituent quark matter (P_Q') , compared to ideal pion gas (P_π) and ideal quark-gluon plasma (P_q') .

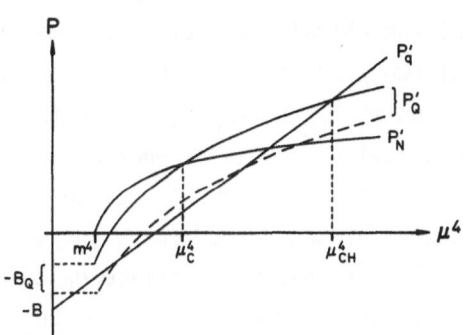

Figure 4 : Constituent quark matter (P_Q') , compared to ideal Fermi gas of hard-core nucleons (P_N') and ideal Fermi gas of massless quarks (P_q') .

The behaviour of P_N' vs. P_q is shown in fig. 2: the hard core repulsion provides a much weaker growth of P_N' at large μ and thus leads to a cross-over,
with

$$\mu_c = (\frac{39\pi^2}{V_N})^{1/3} \cong 7.27 \; V_N^{-1/3} \tag{11}$$

as the critical chemical potential. - Here also a more realistic picture,
with massive nucleons and bag pressure, does not lead to qualitative
changes[*].

We thus note that on a purely phenomenological level, deconfinement is
at $\mu = 0$ basically determined by the bag pressure, while at $T = 0$, it is
the nucleon repulsion which is crucial.

To compare deconfinement and chiral symmetry restoration, we must allow
a third phase: constituent quark matter. Consider a non-interacting gas of
coloured constituent quarks of mass $m_Q \cong \frac{1}{3} m_N$, together with massless pions
as Goldstone bosons; relative to the physical vacuum, there will be a con-
fining bag pressure B_Q, with $|B_Q| < |B|$. For this phase, we have at $\mu = 0$

$$P_Q' = \frac{\pi^2}{90} \; [21 \; \frac{P(m_Q,T)}{P(0,T)} + 3] \; T^4 - B_Q \; , \tag{12}$$

where $P(m_Q,T)$ is the pressure of an ideal gas of massive fermions with one
intrinsic degree of freedom. In fig. 3, we compare eq. (12) to the ideal pion
pressure (1) and that of the chirally symmetric plasma of massless quarks.
There are now in general two transitions: deconfinement at $T_c = f(B_Q,m_Q)$ and
chiral symmetry restoration at $T_{CH} = f(B_Q,B,m_Q)$. Whether the intermediate
constituent quark phase actually occurs, or whether it leads to a pressure
below P_π in the hadronic regime and below P_q' once the plasma pressure crosses
P_π - the behaviour indicated by the dashed curve in fig. 3 - depends on the
actual values of the parameters involved. Lattice calculations[5-8] have led
to $T_c \cong T_{CH}$ and would thus support the latter case. On a phenomenological
level, the question is studied in detail in ref. 9).

At $T = 0$, we must add to the properties of constituent quark matter a
hard core baryonic repulsion between the quarks, characterized by an intrinsic constituent quark volume V_Q ($<\frac{1}{3} V_N$). The resulting pressure then is

_____

[*] The nucleon mass m_N does, however, determine a lower bound for the bag
pressure, if there is to be a cross-over[4]: from P_q ($\mu=m_N$) - B $\geq$ 0 we
obtain $B^{1/4} \geq (162 \; \pi^2)^{-1/4} \; m_N \cong 0.158 \; m_N \cong 148$ MeV.

$$P_Q' = \frac{P_Q(m_Q, \mu)}{1 + n_Q V_Q} - B_Q \quad ; \tag{13}$$

here

$$P_Q(m_Q, \mu) = \frac{m_Q^4}{2\pi^2} \left\{ \frac{\mu}{3m_Q} \left(\frac{\mu^2}{9m_Q^2} - 1 \right)^{1/2} \left(\frac{\mu^2}{9m_Q^2} - \frac{5}{2} \right) \right. \tag{14}$$

$$\left. + \frac{3}{2} \ln \left[\frac{\mu}{3m_Q} + \left(\frac{\mu^2}{9m_Q^2} - 1 \right)^{1/2} \right] \right\}$$

denotes the pressure of an ideal gas of massive fermions[4] with two spin, two flavour and three colour degrees of freedom. Similarly,

$$n_Q = \frac{2}{27\pi^2} (\mu^2 - 9m_Q^2)^{3/2} \tag{15}$$

is the baryon number density for the constituent quark system; here, as above, μ is the baryonic chemical potential. - In fig. 4, we compare the pressure of constituent quark matter, eq. (13), with that of the plasma of massless quarks and with that of nuclear matter. For the sake of consistency, we have now included the nucleon mass in calculating P_N, and the bag pressure B in P_q. Again we obtain in general two transitions: deconfinement at μ_c and chiral symmetry restoration at μ_{CH}. The crucial question for lattice studies thus is whether these phenomena are indeed distinct, or if they occur at the same value of the baryonic chemical potential.

III. LATTICE QCD at $\mu \neq 0$

The starting point for statistical QCD is the partition function

$$Z(T, \mu) = \mathrm{Tr} \{ e^{-(H - \mu N)/T} \} \quad , \tag{16}$$

where H is the Hamiltonian, μ the baryonic chemical potential, and N the net baryon number of the system. On an asymmetric but isotropic Euclidean lattice with N_σ (N_τ) spatial (temporal) lattice sites, the partition function becomes

$$Z(N_\sigma, N_\tau; g^2; \mu) = \int \prod_{\text{links}} dU \, e^{-S_G} \{ \det Q \}^{N_f} \tag{17}$$

Here

$$S_G(U) = \frac{6}{g^2} \sum_{\mathbb{P}} (1 - \frac{1}{3} \mathrm{Re} \, \mathrm{Tr} \, UUUU) \tag{18}$$

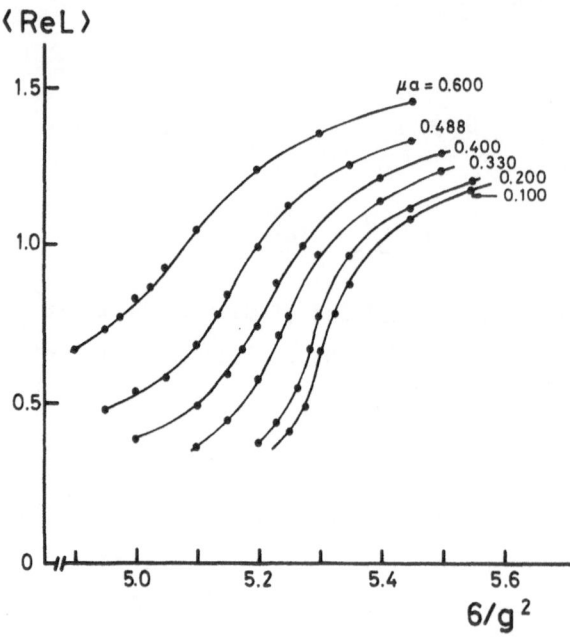

Figure 5 : < Re L > vs. $6/g^2$ for different μa . Curves are only to
guide the eye; naive statistical errors are smaller than the
data points.

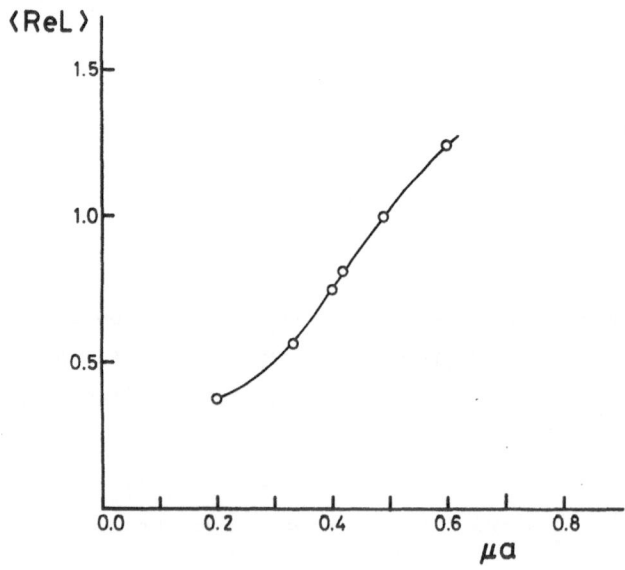

Figure 6 : < Re L > vs. μa at $6/g^2$ = 5.2 ; the curve is only to guide
the eye.

gives the gauge field action as plaquette sum, with U denoting the gauge group elements and g^2 the bare coupling on the lattice. The fermion determinant, det Q, results from the integration of the quark fields, with

$$S_F = \sum_f \bar{\psi}_f \, Q \, \psi_f \tag{19}$$

as quark action; we consider here N_f massless quark species. In Wilson's formulation[10], Q has the form

$$Q = 1 - \varkappa \sum_{\nu=0}^{3} M_\nu \equiv 1 - \varkappa M \tag{20}$$

with

$$(M_\nu)_{nm} = \begin{cases} (1-\gamma_\nu) \, U_{nm} \, \delta_{n,m-\hat\nu} + (1+\gamma_\nu) \, U^+_{mn} \, \delta_{n,m+\hat\nu} \,, & \nu = 1,2,3 \quad\text{(21a)} \\ (1-\gamma_0) \, U_{nm} \, \delta_{n,m-\hat\nu} \, e^{\mu a} + (1+\gamma_0) \, U^+_{mn} \, \delta_{n,m+\hat\nu} \, e^{-\mu a} \,, & \nu = 0 \quad\text{(21b)} \end{cases}$$

Here U_{nm} is the group variable associated to the link between two adjacent sites n and m; n+$\hat\nu$ denotes the site obtained by a unit shift in the ν direction. The strength of the quark interaction is characterized by the hopping parameter $\varkappa(g^2)$; $a(g^2)$ denotes the lattice spacing. For the sake of simplicity, we shall here write all formulae for equal couplings and equal lattice spacings in space and temperature directions. In actual calculations, they are of course set equal only after all operations (differentiation) are carried out.

The introduction of the chemical potential in eq. (21) follows the prescription of ref. 2 and 11; a more general form is discussed in ref. 12. A common feature of all forms is that with $\mu \neq 0$, the U and U^+ terms in eq. (21b) are no longer hermitean conjugates, and as a consequence, det Q becomes complex. Note, however, that Z remains real, since both $\int dU$ and $\int dU^+$ cover the complete group space.

In the hopping parameter expansion, we obtain for the quark action

$$S_F \equiv N_f \ln \det(1 - \varkappa M) = -N_f \, \mathrm{Tr} \sum_{\ell=1}^{\infty} \frac{\varkappa^\ell}{\ell} M^\ell \,, \tag{22}$$

which gives for $N_\tau \leq 4$ as leading terms

$$S_F = -2N_f(2\varkappa)^{N_\tau} \sum_{\text{sites } x} \{ L_x \, e^{\beta\mu'} + L^*_x \, e^{-\beta\mu'} \}$$
$$- 16N_f \, \varkappa^4 \sum_{\mathbb{P}} \mathrm{Re} \, \mathrm{Tr} \, UUUU + O(\varkappa^5) \,, \tag{23}$$

with $\beta \equiv N_\tau a$ for the inverse temperature and

$$L_x = \text{Tr} \prod_{\tau=1}^{N_\tau} U_{\vec{x};\tau,\tau+1} \qquad (24)$$

for the thermal Wilson loop at spatial site $\vec{x}$; the sum in the first term of eq. (23) runs over all such sites. It gives in this approximation the main quark contribution, since the second term amounts only to the shift

$$6/g^2 \rightarrow (6/g^2 + 48N_f \varkappa^4) \qquad (25)$$

in the gauge field action. Writing the first term of the quark action as

$$S_F(L) = -4N_f(2\varkappa)^{N_\tau} \sum_x \{\text{Re } L_x \cosh\beta\mu + i \text{ Im } L_x \sinh\beta\mu\} \qquad (26)$$

we see explicitly that it is complex for $\mu \neq 0$. From

$$\text{Im } Z = \int \prod dU \, e^{-S_G - \text{Re } S_F} \sin [-4N_f(2\varkappa)^{N_\tau} \sum_x \text{Im } L_x \sinh\beta\mu] \qquad (27)$$

we also have explicit, by changing variables $U \rightarrow U^+$, that Im $Z = 0$ and hence Z real. In this order of the hopping parameter expansion we therefore obtain for the partition function

$$Z(N_\sigma, N_\tau; g^2, \mu) = \int \prod dU \, e^{-S_G' - \text{Re } S_F(L)} \cos [\text{Im } S_F(L)] \quad , \qquad (28)$$

where S_G' denotes the gauge action with the shift (25) and $S_F(L)$ is given by eq. (26).

Using the form (28), we now define the thermodynamic average of any quantity $f(U)$ in the usual way

$$<f> \equiv \int \prod dU \, e^{-S_G'-\text{Re}S_F(L)} f \cos(\text{Im }S_F(L)) / \int \prod dU \, e^{-S_G'-\text{Re}S_F(L)} \cos(\text{Im}S_F(L)). \qquad (29)$$

From this, we see immediately that

$$< \text{Im } L_x > = 0 \, \forall \, \vec{x} \quad , \qquad (30)$$

$$< \text{Im } S_F(L) > \sim < \sum_x \text{Im } L_x > = 0 \quad , \qquad (31)$$

since Im L_x changes sign under the transformation $U \rightarrow U^+$. Consider now $<\text{Im } L_x>$; the integration over U, according to eq.(29), gives us the average over configurations. On the other hand, $\sum_x \text{Im } L_x$ is the lattice average for a given configuration. For large enough lattices and sufficiently many confi-

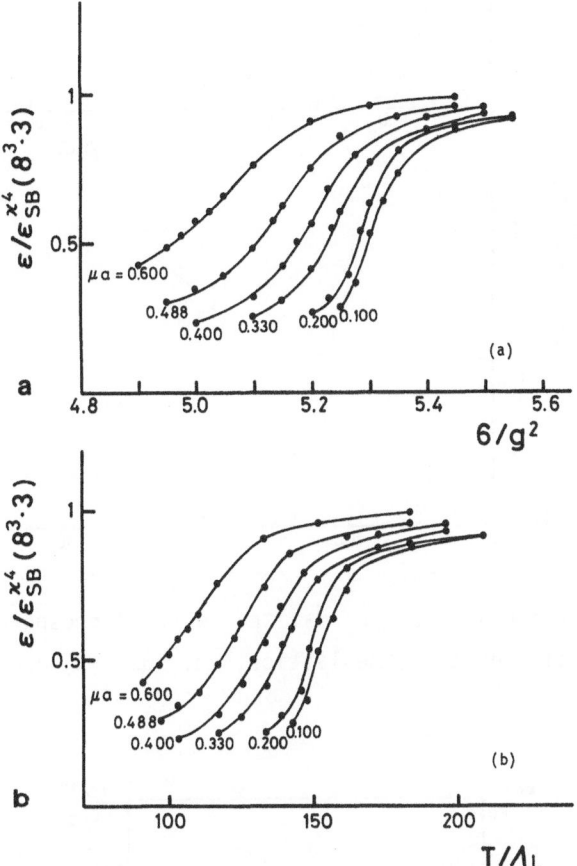

Figure 7 : Energy density ε , normalized to the corresponding $\varkappa^4$ ideal gas value on the same lattice, vs. $6/g^2$ (a) and vs. temperature T/Λ_L (b).

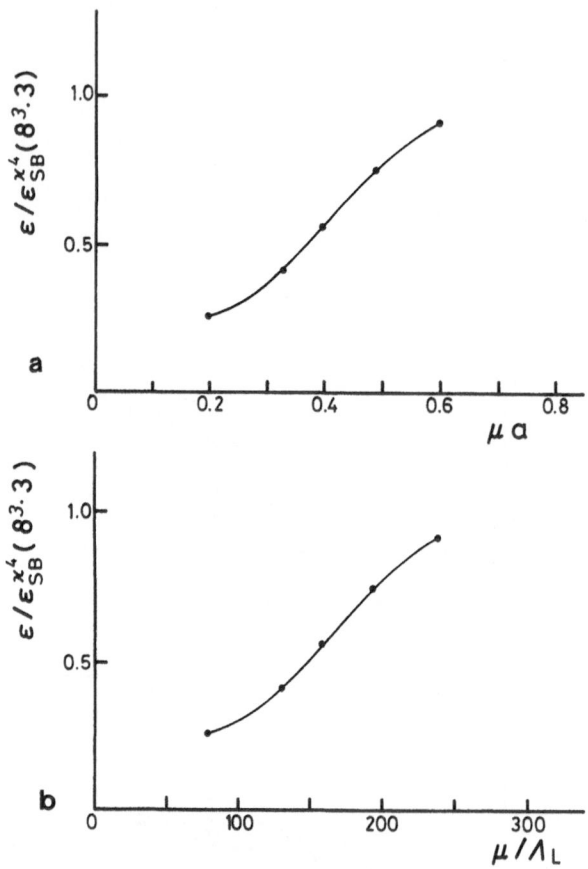

Figure 8 : Energy density ε , normalized to the corresponding $\varkappa^4$ ideal gas value on the same lattice, vs. μa (a) and vs. μ/Λ_L (b).

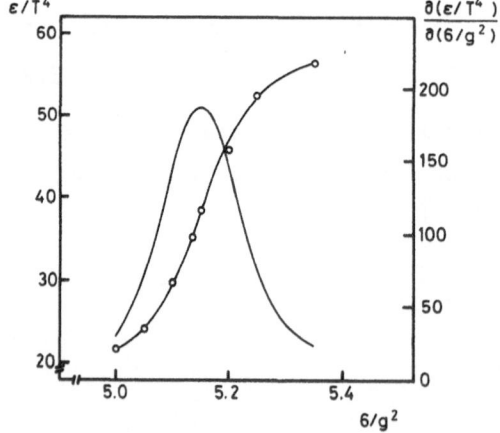

Figure 9 : Energy density and its derivative ($\sim$ specific heat) vs. $6/g^2$ at $\mu a = 0.488$.

gurations we expect these two averages to agree, if we are not at a critical point: we can then imagine the large lattice to be obtained by combining sufficiently many equilibrium configurations on a smaller lattice. These arguments have led to the approximation of "partial quenching", in which we set $Im\ S_F = 0$ everywhere[3]. In this case, the usual Monte Carlo evaluation can be carried out with $exp\ \{-S'_G - Re\ S_F(L)\}$ as weight, and all results to be presented here are obtained in this way. Thus now, with partial quenching,

$$<f>_R = \int \Pi\ dU\ e^{-S'_G - Re\ S_F(L)}\ f\ /\ \int \Pi\ dU\ e^{-S'_G - Re\ S_F(L)} \qquad (32)$$

defines the thermodynamic average.

We had initially considered it possible to test this partial quenching by calculating

$$<f> = <f\ cos(Im\ S_F(L))>_R\ /\ <cos(Im\ S_F(L))>_R \qquad , \qquad (33)$$

i.e., by including $cos(Im\ S_F(L))$ as part of the observable whose average is to be calculated. It turns out that this procedure is not feasible, for the following reason. On a $8^3\ x3$ lattice, we find even after 30000 lattice sweeps, that $<\sum_x Im\ L_x>$ is of order unity, rather than zero, as required in eq. (31). Hence a Monte Carlo evaluation of eq. (33) without importance sampling to assure $<\sum_x Im\ L_x> = 0$ cannot be expected to give reasonable results, since the values of $cos(Im\ S_F(L))$ obtained with $exp\ \{-S'_G - Re\ S_F(L)\}$ as weight for the Metropolis algorithm fluctuate wildly.

If we ignore these difficulties and just calculate e.g. $<Re\ L>$ according to eq. (33) with $exp\ \{-S'_G - Re\ S_F(L)\}$ as Monte Carlo weight, we obtain $<Re\ L> \simeq <Re\ L>_R$, together with $<cos\ (Im\ S_F(L))> \cong 0$: a smoothly varying function, such as $Re\ L$, becomes uncorrelated from the randomly fluctuating $cos\ (Im\ S_F(L))$, and hence can in good approximation be taken outside of the integral. Thus the agreement between $<Re\ L>$ and $<Re\ L>_R$ is here simply a consequence of the fluctuations of $cos\ (Im\ S_F(L))$. In this situation, partial quenching appears to be the most reasonable procedure to follow. - As all averages from here on are defined by eq. (32), we shall now drop the subscript R on the averages $<\ >_R$.

To evaluate thermodynamic observables as functions of μ and T, we need explicit expressions for $\varkappa(g^2)$ and $a(g^2)$. For the hopping parameter, we shall use the weak coupling form[13]

$$\varkappa(g^2) = \frac{1}{8}\ [1 + 0.11\ g^2 + O(g^4)] \qquad , \qquad (34)$$

29

and for the lattice spacing the renormalization group relation with $N_f = 2$

$$a(g^2) \Lambda_L = \exp \{ - \frac{4\pi^2}{29} (\frac{6}{g^2}) + \frac{345}{29^2} \log [\frac{8\pi^2}{29} (\frac{6}{g^2})] \} \quad . \quad (35)$$

In both cases we expect some deviations at the g^2 values actually used; but these relations should suffice to give us at least a reasonable qualitative impression of the resulting critical behaviour.

IV. NUMERICAL RESULTS

Our evaluation was performed on an $8^3 \times 3$ lattice with $N_f = 2$. We have included terms up to κ^4 in the hopping parameter expansion (23). For each g^2 value, we carried out about 3000 - 4000 lattice sweeps. Using these results, we have studied the T and μ behaviour of the thermal Wilson loop < Re L > as deconfinement measure, of $\langle \overline{\psi}\psi \rangle$ as chiral symmetry measure, and of the overall energy density ε.

In our study, we have considered that baryonic chemical potential in the range $0 \leq \mu a \leq 0.6$, i.e., up to about $\mu \sim 300 - 400$ MeV. The reason for stopping here is given by the truncation of the hopping parameter expansion: increasing μ has a similar effect as increasing κ and hence necessitates the inclusion of more terms in eq. (22). To obtain some idea of the error made by including only terms up to order κ^4, we have calculated the energy density of an ideal Fermi gas on an $8^3 \times 3$ lattice for various μ values and compared the results with those given by the hopping parameter expansion up to κ^4 for this quantity[14]. The ratio

$$\varepsilon_F^{\kappa^4}(8^3 \times 3) / \varepsilon_F^{full}(8^3 \times 3) \qquad (36)$$

varies from 0.96 to 1.09 as a is increased from 0 to 0.6; in the μ range considered, the truncation error thus is 10% or less for an ideal Fermi gas. For larger μ, the error increases; we have therefore only gone up to $\mu a = 0.6$. The full result obtained on an $8^3 \times 3$ lattice is of course not identical with the ideal gas value in the continuum[15]. However, up to $\mu a = 0.6$ the difference between the continuum value and the lattice results with $N_\tau = 3$ is essentially independent of μ.

Let us note at this point one of the disadvantages of the hopping parameter approach: the truncation error is evidently N_τ dependent, and hence this approach is not very suitable for studying the scaling behaviour of thermodynamic observables.

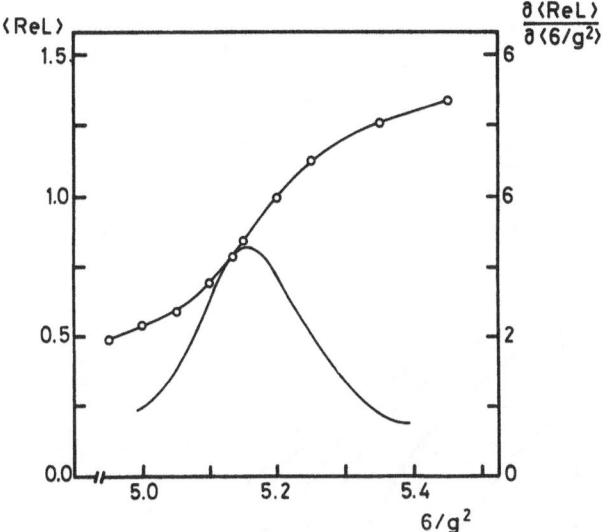

Figure 10 : < Re L > and its derivative vs. $6/g^2$ at $\mu a = 0.488$.

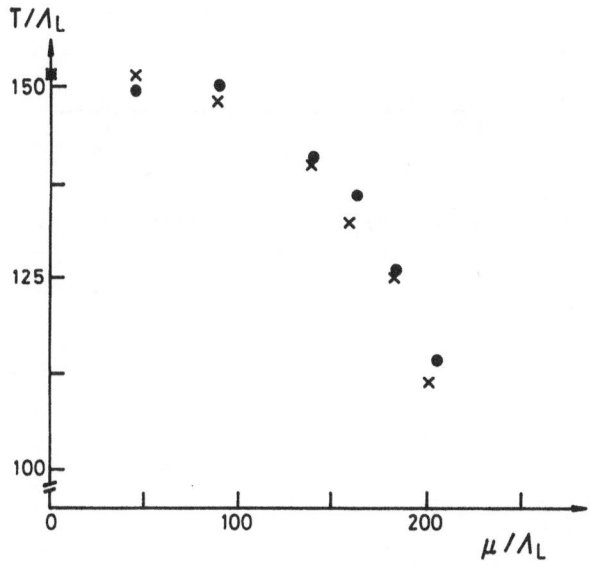

Figure 11 : Phase diagram for deconfinement (x) and chiral symmetry restoration (•).

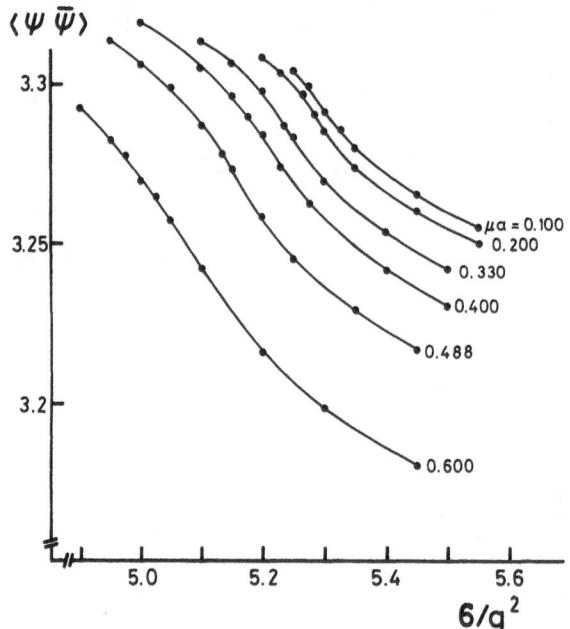

Figure 12 : $<\bar{\psi}\psi>$ vs. $6/g^2$ at different μa .

In fig. 5, we now show the deconfinement measure $<$ Re L $>$ as function of $6/g^2$ for different μa values. There is a clear shift of the deconfinement point to lower $6/g^2$, i.e., to lower temperatures, as the baryonic chemical potential increases. At the same time, the change in regimes becomes less abrupt with growing μa. - In fig. 6, we show the onset of deconfinement obtained at fixed temperature by increasing μa. It is seen that a variation of μ results in a behaviour quite similar to that obtained for a variation in T, so that deconfinement can indeed be induced either way.

The behaviour of the overall energy density,

$$\varepsilon \equiv \{T^2(\partial \ln Z / \partial T) + \mu \, T(\partial \ln Z / \partial \mu)\} / V \qquad (37)$$

calculated for different μa values as function of coupling $6/g^2$ and temperature T/Λ_L, is shown in fig. 7. It is normalized here to the value $\varepsilon_{SB}^{K^4}(8^3 \times 3)$ for an ideal gas of quarks and gluons, also calculated on an $8^3 \times 3$ lattice in 4th order hopping parameter expansion. The lattice evaluation procedure of both ε and $\varepsilon_{SB}^{K^4}(8^3 \times 3)$ is described in ref. 16. - We note in fig. 7 again a rapid deconfinement transition, becoming slightly "softer" with increasing μa. In fig. 8, we show the corresponding behaviour as function of μ at a fixed value of T.

We can now use either the ε results or those for $<$ Re L $>$ to determine the transition parameters. At T_c, $c_V \sim (\partial \varepsilon / \partial T)$ should become singular; $<$Re L $>$ should become exponentially small there. In fig. 9, we show as an example the behaviour of ε/T^4 vs. $6/g^2$ together with its derivative. The critical couplings thus obtained, together with the resulting critical temperature, are listed in table I. Also listed there are the corresponding points of maximum variation of $<$ Re L $>$, and the temperature obtained from them; an illustration of this functional behaviour is shown in fig. 10. It is seen that the two determinations of $T_c(\mu)$ agree quite well, leading to the variation of T_c with μ as shown in fig. 11. At the highest value of the baryonic chemical potential studied here (μ/T ≈ 1.8 or $\mu/\Lambda_L \cong 200$), the critical temperature has dropped by about 25%:

$$T_c \, (\mu = 0) \, / \, T_c \, (\mu / \Lambda_L \simeq 200) \approx 0.74 \qquad . \qquad (38)$$

Finally, we want to consider chiral symmetry restoration and its relation

Table I : Critical parameters for deconfinement and chiral symmetry restoration, as obtained on an $8^3 \times 3$ lattice

μa	μ/Λ_L	$6/g_c^2$ (from ϵ)	T_c/Λ_L (from ϵ)	$6/g_c^2$ (from ReL)	T_c/Λ_L (from ReL)	$6/g_{CH}^2$ (from $\bar{\psi}\psi$)	T_{CH}/Λ_L (from $\bar{\psi}\psi$)
0.1	45	5.299	151	5.288	149	5.290	150
0.2	89	5.282	148	5.294	150	5.292	150
0.33	139	5.239	140	5.239	140	5.243	141
0.4	161	5.194	132	5.214	136	5.218	136
0.488	184	5.151	125	5.158	126	5.156	126
0.6	203	5.063	112	5.077	114	5.081	114

to deconfinement. As is known, the Wilson fermion formulation is not ideal for this purpose, since the chiral symmetry measure $<\bar{\psi}\psi>$ never vanishes on a finite lattice. Nevertheless, for $\mu = 0$ it is found to show a rapid variation presumably related to the onset of chiral symmetry restoration[16], and it therefore appears meaningful to study the μ-dependence of this variation. The results are shown in fig. 12; we see again a clear shift to lower turning point values of $6/g^2$ with increasing μa . Defining again the critical parameter as that given by the point of maximum variation, we obtain for $<\bar{\psi}\psi>$ the values shown in table I. They are seen to agree very well with those obtained for the deconfinement point. We therefore conclude that our results provide up to $\mu a = 0.6$ a common point of deconfinement and chiral symmetry restoration. In fig. 11, we have the resulting phase diagram for both deconfinement and chiral symmetry restoration, as it emerges form the transition parameters listed in table I.

REFERENCES

1) See e.g. H. Satz, Ann. Rev. of Nucl. and Part. Sci. 35 : 245 (1985)

2) J. Kogut et al., Nucl. Phys. B225 [FS9] : 93 (1983)

3) J. Engels and H. Satz, Phys. Lett. 159B : 151 (1985)

4) V.V. Dixit and E. Suhonen, Z. Phys. C18 : 355 (1983)

5) J. Polonyi et al., Phys. Rev. Lett. 53 : 644 (1984)

6) R.V. Gavai, M. Lev. and B. Petersson, Phys. Lett. 149B : 492 (1984)

7) F. Fucito, C. Rebbi and S. Solomon, Cal. Tech. Report CALT-68-1127 (1984)

8) T. Celik, J. Engels and H. Satz, Nucl. Phys. B256 : 670 (1985)

9) J. Cleymans, K. Redlich, H. Satz and E. Suhonen, in preparation

10) K. Wilson, in: New Phenomena in Subnuclear Physics, A. Zichichi, ed., Plenum Press, New York (1977)

11) P. Hasenfratz and F. Karsch, Phys. Lett. 125B : 308 (1983)

12) R.V. Gavai, Phys. Rev. D32 : 519 (1985)

13) N. Kawamoto, Nucl. Phys. B190 [FS3] : 6171 (1981)

14) See e.g. F. Karsch, in: Quark Matter Formation and Heary Ion Collisions, M. Jacob and H. Satz, ed., World Scientific, Singapore (1982)

15) R.V. Gavai and A. Ostendorf, Phys. Lett. 132B : 137 (1983)

16) T. Celik, J. Engels and H. Satz, Nucl. Phys. B256 : 670 (1985)

MONTE CARLO RENORMALIZATION GROUP: A REVIEW‡

Rajan Gupta†

MS-B276, Los Alamos National Laboratory
Los Alamos, N.M. 87545

ABSTRACT

The logic and the methods of Monte Carlo Renormalization Group ($MCRG$) are reviewed. A status report of results for 4-dimensional lattice gauge theories derived using $MCRG$ is presented. Existing methods for calculating the improved action are reviewed and evaluated. The Gupta-Cordery improved $MCRG$ method is described and compared with the standard one.

The development of Monte Carlo Renormalization group method ($MCRG$) was essentially complete in 1979 with the work of Wilson[1], Swendsen[2] and Shenker and Tobochnik[3]. Prior to this Ma[4] and Kadanoff[5] had provided key ingredients. The method is therefore relatively new, furthermore its application to field theories has been carried out only since 1982. In this short period there has been considerable activity and I shall review the methodology and summarize the status with emphasis on 4-dimensional gauge theories. There already exists extensive literature on $MCRG$ and I direct the reader to it[1,3,6,7] for details and for a wider exposure. Similarly, the reviews[8,9] are a good starting point for background on Lattice Gauge Theories and on spin systems. The topics I shall cover are

1) Introduction to $MCRG$ and its methodology.
2) Renormalization Group Transformations for $d = 4$ lattice gauge theories.
3) U(1) Lattice Gauge theory.
4) β-function and Scaling for SU(3) Lattice Gauge Theory.
5) Improved Actions and Methods to calculate them.
6) Improved Monte Carlo Renormalization Group.
7) Effective Field Theories.

The main results in QCD from $MCRG$ are the determination of the β-function

---

‡ Invited Talk given at the Nov. 1985 Wuppertal Conference on: *Lattice Gauge Theories —— A Challange in Large Scale Computing.*

† *J. Robert Oppenheimer Fellow*

and the consequent prediction for the value of the coupling at which asymptotic scaling sets in and second an estimate of the improved gauge action[10]. These results are not spectacular in the sense of confirming that QCD is the correct theory of strong interactions, however they have led to a deeper understanding of the lattice theory and provided a quantitative estimate of the approach to the continuum limit. I shall attempt to show that this method is as yet in its infancy and should be used to tackle a number of problems.

1) INTRODUCTION TO *MCRG*

Renormalization Group[11,12,13] (RG) is a general framework for studying systems near their critical point where singularities in thermodynamic functions arise from coherence at all length scales. This phenomenon occurs in Statistical Mechanics near and on the critical surface (defined by a divergent correlation length) and in the strong interactions of quarks and gluons. The *MCRG* method was developed to handle this problem of infinitely many coupled degrees of freedom so that sensible results can be obtained from finite computers. There are two central ideas behind *MCRG*: One is to average over these infinitely many degrees of freedom in discreet steps preserving only those which are relevant for the description of the physical quantities of interest. The interaction between these averaged (block) fields is described by an infinite set of couplings that get renormalized at each step. In QCD this discrete reduction is carried out until the correlation length is small enough so that the system can be simulated on a lattice with control over finite size effects. The second is that there are no singularities in the coupling constant space even though the correlation length diverges on the critical surface and that the fixed point is short ranged. Thus even though there are an infinite number of couplings generated under renormalization, only a few short range ones are necessary to simulate the system at a given scale and preserve the long distance physics. Present results suggest that the fixed point for QCD is short ranged.

Standard Monte Carlo: Consider a magnetic system consisting of spins $\{s\}$ on the sites of a $d - dimensional$ lattice L described by a Hamiltonian H with all possible couplings $\{K_\alpha\}$. All thermodynamic quantities can be found from a detailed knowledge of the partition function

$$Z = \sum e^{-H} = \sum e^{K_\alpha S_\alpha} \qquad (1.1)$$

where S_α are the interactions. In Monte Carlo, configurations of spins on the original lattice are generated by the Metropolis[14], heat bath[15], molecular dynamics alias Microcanonical[16] or the Langevin[17,18] algorithm with a Boltzmann distribution $e^{-H} \equiv e^{K_\alpha S_\alpha}$. All thermodynamic quantities are given as simple averages of correlation functions over these weighted configurations. The accuracy of the calculations depend on the size of the statistical sample and the lattice size L used. Both these quantities depend on the largest correlation length ξ in the system. Near the critical temperature, T_c, associated with second order phase transitions, the correlation length and consequently thermodynamic quantities like the specific heat *etc* diverge as functions of $(T-T_c)$ with universal critical exponents that have been calculated for many systems either analytically or by the Monte-Carlo / *MCRG* method. Because of a diverging ξ, long runs are needed to counter the critical slowing down and the lattice size has to maintained at a few times ξ. The problem of critical slowing down is addressed by analyzing update algorithms (Metropolis vs. heat bath vs. Microcanonical vs. Langevin with acceleration techniques like multi-grid[19], fourier acceleration[18,20] etc). The optimum method is, of course, model dependent and has to take care of metastability (local versus global minima) and global excitations like vortices, instantons etc that are not efficiently handled by local changes. This last feature has not received adequate attention. To control the second problem in standard Monte Carlo, effects of a finite lattice especially as $\xi \to \infty$, finite size scaling has been used with success. In this review I shall concentrate on $MCRG$. First I shall describe how universality and scaling are explained by the renormalization group.

The renormalization group transformation (RGT) $H^1 = R(H)$ is an operator defined on the space of coupling constants, $\{K_\alpha\}$. In practice the RGT is a prescription to average spins over a region of size b, the scale factor of the RGT, to produce the block spin which interacts with an effective theory H^1. The two theories H and H^1 describe the same long distance physics but the correlation length in lattice units $\xi \to \frac{\xi}{b}$. If this RGT has a fixed point H^* such that $H^* = R(H^*)$, then clearly the theory is scale invariant there and ξ is either 0 or ∞. An example of a fixed point with $\xi = 0$ is $T = \infty$ and these are trivial. The interesting case is $\xi = \infty$ about which the theory is governed by a single scale ξ. If this fixed point is unstable in 1 direction only (this direction is called the Renormalized Trajectory (RT)), then non-critical H will flow away from H^* along trajectories that asymptotically converge to the RT. Thus the long distance physics of all the trajectories that converge is identical and is controlled by the RT. Similarly, points ϵ away from H^* on the $\infty - 1$ dimension hypersurface at which $\xi = \infty$ (the critical surface) will converge to H^*. The fact that the fixed point with its associated RT control the behavior of all H in the neighborhood of H^* is universality. Next, consider a non-critical H that approaches H^* along the RT. Thermodynamic quantities depend on a single variable *i.e.* distance along the RT. This is scaling. Corrections to scaling occur when H does not lie on the RT. These are governed by the irrelevant eigenvalues of the RGT which give the rate of flow along the critical surface towards H^* and for H not on the RT, the rate of convergence towards it. The relevant eigenvalue gives the rate of flow away from the fixed point (along the unstable direction RT) and is related to the critical exponent ν. This terse exposé ends with a word of caution; all these statements have validity close to H^*.

In the $MCRG$ method, configurations are generated with the Boltzmann factor $e^{K_\alpha S_\alpha}$ as in standard Monte Carlo. The RGT, $P(s^1, s)$, is a prescription for averaging variables over a cell of dimension b. The blocked variables $\{s^1\}$ are defined on the sites of a sublattice L^1 with lattice spacing b times that of L. They interact with undetermined couplings K_α^1, however the configurations are distributed according to the correct Boltzmann factor e^{-H^1} *i.e.*

$$e^{-H^1(s^1)} = \sum P(s^1, s)\, e^{-H(s)} \tag{1.2}$$

so expectation values can be calculated as simple averages. The RGT should satisfy the Kadanoff constraint

$$\sum^1 P(s^1, s) = 1 \tag{1.3}$$

independent of the state $\{s\}$. This guarantees that the two theories H and H^1 have the same partition function. The blocking is done n times to produce configurations with hamiltonians H^n describing the same long distance physics but on increasingly coarser lattices. The fixed point H^*, the RT and the sequence of theories, H^n, generated from a given starting H depend on the RGT. Many different RGT can be used to analyze a given model (determine the universal exponents) and I defer discussion on how to evaluate their efficiency to sections 2 and 5.

1.1) Methods to calculate the critical exponents.

There are two methods to calculate the critical exponents from expectation values calculated as simple averages over configurations. In both there is an implicit assumption that the sequence H^n stays close to H^*. The more popular method is due to Swendsen[2,7] in which the critical exponents are calculated from the eigenvalues of the linearized transformation matrix $T^n_{\alpha\beta}$ which is defined as

$$T^n_{\alpha\beta} = \frac{\partial K^n_\alpha}{\partial K^{n-1}_\beta} = \frac{\partial K^n_\alpha}{\partial \langle S^n_\sigma \rangle} \frac{\partial \langle S^n_\sigma \rangle}{\partial K^{n-1}_\beta} \quad . \tag{1.4}$$

Each of the two terms on the right is a connected 2-point correlation function

$$\frac{\partial \langle S^n_\sigma \rangle}{\partial K^{n-1}_\beta} = \langle S^n_\sigma S^{n-1}_\beta \rangle - \langle S^n_\sigma \rangle \langle S^{n-1}_\beta \rangle . \tag{1.5}$$

and

$$\frac{\partial \langle S^n_\sigma \rangle}{\partial K^n_\beta} = \langle S^n_\sigma S^n_\beta \rangle - \langle S^n_\sigma \rangle \langle S^n_\beta \rangle . \tag{1.6}$$

Here $\langle S^n_\sigma \rangle$ are the expectation values on the nth renormalized lattice and K^n_σ are the corresponding couplings. The exponent ν is found from the leading eigenvalue λ_t of $T^n_{\alpha\beta}$ as

$$\nu = \frac{\ln b}{\ln \lambda_t} \tag{1.7}$$

where b is the scale factor of the RGT. The eigenvalues less than one give exponents that control corrections to scaling. The accuracy of the calculated exponents improves if they are evaluated close to the fixed point. This can be achieved by starting from a critical point and blocking the lattice a sufficient number of times *i.e.* for large n. Thus the convergence is limited by the starting lattice size and how close the starting H^c is to H^*. If H^* can be approximated by a small number of short range couplings (a necessary assumption in the RG), then this method can be improved if the renormalized couplings $\{K^n\}$ are determined starting from a known critical Hamiltonian. These should then be used in the update. A second possibility is to tune the RGT so that the convergence to H^* from a starting H^c is improved. In section 5, I will describe a number of methods to calculate the renormalized couplings. Tuning of the RGT is discussed in section 2.5 and a careful analysis of the accuracy of this method is deferred until section 6.

The second method to calculate the leading relevant exponent is due to Wilson[6]. Consider once again the 2-point connected correlation function (the derivative of an expectation value) $\langle S^i_\alpha S^j_\beta \rangle_c$ with $j > i$. Expand S^i_α in term of the eigenoperators O^i_α of the RGT. Close to H^* the level dependence in O^i_α (equivalently in the expansion coefficients $c^i_{\alpha\beta}$) can be neglected. Then to the leading order

$$\langle S^i_\alpha S^j_\beta \rangle \sim \lambda^{j-i}_t \, c_{\alpha,t} \langle O_t S^j_\beta \rangle \tag{1.8}$$

where λ_t is the leading relevant eigenvalue and corrections are suppressed by $(\frac{\lambda}{\lambda_t})^{j-i}$. Thus for each α and β, the ratio $\frac{\langle S^i_\alpha S^j_\beta \rangle}{\langle S^{i+1}_\alpha S^j_\beta \rangle}$ gives an estimate for the leading eigenvalue

λ_t. The accuracy of the method improves if $j - i$ is large since non-leading terms are suppressed geometrically. So far this method has not been used extensively so its practical accuracy cannot be evaluated.

QCD: At the tree level, the coupling g in QCD does not renormalize and the fixed point is at $g_{bare} = 0$. At 1-loop the leading operator has eigenvalue equal to one, is relevant and the fixed point changes from simple gaussian to being asymptotically free and non-trivial. A special feature of asymptotic freedom is that even when the leading eigenvalue is one there is a flow away from the fixed point at a constant rate. At 2-loop, this operator becomes truely relevant *i.e.* with eigenvalue > 1. Perturbation theory also tells us that leading scaling violations are $\sim \frac{1}{k^2}$, so the second eigenvalue should be $\sim \frac{1}{b^2}$ for a RGT with scale factor b. Present studies[22b] show that the leading eigenvalue is close to 1 and the second near $\frac{1}{b^2}$. However, the statistics are poor and the calculation was done at large g_{bare}. Thus reliable quantitative results are lacking.

1.2: Wilson's method to find a critical point

Consider $MCRG$ simulations L and S with the same starting couplings K_α^0 but on lattice sizes $L = b^n$ and $S = b^{n-1}$. If K_α^0 is critical and after a few blockings the 2 theories are close to H^*, then all correlation functions attain their fixed point values. For non-critical starting H, expand about H^* in the linear approximation

$$\langle L_\alpha^m \rangle - \langle S_\alpha^{m-1} \rangle = \frac{\partial}{\partial K_\beta^0} \{\langle L_\alpha^m \rangle - \langle S_\alpha^{m-1} \rangle\} \Delta K_\beta^0$$

$$= \{\langle L_\alpha^m L_\beta^0 \rangle_c - \langle S_\alpha^{m-1} S_\beta^0 \rangle_c\} \Delta K_\beta^0 \qquad (1.9)$$

to determine ΔK_α^0. To reduce finite size effects the compared expectation values are calculated on the same size lattices. The critical coupling is given by

$$K_\alpha^c = K_\alpha^0 - \Delta K_\alpha^0 \qquad (1.10)$$

and this estimate should be improved iteratively.

On the critical surface the 2-point correlation functions (like in Eq. (1.5) and (1.6)) diverge in the thermodynamic limit. However, their ratio is the rate of change of couplings and these are well behaved. The reason $MCRG$ has better control over finite size effects is that if H^* is short ranged then only short ranged correlation functions need to be evaluated in determining $T_{\alpha\beta}^n$ or in Eq. (1.9). The finite size contributions to the ratios fall off like the couplings *i.e.* exponentially. Thus reliable estimates are obtained from small lattices.

2: RENORMALIZATION GROUP TRANSFORMATIONS IN $d = 4$

It has been mentioned before that there is no unique RGT for a given model. There are at present four different transformations that have been proposed for 4-dimensional lattice gauge theories. In each of them the block link variable is constructed from a sum of paths $\Sigma \equiv \sum paths$. This sum of SU(N) matrices

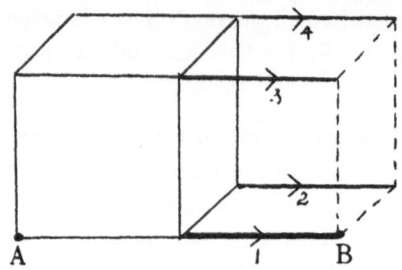

Figure 1: Wilson's b = 2 RGT. Four of the eight paths in a given direction are shown.

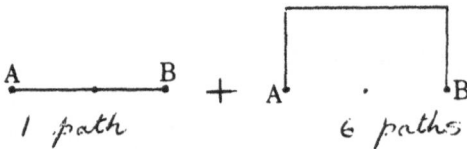

Figure 2a: Swendsen's b = 2 RGT.

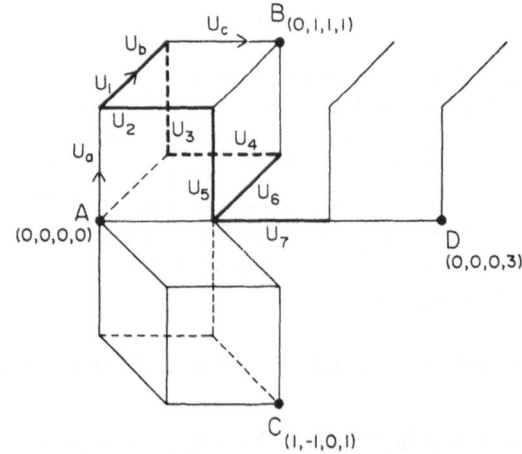

Figure 2b: Generalized Swendsen RGT with parameters α_i that have to be optimized.

4-DIMENSIONAL HYPERCUBIC LATTICE

Figure 3: The geometry of the $\sqrt{3}$ block transformation.

is not an element of SU(N), and the new block link matrix is selected with the distribution

$$P(U_b) = e^{p \, Tr \, U_b \Sigma} . \qquad (2.1)$$

where p is a free parameter to be optimized. The advantage of taking the sum is that such a RGT preserves gauge invariance. The 4 RGT are (in cronological order)

2.1) $b = 2$ **by Wilson**[1]: The geometry of the transformation is shown in Fig. 1. There are 8 links in a given direction of which 4 are shown in the 3-dimensional projection. In this method the gauge has to be fixed on 15 sites other than the block site. This fixing has to take into account the fact that the ends of the 8 links are at different sites. The ansatz Wilson used was to transform the hypercube locally into the Landau gauge. The process of fixing the gauge is slow and a disadvantage of the method. The need for gauge fixing can be avoided by defining 8 paths that run between the block sites and include the same links. This modified construction violates cubic rotational invariance because of the particular choice of the ordering of the paths within the cell. In both forms only $\frac{32}{49}$ degrees of freedom are used in this approximate averaging at each level. This method has not been used since Wilson's preliminary investigation because the next two methods are simpler.

2.2) $b = 2$ **by Swendsen**[21]: The transformation in its initial form is shown in Fig. 2a. The more general version is shown in Fig. 2b where the parameters α_i have to be determined. In this construction all paths start and end at the block sites. Thus no gauge fixing is necessary and arbitrarily complex paths can be included. However calculations show that an optimization of the parameters has to be done to improve the convergence. I shall discuss this tuning later.

2.3) $b = \sqrt{3}$ **by Cordery, Gupta and Novotny**[22]: This transformation is specific to gauge theories in 4-dimensions and is based on the fact that the body diagonals of the 4 positive 3-cubes out of a site are orthogonal and of length $\sqrt{3}$. The geometry is shown in Fig. 3 and under one RGT the new lattice is still hypercubic but rotated with respect to the old basis. Also, the box boundary becomes jagged. This can be undone by a second application of the RGT with different basis vectors. So the original box geometry is recovered after every scale change by a factor of 3. The construction of the paths requires no gauge fixing, all paths are of equal length (no free parameters to be tuned) and $\frac{24}{28}$ degrees of freedom are used at each step. Further, the block cell consists of the block site and its 8 nearest neighbors. This provides an easy and natural way to include complex matter fields and block them simultaneously. It is also better suited to the fermion block diagonalization process of Mutter and Schilling[64] as is explained in section 5.10. In practice, for both SU(2) and SU(3), this RGT has consistently shown good convergence at strong and at weak coupling. It is therefore recommended.

2.4) $b = \sqrt{2}$ **by Callaway and Petronzio**[23]: The construction of paths shown in Fig 4a is based on a planer structure *i.e.* $x - y$ and $z - t$ planes are treated separately at all blocking steps. No gauge fixing is required but only 2 paths are used in the averaging *i.e.* in Eq. (2.1). This drawback of using only 2 planar paths can be improved by including nonplanar paths as shown in Fig. 4b. Because this RGT has the advantage that $b = \sqrt{2}$ is the smallest scale factor possible, a serious test should be made.

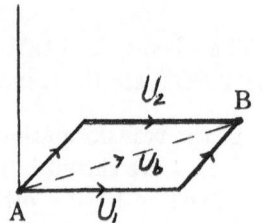

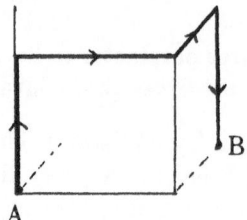

Figure 4a: The two paths in the
b = $\sqrt{2}$ RGT.

Figure 4b: Additional 4 link paths.

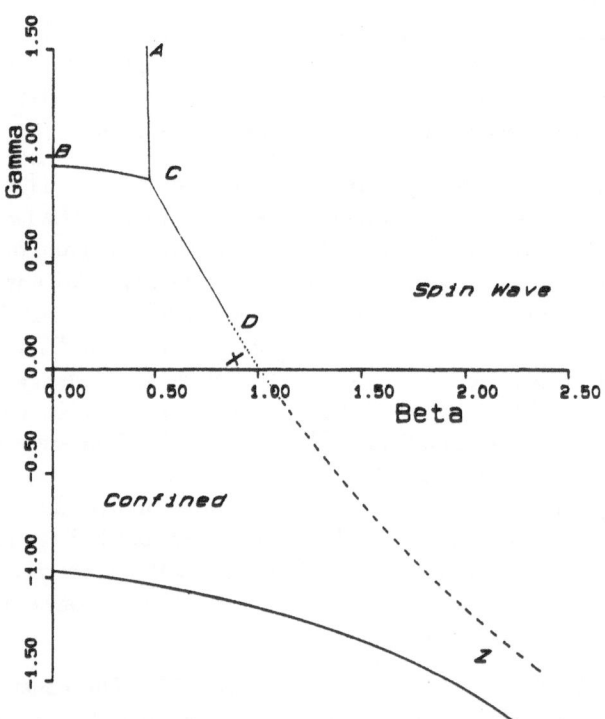

Figure 5: The phase diagram of U(1) gauge theory. The location of the
TCP is somewhere near the dotted line DX.

2.5) Optimization of the RGT: In addition to the freedom of the choice of the RGT, there are the free parameters p and α_i. Hasenfratz et al.[24] have shown that the convergence of the original $b = 2$ Swendsen transformation is improved if p is tuned. I will give a qualitative description of how this works. Consider a set of RGT that are a function of the continuous parameter p i.e. R_p. Starting from a given point H, the blocked theories generated are described by $H^1(p)$. They all have the same long distance behavior as can be checked by measuring expectation values of large Wilson loops. In fact there is an effective Wilson action H_{eff} which will have the same long distance behavior. The short distance behavior of $H(p)$ will be different and for some values of p, the $\langle plaq \rangle_p$ will be larger than the $\langle plaq \rangle_w$ corresponding to H_{eff}. I have checked that this is the case for the original Swendsen transformation when $p = \infty$ and $g^2 < 1$. Lowering p reduces the blocked $\langle plaq \rangle_p$, making it agree better with H_{eff}. Thus, the tuning makes the short and long distance behavior correspond better to the same approximate H_{eff}. This improves the matching (using small loops) in the 2-lattice method to calculate the β-function. Hasenfratz et al.[24] estimate p using perturbation theory and by Monte Carlo using the criterion of early matching of block expectation values in Wilson's two lattice method. They found that the best value at $\frac{6}{g^2} = 6.$ given by Monte Carlo (~ 35) does not agree with the value found using perturbation theory (~ 15). So as of now this optimization is still by trial. Also, p_{opt} depends on the coupling g. This implies that the RT cannot be pulled close to the Wilson axis globally by this optimization. So the usefulness of such optimization is limited to the β-function calculation. The parameters α_i can similarly be optimized using the same improvement criterion.

Gupta and Patel[22] used $p = \infty$ in the $\sqrt{3}$ RGT. This is equivalent to choosing the matrix U such that $Tr\, U \sum$ is maximized (the δ-function construction). They find that even with this choice the small block Wilson loops are more disordered than for an H_{eff} determined using large loops. Thus lowering p would not help. The $\sqrt{3}$ RGT has shown good convergence properties and provided reliable results with $p = \infty$.

The freedom to choose the RGT and further tune the parameters α_i and p leads to the question: What are the criteria by which to decide what is the best RGT? I will first address the question -- what is the effect of changing the RGT on the fixed point and on the RT? Conjecture[25]: Changing the RGT moves the fixed point on the critical surface but only along redundant directions. A simple argument is as follows: Consider two different RGT, R_1 and R_2, and their associated fixed points H_1^* and H_2^*. There are no non-analytic corrections to scaling at either fixed points and the associated RT. If these two points are distinct, then under R_1 H_2^* flows to H_1^*. Consequently there are no scaling violations along the flow. This is by definition a redundant direction. This implies that the associated RT differ by redundant operators.

The presence of redundant operators does not effect the physics, however it can obscure results. The redundant eigenvalues are not physical, depend on the RGT, and can be relevant or irrelevant. If a relevant redundant operator is present then the flows will not converge to the H^* or to the RT. Thus it is desirable to pick a RGT for which the redundant eigenvalues are small. Similarly, the coefficients of the leading irrelevant operators should be reduced. To some extent the irrelevant basis vectors are a function of the position of H^*, so it is possible to simultaneously reduce the two coefficients. In QCD, there is an additional freedom -- all possible

Wilson loops form an overcomplete set. Therefore, in order to tune the RGT and to find an efficient improved action, it is necessary to determine the operators that can be eliminated because of the overcompleteness and the redundant combinations.

Swendsen[26] has conjectured that the fixed point can be moved anywhere on the critical surface by tuning the RGT. In particular, if the simulation point is made H^*, then that RGT is optimal. There is some support for this in spin systems, where by adding terms to the RGT, one can successively kill terms in the renormalized hamiltonian. There are two things to check here: first whether the coefficients of the RGT terms fall off like the couplings, *i.e.* exponentially, and second whether the long range untuned couplings continue to fall off at least as fast as before. The quantity to optimize is the update complexity (embodied in the RGT or the hamiltonian) versus the decrease in the coefficient of the leading irrelevant operator. Swendsen[26] found that the eigenvalues for the $d = 3$ Ising model are significantly improved with a tuned 10 term RGT. However, he did not compare it with a simulation that used a 10 term truncated renormalized hamiltonian close to the H^* for a simple RGT. There is one additional anomaly in this approach: Tuning the RGT improved the thermal exponent but the results for the magnetic exponent deteriorated in quality. This is surprising because the fixed point is at zero odd couplings and these remain unchanged in tuning the RGT. The previous conjectures are in conflict and the results are ambiguous. Consequently, this subject is being explored[69] further.

The criterion for an optimum RGT is to make the H^* and the RT as short ranged as possible. In critical phenomena, the improvement can be quantified by measuring the convergence of the exponents as a function of the blocking level. In QCD we are interested in continuum mass-ratios *etc*. These have so far been hard to measure so the improvement cannot be judged. The behavior of the RT for QCD is discussed at the end of section 5. For the moment let me conclude this section by: The the question of how best to optimize $MCRG$ has not been adequately answered and is under investigation.

3: U(1) LATTICE GAUGE THEORY

The phase diagram of the theory defined by the action

$$S = \beta \sum \cos\Theta_{\mu\nu} + \gamma \sum \cos 2\dot{\Theta}_{\mu\nu} \qquad (3.1)$$

where β (γ) is the charge 1 (charge 2) coupling is known to have a phase boundary separating the confining (strong-coupling) phase from the spin-wave (QED) phase[27,28,29]. The order of the transition along the boundary DXZ in Fig. 5 is not known. In particular it is not known if the gradually weakening first order transition along CD ends in a tricritical point, and if so what is its location. Evertz *et al.*[28] claim that the location of the TCP is at $\beta = 1.09\pm0.04$ and $\gamma = -0.11\pm0.05$ on basis of a scaling analysis of the discontinuity in the energy ΔE. The mechanism driving the transition are topological excitations[30,31], *i.e.* closed loops of monopoles, whose density is observed to change at the transition[32,33]. This change in density is caused by a growth in the size of the largest monopole loop which begins to span the finite lattices used in the calculations[32,34]. Thus, the usual difficulty of finite size effects near a TCP in determining the location of the TCP by an extrapolation of the latent heat ΔE along the phase boundary is here compounded by the presence of monopole current loops that are closed due to the lattice periodicity[32,34]. These contribute a fake piece to the ΔE which makes the extrapolation unreliable. One solution is to calculate and then subtract the contribution of these loops from ΔE before making the extrapolation. The more reliable method is $MCRG$ and in particular the 2-lattice method discussed in section 1.2 should be used to locate the TCP. A word of caution for the U(1) model when using this method: There is a large shift in the critical coupling as a function of the lattice size[32] and consequently in the contribution of the fake monopole loops. One should therefore use a starting coupling for which both lattice simulations are on the same side of the transition.

The present status is that in a $MCRG$ calculation done along the Wilson axis[32] only one relevant exponent was found using the $\sqrt{3}$ RGT. Furthermore, the value of the exponent showed a variation with β. At $\beta = 1.0075$, $\nu \approx 0.32$ and this value changes to $\nu \approx 0.43$ (or even the classical value 0.5) at $\beta = 1.01$. One explanation is that the TCP lies above the Wilson axis and in simulations along the Wilson axis one first measures the tricritical exponent and then the critical one. The same conclusion is also reached in a recent $b = 2$ $MCRG$ study[35]. Therefore the location of the TCP is still an open question.

The interest in this model (goal of MC calculations) is to know if there exists a fixed point at which a non-trivial field theory can be defined. To settle this important question requires considerably more work.

4: β-FUNCTION AND SCALING FOR SU(3) LATTICE GAUGE THEORY

The non-perturbative β-function tells us how the lattice spacing goes to zero as $g_{bare} \to 0$. Since on the lattice all dimensionful quantities, like masses, are measured in units of the lattice spacing a, we need to know how a scales in order to take the continuum limit. One option is to use the 2-loop perturbative result provided it is demonstrated that this is valid at values of g_{bare} where the calculations are done. The other is to measure the non-perturbative β-function. In case there is only asymptotic scaling, this calculation is still necessary since it provides the value of g_{bare} at which such scaling sets in.

There are two methods for calculating the non-perturbative β-function directly.

4.1) $MCRG$ using Wilson's 2 lattice method[1,3]: There are 2 groups who have used this method for SU(3); one with $b = \sqrt{3}$ RGT [37] and the second[24] with $b = 2$ proposed by Swendsen[21]. The outline of the method is: First a system of size $L = (b^n)^d$ is simulated with couplings K_α^A and the expectation values of Wilson loops are calculated on the original lattice and the n block lattices. A second system of size $S = (b^{n-1})^d$ is then simulated with couplings K_α^B (chosen judiciously) and again the expectation values are calculated on the n lattices. The expectation values from the two simulations are compared with the ones from the larger lattice L blocked one more time $i.e.$ L^m with S^{m-1}. Finite size effects are minimized since the comparison is on approximately the same physical size lattices. The couplings K_α^B are adjusted (which requires a new simulation) until there is matching at the last, n^{th}, level. In practice it is sufficient to do two simulations S_1 and S_2 which bracket L and then use interpolation. The test for convergence of the two theories L^m and S^{m-1} is that the expectations values should match simultaneously at the last few levels. This situation is shown in the coupling constant space in Fig. 6. At matching, the correlation length at K_α^A is larger than at K_α^B by the scale factor b. Thus if the starting trajectory is taken to be the Wilson axis (or any 1 parameter line) then the value of the β-function, $\Delta\beta$, for a scale change b is $K^A - K^B$.

Under the assumption that the fixed point action is local, and that at any scale a few short range couplings are sufficient to characterize the action, matching the expectation values of a few small Wilson loops is sufficient to guarantee that the two actions are equal. Recall that there is a one to one correspondence between the value of the couplings and the expectation values. Also note that finite size effects in expectation values are irrelevant once there is matching because then the two theories flow along a common trajectory under a RGT and continue to match. Thus it is sufficient to require that matching first take place on lattices which are large enough to accommodate the important couplings. Thereafter, the check can be on a 1^4 lattice too! It is the range of the couplings that controls finite size effects in $MCRG$ and not the correlation length and this range falls off exponentially even on the critical surface. This is why $MCRG$ has better control over finite size effects and is a powerful method.

For the simple plaquette SU(3) action with $K_F \equiv \frac{6}{g^2}$, asymptotic scaling is defined by the 2-loop perturbative β-function,

$$\frac{\partial(g^{-2})}{\partial(lna)} = -\frac{11}{8\pi^2} - \frac{51}{64\pi^4}g^2 + \cdots . \tag{4.1}$$

The quantity calculated using $MCRG$ is,

$$\Delta\beta = -\frac{\partial(6g^{-2})}{\partial(lna)} \cdot lnb , \tag{4.2}$$

i.e., the discrete β-function at K_F evaluated for a scale change b.

The results for the $b = \sqrt{3}$ calculation[37] are shown in Table 1, while those for $b = 2$ are shown[24] in Table 2. There is clear evidence of a dip at $\frac{6}{g^2} \sim 6.0$ which is caused by the end point of the phase transition line in the fundamental-adjoint coupling space. The conclusion of these calculations is that there is no asymptotic scaling below $\frac{6}{g^2} = 6.1$.

4.2) Loop ratio method[38,24]: This method is simpler as it uses expectation values of Wilson loops calculated in standard Monte Carlo. Thus it can be used for gauge theory with dynamical fermions while method 4.1 cannot until one learns how to block fermions. The ratios of Wilson loops that cancel the perimeter and corner terms

$$R(i,j,k,l) = \frac{W(k,l)}{W(i,j)} \qquad where \ \ i+j = k+l . \tag{4.3}$$

satisfy an approximate homogeneous renormalization group equation

$$R(2i, 2j, 2k, 2l, g_a, 2L) = R(i, j, k, l, g_b, L) . \tag{4.4}$$

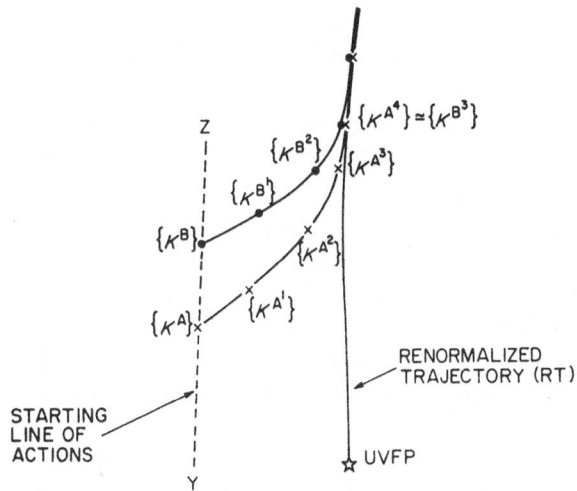

Figure 6: The evolution of actions under the renormalization group transformation. The two actions $\{K^A\}$ and $\{K^B\}$ have the same long-distance behavior and their lattice correlation lengths are related by the scale transformation factor b.

Table 1 : The values of $\Delta\beta$ for $b=\sqrt{3}$ at different levels of matching for different values of the couplings [37]. The matching K_F on $(3\sqrt{3})^4$ were determined by linear interpolation and the errors are based on a 1σ fit. Also shown are the values of $\Delta\beta$ corresponding to asymptotic scaling.

SU(3)			
9^4	$\Delta\beta$ for $b=\sqrt{3}$ from matching on		2-loop
K_F	3^4	$(\sqrt{3})^4$	1^4	$\Delta\beta$
6.0	.337(5)	.323(5)	.308(6)	.489
6.125	.387(5)	.376(5)	.351(6)	.488
6.25	.421(4)	.424(5)	.401(5)	.488
6.35	.431(4)	.452(5)	.445(9)	.487
6.45	.432(4)	.464(6)	.423(12)	.487
6.5	.435(4)	.464(6)	.449(15)	.487
6.75	.430(4)	.485(5)	.443(9)	.485
7.0	.422(7)	.503(11)	.488(20)	.484

Thus using Monte Carlo data for ratios calculated on 2 lattices of size $2L$ and L, with couplings g_a and g_b respectively, the desired function $\Delta\beta$ for $b = 2$ can be calculated. Caveats: Eq. (4.4) becomes correct only as $i \to \infty$, otherwise there are corrections due to lattice artifacts. The quality of results for large $i, j, \ldots$ are limited by statistics. The reliability of the results therefore depends on obtaining the same $\Delta\beta$ for $i = 1, 2, 3, 4, \ldots$.

The contribution of lattice artifacts can be reduced in perturbation theory. For this consider Eq. (4.4) for a linear combination of loop ratios with coefficients α_i. To determine the α_i, use the expectations values of loops calculated in perturbation theory and require that $\Delta\beta = 0$ (tree-level), 0.579 (1-loop) .... Then go back and use the monte carlo data for Wilson loops to calculate $\Delta\beta$. The drawback of this approach is that if two (or more) ratios of different scale, $i = 1$ and 4 say, are used then the difference in statistical errors is a problem. Otherwise, at weak coupling each ratio roughly satisfies Eq. (4.4) and there is a loss of sensitivity in determining α_i. At strong coupling, perturbation theory calculation/improvement of α_i breaks down. So one can, at best, expect a window where reliable results are obtained. Hasenfratz et $al.$[24] claim this is true for $\frac{6}{g^2}$ in the range [6,6.6]. Their results are in agreement with their $b = 2$ $MCRG$ results as shown in Table 2. It has been observed by Gutbrod[39] in SU(2) that stability with respect to loop size is reached slowly. Therefore, one has to be cautious of apparent convergence.

4.3) Results: It is hard to compare the results of the $b = \sqrt{3}$ study directly with the $b = 2$ ones because of the different scale factor of the RGT. Petcher[40] has carried out the following analysis: he fits the $b = \sqrt{3}$ data to a smooth function which had the correct asymptotic value built in. This function can then be used to determine the discrete change $\Delta\beta$ in the couplings for any other scale factor b. In Fig. 7 the smooth function found from the $\sqrt{3}$ data rescaled to $b = 2$ is compared with the $b = 2$ $MCRG$ data.

Next we would like to check if the $\Delta\beta$ calculated from MC determinations of different physical observables are identical and agree with the $MCRG$ calculations. This comparison tests two things, first whether there exists scaling (constant mass ratios) before (larger g) asymptotic scaling and second whether the MC measurements are reliable. The lattice value of a mass ma calculated at two values of the coupling, $\frac{6}{g_1^2}$ and $\frac{6}{g_2^2}$, gives the $\Delta\beta$ for a scale change $\frac{a_1}{a_2}$. Unfortunately the values of couplings cannot be selected to give the $\Delta\beta$ for a given constant scale change. This again introduces the problem of rescaling data thereby preventing a definite statement on scaling. In Fig. 8 we have only used pairs of data points with a scale factor close to $\sqrt{3}$. At $\frac{6}{g^2}$=6.0, the 0^{++} glueball mass[41], string tension σ[42] and the deconfinement temperature T_c [43] represent scales of 2,5 and 8 lattice units respectively. Thus identical $\Delta\beta$ would be a reasonable test of scaling. Bearing in mind the problem of rescaling data, the only significant statement is that the behavior of the glueball mass is different.

The onset of asymptotic scaling has also been checked by plotting $\frac{ma}{\Lambda a}$ where m is the deconfinement temperature T_c and Λ is the 2-loop perturbative scale. Kuti et $al.$[43] found that for $N_r = 10, 12, 14$ this ratio is constant and different from the value at $N_r \leq 8$. From this they deduce that there is asymptotic scaling for $\frac{6}{g^2} > 6.15$. The only drawback of this method is the reliance on Λ calculated in 2-loops to define asymptotic scaling. There could be corrections $i.e.$ $(1 + O(g^2))$ terms,

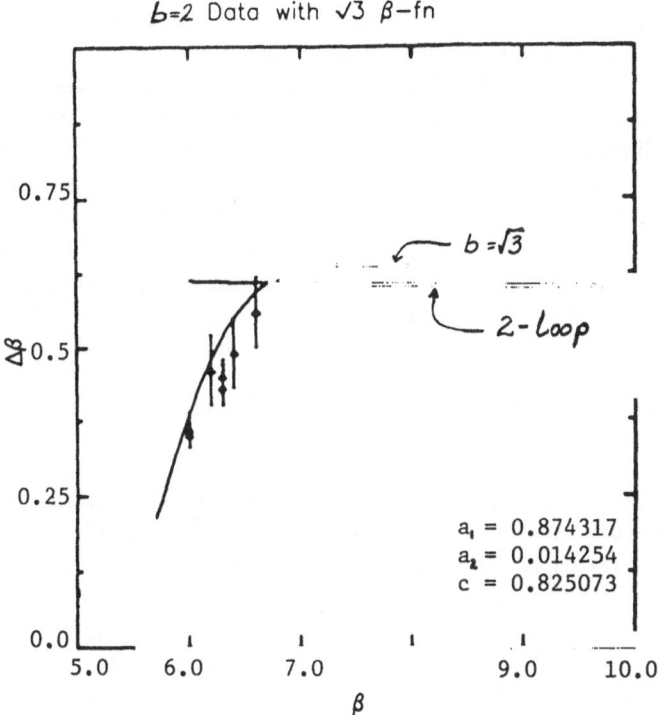

Figure 7: The smooth curve is a fit to the b = √3 data. The data points are for b = 2. (Courtesy of D. Petcher).

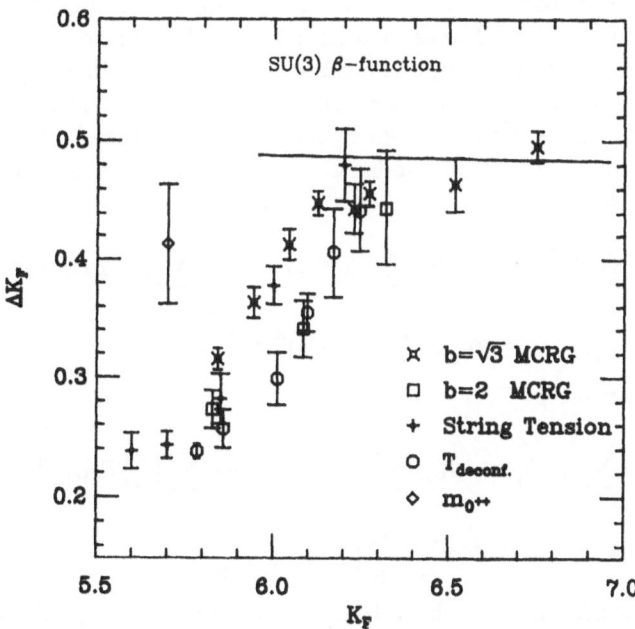

Figure 8: The discrete β-function for a scale change of √3.

that are large for $g \sim 1$. Thus these calculations should be used as a guide and the goal should always be to attain constant mass-ratios.

To conclude this section; $MCRG$ calculations have not yielded any physical results so far, but they have provided us with a definitive statement on the approach to the continuum limit. This is non-trivial. The present MC determination of σ and the glueball masses need improvement before a definite statement of scaling can be made. The largest lattice calculation of σ by de Forcrand[44] show deviations from asymptotic scaling $i.e.$ $\sqrt{\sigma} = 92$ (79) Λ_L at $\frac{6}{g^2} = 6.0$ (6.3). Since these calculations have already taxed the power of a Cray XMP-48, it leads us to the question whether improved actions can help. This is discussed next.

5: DETERMINATION OF THE IMPROVED ACTION

The advantage of using an improved action in MC simulations is to reduce the effect of operators that lead to scaling violations. In QCD this means that corrections to mass-ratios determined from small lattices can be reduced. Second, we want to avoid regions near singularities where continuum mass-ratios are violated. A known example is the end point of the phase structure in the fundamental-adjoint plane. There are, to the best of my knowledge, 11 methods in existence to calculate the renormalized couplings. All, except for those using perturbation theory (and therefore only valid near $g = 0$ where scheme dependence is negligible), are based on $MCRG$. In fact, since the fixed point and the Renormalized trajectory is a function of the RGT, an improved action is content-free unless the RGT is specified.

I shall briefly describe the methods, state their advantages and disadvantages and mention results obtained with them. The generic problem of systematic errors in the estimate of the couplings due to a truncation in the number of couplings kept in the analysis will be referred to as "truncation errors". This is a serious drawback because the errors can be very large and there is no way of estimating them without a second long simulation. In order to consider this truncated ansatz to be the best 'fit', a criterion to judge the improvement has to be established. This is discussed after a brief description of the methods. To fix the notation, the pure gauge SU(2) action is written as

$$S \;=\; K_F \sum TrU_p + K_{6p} \sum TrU_{6p} + K_A \sum \{\frac{4}{3}(TrU_p)^2 - \frac{1}{3}\}$$

$$+ K_{\frac{3}{2}} \sum \{2(TrU_p)^3 - TrU_p\} \tag{5.1}$$

while the SU(3) action is

$$S \;=\; Re[K_F \sum TrU_p \;+\; K_{6p} \sum TrU_{6p} \;+\; K_6 \sum \{\frac{3}{2}(TrU_p)^2 - \frac{1}{2}TrU_p\}$$

$$+ K_A \sum \{\frac{9}{8} \mid TrU_p \mid^2 - \frac{1}{8}\}] \;, \tag{5.2}$$

Here the higher representations have been constructed from U_p, all the traces are normalized to unity and the sums are over all sites and positive orientations of the loops.

Table 2 :   The values of $\Delta\beta$ for a scale change of $b=2$. The results are from Hasenfratz et al.[24].

$\Delta\beta$ for SU(3) (16^4 matched with 8^4)		
K_F	$b=2$ MCRG method	$b=2$ 1-loop Ratio method
6.0	0.35(2)	0.36(3)
6.3	0.43(3)	0.45(3)
6.6	0.55(9)	

Table 3 : Projection of the renormalised SU(2) action onto the $[K_F, K_A, K_{3/2}, K_{6p}]$ space for several starting actions. For each starting action, the first row shows the couplings after one $b=\sqrt{3}$ RGT on starting lattices of size 9^4 calculated by the 2-lattice method [22b]. The second row shows the couplings after two RGT on starting lattices of size 18^4 calculated using the microcanonical demon method [63]. The last set, $K_F=4.35$, is on the MK trajectory Eq. (5.6).

Initial Action K_F	K_F	$\dfrac{K_A}{K_F}$	$\dfrac{K_{3/2}}{K_F}$	$\dfrac{K_{6p}}{K_F}$
2.50 (W)	2.571(05)	–0.195(01)	0.043(01)	–0.0036(003)
	2.058(06)	–0.186(06)	0.038(03)	–0.010(02)
2.75 (W)	3.16(1)	–0.199(03)	0.042(02)	–0.0208(015)
	2.815(35)	–0.214(11)	0.044(06)	–0.019(04)
3.00 (W)	3.69(1)	–0.190(04)	0.040(02)	–0.0314(007)
	3.469(48)	–0.211(12)	0.039(04)	–0.032(03)
3.25 (W)	4.12(2)	–0.160(05)	0.025(03)	–0.0374(004)
	4.003(37)	–0.182(10)	0.032(06)	–0.040(03)
3.50 (W)	4.71(2)	–0.168(05)	0.028(03)	–0.0402(004)
	4.396(67)	–0.150(15)	0.007(06)	–0.049(02)
4.35 (MK)	3.42(1)	–0.211(02)	0.044(01)	–0.0268(011)
	3.098(33)	–0.235(12)	0.055(04)	–0.029(03)

5.1) Symanzik Program[45]: This is a perturbation theory method to remove all $O(a^2)$ corrections in physical observables. At the tree level, at 1-loop[46] and in the leading log[47] analysis, the $O(a^2)$ corrections are removed by including the 6-link planar loop with strength

$$\frac{K_{6p}}{K_F} = -0.05 \quad . \tag{5.3}$$

There have been some SU(3) calculations[36] done with this action, but they are inconclusive and no statement for an improvement in mass-ratios can be made.

5.2) Block Spin Renormalization Group (perturbation theory): The first work in this direction is by Wilson[1] who wrote down the ansatz

$$\frac{K_{6p}}{K_F} = -0.0576 \quad , \quad \frac{K_{6t}}{K_F} = -0.0388 \quad , \tag{5.4}$$

where K_{6t} is the twisted 6-link coupling. No calculation of physical observables has been done with this action. The group of Iwasaki et al.[48] have made a large independent effort in this direction of improvement. They find that near $g = 0$ the action after 3 RGT can be approximated by including the 6-link planar loop with strength

$$\frac{K_{6p}}{K_F} = -\frac{0.331}{3.648} \quad . \tag{5.5}$$

They show that for both the Wilson ansatz, Eq. (5.4), and for this action instantons are stable on the lattice. Since this is not true of the simple plaquette action, they regard it as another criterion for improvement. They have recently calculated the string tension and the hadron masses in the quenched approximation using the improved gauge action of Eq. (5.5) and the standard Wilson action for the quark propagator on a $12^3 \times 24$ lattice at an effective $\frac{6}{g^2} \sim 5.9$. Their results for mass ratios are impressive. A comparison with an equivalent calculation on the Wilson axis is limited because a number of parameters are different. The accuracy of their results warrants more attention.

5.3) Migdal-Kadanoff Recursion Technique: This calculation[49] is limited to the plaquette in the fundamental and higher representations. The integration over links is done by expanding the action in terms of the characters and then using the recursion formula. In the improved action, the effect of the singularity in the fundamental-adjoint plane is reduced but the leading irrelevant coupling K_{6p} is not included. For SU(2)[49], the convergence in the character expansion was good, the recursion was stable on keeping 20 characters. The improved action is dominated by the spin 1 and 3/2 representations, and the K-M improved trajectory was approximated by

$$\frac{K_A}{K_F} = -0.24 \quad . \tag{5.6}$$

It was later shown by Bitar et al.[50] that the heat Kernal action works very well in the recursion scheme and in fact is the solution in the perturbative limit. For a SU(2) calculation of the β-function along the K-M improved trajectory $K_A = -0.24 K_F$, and for an analysis of the improved action see Ref. 22b.

5.3b) Phenomenological (Lines Of Constant String Tension): The continuum limit is taken along directions perpendicular to the lines of constant string tension in the negative fundamental-adjoint plane. Rebbi[52] *et al.* have measured the $q\bar{q}$ potential, while Samuel[53] has spearheaded a calculation with scalar quarks. The effective coupling for comparison on the Wilson axis is defined by using the large N resummation technique[53a]. Since no direct comparison has been made it is hard to state if better mass ratios are obtained.

5.4) Swendsen's method[54] using the Callen representation: The block expectations values of Wilson loops are calculated in two ways. First as simple averages over block configurations, and second using the Callen representation[55] with a guess for the block couplings. From these two estimates, the block couplings are determined iteratively. The method is fast and easy to implement. It does have undetermined truncation errors. Lang[56] has used this method to show that the quartic coupling $\lambda\phi^4$ in the self-interacting scalar field theory renormalizes to zero. Recently Burkitt[35] has used it to map the flow of the action under the $b = 2$ *RGT* (section 2.2) for the U(1) model. From a difference in the flows he can estimate the transition point on the Wilson axis. It would be instructive to extend the U(1) analysis to $\pm\gamma$ coupling values along the phase transition line and check if there is a qualitative change at the *TCP*.

5.5) Callaway-Petronzio-Wilson[57,58] method of fixed block spins: This method is useful for discrete spin systems like the Ising model and models in the same universality class. A *MCRG* calculation is modified by fixing all the block spins except one such that only a controllable few block interactions are non-zero. The system is simulated with the *RGT* used as an additional weight in the Metropolis algorithm. The ratio of probability of this unfixed spin being up to it being down is equal to a determined function of a certain number (depending on how many block interactions are non-zero) of block couplings. By using different configurations of fixed block spins a system of linear equations is set up from which the block couplings are determined. The drawback of this method, even for the Ising model, is that it is hard to set up the block spins so that only a few (≈ 10) block interactions are nonzero. Wilson showed that this can be done if one uses the lattice gas representation *i.e.* 0 or 1 for spin values. The couplings in the ± 1 representation are then given by an expansion in the lattice gas couplings. The second improvement due to Wilson is that instead of a MC determination of the ratio of probabilities, the exact result can be obtained in the transfer matrix formalism. In the $d=2$ Ising model, the convergence of the ± 1 couplings in terms of the lattice gas couplings is slow[58]. About a 1000 lattice gas couplings were necessary for an accuracy of $\sim 10^{-4}$. However the calculation is non-statistical and very fast.

5.6) Character Expansion method of Bitar[59]: I will describe this method with a restriction to simple plaquette actions. The character expansion for the action is $S = \sum_p \sum_r K_r \chi_r(U_p)$ where χ_r is the character in the rth representation and K_r is the corresponding coupling. Similarly the Boltzmann factor F_p for each plaquette p can be expanded in a character expansion $F_p = \sum_r d_r f_r \chi_r(U_p)$ where d_r is the dimension and f_r the coefficient for r^{th} representation. The couplings K_r are given by

$$K_r = \int d(U_p) \, lnF_p(U_p) \, \chi_r(U_p) \, . \tag{5.7}$$

The novel idea is that the ratio $\frac{d_r f_r}{f_1}$ can be calculated as a ratio of expectation values over block configurations. From this the Boltzmann factor F_p and consequently K_r can be determined. The method is sensitive to the convergence of the character expansion *i.e.* the number of terms in r needed to determine F_p accurately. After this there are no truncation errors in determining K_r. The method grows in complexity if larger loops are to be included in the analysis. The first results[59] for the simple plaquette action in SU(2) are encouraging.

5.7) The Schwinger-Dyson Equation method of Falconi et al.[60]: In this method the lattice Schwinger-Dyson equations (equations of motion for expectation values of n-point functions) are used to write down a set of inhomogeneous linear equations for the couplings. The coefficients and the inhomogeneous term are given in terms of expectation values of n-point functions. In deriving these equations the action has to be truncated to the subspace of couplings to be determined. Thus the method has truncation errors. Preliminary results for the O(3) non-linear σ-model in $d = 2$ are encouraging.

5.8) 2-Lattice $MCRG$ method[61,6]: The calculation steps are the same as Wilson's 2-Lattice method to determine the β-function. The method consists of expanding the block expectation values (with unknown couplings) about those from a simulation with known couplings. Keeping just the linear term in the expansion gives the difference between the two sets of couplings. The main advantage is that this comes free with the calculation of the β-function. The method has a statistical drawback that it requires two different simulations so there is no possibility of cancellation of statistical errors. Also, far from the RT, only the first renormalized couplings can be determined accurately. There exist extensive calculations for both the SU(2) and the SU(3) models using the $\sqrt{3}\ RGT$. The estimate for the improved action in a 4-parameter space for SU(2) is[22]

$$\frac{K_{6p}}{K_F} = -0.06 \quad , \quad \frac{K_A}{K_F} = -0.19 \quad , \quad \frac{K_{\frac{3}{2}}}{K_F} = 0.03 \tag{5.8}$$

and for SU(3) is[37]

$$\frac{K_{6p}}{K_F} = -0.04 \quad , \quad \frac{K_8}{K_F} = -0.12 \quad , \quad \frac{K_6}{K_F} = -0.12 \ . \tag{5.9}$$

The truncation errors are known to be large and the reliability of the results is being tested by using the estimated improved action in the update and repeating the calculation of the β-function and the improved action. Also the hadron spectrum is being calculated to test if better mass-ratios are obtained. A detailed comparison of the results for the renormalized action is made with the microcanonical method discussed next.

5.9) Microcanonical (Creutz's Demon) Method[62]: This method is very efficient if from a previous $MCRG$ calculation expectation values of n block Wilson loops at each of the l block levels are determined. To determine the corresponding couplings at the lth level, a microcanonical simulation is then done (on a same size lattice as on which the block expectation values were calculated) with the corresponding n energies fixed and with one demon per interaction. The desired n couplings are then determined from the distribution of demon energies. P. Stolorz[63] at Caltech used the block expectations values obtained after two applications of the $\sqrt{3}$ RGT for SU(2). From these he obtained the second, ($l = 2$), renormalized action in a truncated coupling constant space (four couplings of Eq(5.1)). The results are shown in Table 3 and compared with the first renormalized couplings obtained from the 2-Lattice $MCRG$ method described above. The results show a rapid convergence of the action to the RT consistent with the estimates given in Eqs (5.8). This is evidence that the $\sqrt{3}$ RGT transformation has good convergence properties after two steps. In this calculation it was easy to thermalize the four energies. The simulation is faster than the 2-Lattice method and has better statistical properties. Also the block couplings at all levels can be determined once the block expectation values are known. The truncation errors are the same as in the 2-Lattice method.

5.10) Block Diagonalization method of Mutter and Schilling[64]: This is at present the only method that attempts to improve both the gauge and the fermion action. The main idea is that quark propagators are calculated on blocked gauge configurations using a blocked fermion action. The blocked fermion action is calculated as follows: Let the starting action be the Wilson action

$$\Psi \, M \, \Psi \quad , \tag{5.10}$$

where M is the interaction matrix. The lattice is now divided into blocks which for the $\sqrt{3}$ RGT contain 9 sites each. The site action is then cast into a block action

$$\Xi \, \Gamma \, \Xi \tag{5.11}$$

where Ξ is a 9 component Dirac fermion field and Γ is the interaction matrix set up to reproduce Eq. (5.10). The mass term part of Γ, Γ_m, is now diagonalized to provide the non-interaction fermion basis vectors. For the $\sqrt{3}$ RGT, the 9 eigenvalues of Γ_m are 0 and 8 degenerate ones with value $\frac{9}{a}$. The light mode alone is kept on the blocked lattice. The interaction between the light and heavy modes is calculated in perturbation theory and these terms are added to the Wilson action to give the improved fermion coupling matrix. It is to be noted that this fermion diagonalization is only approximate. Thus lattice masses will not a priori change by the scale factor b between the original and the blocked lattice. It is therefore necessary to first check how good the transformation is in preserving mass-ratios of the unblocked system. The results on a twice blocked set of configurations using $b = 2$ are encouraging[65]. Results of a test of preservation of mass ratios under blocking should be available soon for both the $b = 2$ and $b = \sqrt{3}$ RGT. At this point it is worth mentioning that the following advantages were observed in the diagonalization process for the $\sqrt{3}$ RGT in comparison to $b = 2$.

(a) The separation between the light modes $m \sim 0$ and the heavy modes is better *i.e.* $\frac{9}{a}$ versus $\frac{2}{a}$, so the perturbative corrections are more reliable.

(b) Rotational invariance is not broken as is in the $b = 2$ transformation.

(c) No closed gauge loops which manifest themselves as additional contact terms in the fermion operators arise. This implies that the value of the Wilson parameter r does not get modified and κ_c would remain the same on the blocked lattice for the Wilson fermions if the exact fermion coupling matrix was used in the calculations.

(d) The blocking of gauge links is the same as defined in section 2.3.

Discussion: There are some features of the improved action that seem common to the various analysis done. The details will certainly depend on the specific RGT.

(a) The leading irrelevant operator is dominated by K_{6p}, the $6-link$ planar Wilson loop. Thus a RGT that kills it is an improvement.

(b) From the $\sqrt{3}$ RGT analysis, one gets an estimate of $\frac{K_A}{K_F} \sim \frac{K_6}{K_F} \sim -0.12$. Thus near $\frac{6}{g^2} = 6.$, the phase structure in the $\{K_F, K_A\}$ plane is avoided. This is necessary because in the vicinity of the end point of the phase structure universality is violated.

(c) The RT for the $b = \sqrt{3}$ RGT shows significant deviations from linearity in the region accessible to Monte Carlo. The ratios given in Eqs. (5.8) and (5.9) are an estimate of the asymptotic behavior.

(d) The RT out of the fixed point is local $i.e.$ dominated by small loops. The Wilson axis is tangent to the strong coupling RT at the trivial fixed point at $K_\alpha = 0$. The change from the weak coupling RT to flow close to the Wilson axis takes place in the region where current Monte Carlo calculations have been done $i.e.$ between 5.7 and 6.5. This feature needs to be investigated since current mass-ratios show a behavior that is in between strong coupling and the expected continuum one.

It is still necessary to evaluate whether constant mass-ratios in the quenched approximation are obtained earlier with an improved action. The results have to justify the factor of ~ 5 by which the gauge update slows down when the above four couplings are used. The key lies in improving the fermion sector. For dynamical quarks, the gauge update is a small fraction of the update time. So, an investment in improving the action is justified.

6: IMPROVED MONTE CARLO RENORMALIZATION GROUP[66]

I shall review the Gupta-Cordery Monte Carlo Renormalization Group method ($IMCRG$) in some detail. In this method the Renormalized Hamiltonian and the Linearized Transformation Matrix, LTM, are determined without any truncation errors. There are no long time correlations even on the critical surface and the block n-point correlation functions like $\langle S_\alpha^1 S_\beta^1 \rangle - \langle S_\alpha^1 \rangle \langle S_\beta^1 \rangle$ are calculable numbers. Also, the method allows a careful error analysis in the determination of the renormalized couplings and in the LTM.

In the $IMCRG$ method the configurations $\{s\}$ are generated with the weight

$$P(s^1, s)e^{-H(s)+H^g(s^1)} \tag{6.1}$$

where H^g is a guess for H^1. Note that both the site and block spins are used in the update of the site spins. In analogue to Eq. (1.2), the distribution of the block spins is given by

$$e^{-H^1(s^1)+H^g(s^1)} = \sum P(s^1, s)e^{-H(s)+H^g(s^1)} \ . \tag{6.2}$$

If $H^g = H^1$, then the block spins are completely uncorrelated and the calculation of the n-point functions on the block lattice is trivial.

$$\langle S_\alpha^1 \rangle = 0 \qquad \langle S_\alpha^1 S_\beta^1 \rangle = n_\alpha \delta_{\alpha\beta} \qquad \cdots \tag{6.3}$$

where for the Ising model (and most other models) the integer n_α is simply a product of the number of sites times the multiplicity of interaction type S_α. When $H^g \neq H^1$, then to first order

$$\langle S_\alpha^1 \rangle = \langle S_\alpha^1 S_\beta^1 \rangle_{H^g=H^1} (K^1 - K^g)_\beta \qquad (6.4)$$

and using Eq. (6.3), the renormalized couplings $\{K_\alpha^1\}$ are determined with no truncation errors as

$$K_\alpha^1 = K_\alpha^g + \frac{\langle S_\alpha^1 \rangle}{n_\alpha} . \qquad (6.5)$$

This procedure can be iterated $--$ use H^{n-1} as the spin H in Eq. (6.1) to find H^n. If the irrelevant eigenvalues are small, then after two or three repetitions of the RGT, the sequence H^n converges to the fixed point Hamiltonian H^* which is assumed to be short ranged. For the $d = 2$ Ising model, the method has been shown to be extremely stable[67]. The only limitations of this method are the linearity approximation, Eq. (6.4), (this is trivially handled by iterating H^g) and the use of a truncated H^{n-1} for the spin Hamiltonian in the update to find H^n. The second limitation can be overcome and the solution is straightforward: In Eq. (6.1) use H^g as the guess for H^n. The update now involves the original spins and all block spins up to the nth level in the Boltzmann weight

$$P(s^n, s^{n-1}) \ldots\ldots P(s^1, s)e^{-H(s)+H^g(s^n)} . \qquad (6.6)$$

The four Eqs. (6.2-6.5) are unchanged except that the *level* superscipt is replaced by n, *i.e.* the nth *level* block-block correlation matrix is diagonal and given by Eq. (6.3). With this modification, the H^n is calculated directly. The limitation on n is the size of the starting lattice. Such a check is necessary because errors in long range couplings due to finite statistics and the effects of a truncation in the spin H^{n-1} get magnified and the system rapidly flows away from the fixed point.

The calculation of the LTM proceeds exactly as in the standard $MCRG$ *i.e.* Eqs. (1.4) to (1.6). However, in the limit $H^g = H^1$, the block-block correlation matrix is diagonal and given by Eq. (6.3). Thus it has no truncation errors, can be inverted with impunity and the final LTM elements are also free of all truncation errors. The only error is in finding the eigenvalues from a truncated matrix. These errors can be estimated and the results improved as explained below.

$IMCRG$ is therefore more complicated than $MCRG$ and requires a simultaneous calculation of a many term $H(s)$ and $H^g(s^1)$ at update. However, the system does not have critical slowing down. Secondly, the correlation length ξ can always be made of $O(1)$, so finite size effects are dominated by the range of interactions, which by assumption of a short range H^* fall off exponentially. Thus, critical phenomenon can be studied on small lattices and with no hidden sweep to sweep correlations that invalidate the statistical accuracy of the results.

6.1: Truncation Errors In The LTM

Consider the matrix equation for T in block form

$$\begin{pmatrix} D_{11} & D_{12} \\ D_{21} & D_{22} \end{pmatrix} \begin{pmatrix} T_{11} & T_{12} \\ T_{21} & T_{22} \end{pmatrix} = \begin{pmatrix} U_{11} & U_{12} \\ U_{21} & U_{22} \end{pmatrix} \qquad (6.7)$$

where D_{11} and U_{11} are the 2 derivative matrices calculated in some truncated space of operators that are considered dominant. The elements of the sub-matrix T_{11} will have no truncation errors provided we can calculate

$$T_{11} = D_{11}^{-1} \{U_{11} - D_{12}T_{21}\} . \qquad (6.8)$$

In the $IMCRG$ method the matrix D is diagonal and known, so D_{12} is 0. Thus elements of T_{11} determined from U_{11} have no truncation errors. The errors in the eigenvalues and eigenvectors arise solely from diagonalizing T_{11} rather than the full matrix T. Calculations in the $d = 2$ Ising model have shown that these errors are large, *i.e.* of order 10%, if all operators of a given range are not included. An open problem right now is a robust criterion for classifying operators into sets such that including successive sets decreases the truncation error geometrically by a large factor.

The errors arising from using a sub-matrix T_{11} can be reduced significantly by diagonalizing

$$T_{11} + T_{11}^{-1}T_{12}T_{21} = D_{11}^{-1} U_{11} + \{-D_{11}^{-1}D_{12} + T_{11}^{-1}T_{12}\} T_{21} \qquad (6.9)$$

as shown by Shankar, Gupta and Murthy[68]. The correction term $T_{11}^{-1}T_{12}T_{21}$ is the 2nd order perturbation result valid for all eigenvalues that are large compared to those of T_{22}. This correction matrix can be calculated in $IMCRG$ from $(T^2)_{11} - (T_{11})^2$. I am here overlooking the errors due to the RG flow, because of which T^2 is evaluated at a different point than T. Another aspect of these errors is their behavior as a function of how close to H^* the calculation is done. For the $d = 2$ Ising model we[66,69] find that the truncation errors in the relevant eigenvalues are large. Adding more operators does not monotonically decrease the error. The fluctuations can be as large as 2% even after the 20 largest operators are included in T_{11}.

In standard $MCRG$, the calculations with $T_{11} = D_{11}^{-1}U_{11}$ have shown good convergence once few operators, $O(5 - 10)$, are included in T_{11}. The reason for this is an approximate cancellation of a term ignored and the correction term. Using Eq. (6.7), ignoring terms with T_{22} and approximating $T_{11} = D_{11}^{-1}U_{11}$ we get

$$-D_{11}^{-1}D_{12} + T_{11}^{-1}T_{12} \sim -D_{11}^{-1}D_{12} + U_{11}^{-1}U_{12} .$$

Further, these derivative matrices are roughly proportional, *i.e.* $U \sim \lambda_t D$ and the corrections fall off as the ratio of non-leading eigenvalues to the leading one λ_t. This follows from the arguments of section 1.1 and can be checked by expanding operators in term of eigenoperators. Thus Swendsen[7] by calculating just $D_{11}^{-1}U_{11}$ and ignoring all truncation problems was effectively incorporating a large part of the perturbative correction piece. This explains his success. Shankar[70] has found a correction term to further decrease the truncation effects in $MCRG$. However, given the assumptions, the flow under a RG and the success of the procedure as it exists, an improvement may be hard to evaluate.

Thus, at present the best way to get accurate results is to use $IMCRG$ to calculate the Renormalized couplings and Swendsen's $MCRG$ method to calculate the eigenvalues. The topics that need more work are the accuracy of perturbative improvement in $IMCRG$, the classification of interactions into complete sets and a quantitative understanding of the tuning of the RGT.

Let me also summarize some of the other results obtained from the study of the $d = 2$ Ising model.

[1] The LTM has elements that grow along rows and fall along columns[68], therefore it can be arranged to look like

$$\begin{pmatrix} A & B \\ \varepsilon & D \end{pmatrix} \qquad (6.10)$$

with A the minimal truncated $n \times n$ block matrix that should be calculated. The case $\varepsilon = 0$ is simple; there are no truncation errors in either method and diagonalizing A gives the n largest eigenvalues. Otherwise for $IMCRG$ the truncation error depends on the dot product of terms in ε and B. The requirement of absolute convergence in the dot product only guarantees that this product is finite but it may be arbitrarily large i.e. $O(1)$. Therefore for each model, a careful study of the signs and magnitude of the elements in ε as a function of the RGT becomes necessary. This is being done at Cornell[69].

[2] The leading left eigenvector is normal to the critical surface[68]. Its elements give an estimate of the growth in the elements along the rows of the LTM.

[3] Using H^0 as the known nearest-neighbor critical point $K^c_{nn} = 0.44068$, the $IMCRG$ results[67] for H^1 are independent (within statistical accuracy) of finite size effects for lattice sizes 16, 32, 64 and 128.

[4] The results for H^n converged provided the couplings in H^g were correct to $O(10^{-3})$. This initial accuracy can be achieved[67] with a few thousand sweeps on a 128^2 lattice.

[5] The statistical errors in $IMCRG$ can be evaluated very reliably[67]. Detailed binning analysis showed that each sweep is approximately independent and an accuracy of 10^{-5} is obtained in all couplings with $\sim 2 \cdot 10^6$ sweeps on a 64^2 lattice. This could be achieved with 3000 Vax 11/780 hours.

To conclude, I believe that $IMCRG$ provides a complete framework to analyze the critical behavior of spin and gauge models. With the increased availability of supercomputer time we shall have very accurate and reliable results.

7: EFFECTIVE FIELD THEORIES

The point of effective field theories is that physical phenomena at some given length scale can be described by some effective/composite degrees of freedom. The couplings between these variables are determined by the underlying microscopic theory. Thus we would like to know these effective degrees of freedom and the couplings. So far the discussion of $MCRG$ has focused on the change of scale without a change of variables. To make full use of its power, a transformation of variables at the appropriate scale should be added i.e. in addition to a RGT that just averages over degrees of freedom, consider a change from the microscopic theory to an effective theory with new variables at some give length scale. These variables can be composite (as is the case in going from QCD to a theory where the degrees of freedom are hadrons) or represent a freezing as in SU(2) at high temperatures where the interaction between the Wilson lines is described by an effective $d = 3$ Ising model. Here one transforms from link variables to Wilson lines to Ising spins.

Once the effective theory has been constructed, it is important to know the universality class to which it belongs. This would provide a detailed knowledge of the critical/long distance behavior. Little work has been done in actually exploring universality classes by mapping flows that incorporate a change of variables.

The way to do this in standard MC is to define the composite degrees of freedom and their n-point functions in terms of the microscopic variables. From the expectation values of these n-point correlation functions calculated as simple averages, the corresponding couplings can then be determined by a Microcannonical simulation as described in section 5.9. One such calculation is by Ogilvie and Gocksch[70] in which they determine the nearest neighbor couplings between the Wilson lines in SU(2).

In $MCRG$, the transformation from the microscopic degrees of freedom to the composite variables is made on the original lattice (same as in MC). The RGT is defined on the composite variables and the critical exponents of the effective theory are calculated from the LTM. The couplings can be determined by one or more of the methods of section 5. This process also maps the universality class. Similarly, $IMCRG$ can be used provided H^g is a guessed hamiltonian for the effective theory. This subject is being actively pursued in collaboration with A. Patel, C. Umrigar and K. G. Wilson and we hope it will blossom.

ACKNOWLEDGEMENTS

I would like to thank my collaborators R. Cordery, G. Guralnik, G. Kilcup, G. Murthy, M. Novotny, R. Shankar, S. Sharpe, C. Umrigar, K. Wilson and especially A. Patel for the work presented here and for many long and fruitful discussions. It is a pleasure to acknowledge the hospitality of K. H. Mutter and savor the experience of watching him attend to details necessary to make this an exciting conference.

REFERENCES

[1] K. G. Wilson, in *Recent Developments in Gauge Theories*, Cargese (1979), eds. G. t' Hooft, *et al.* (Plenum, New York, 1980).

[2] R. H. Swendsen, Phys. Rev. Lett. **42**, (1979) 859.

[3] S. H. Shenker and J. Tobochnik, Phys. Rev. **B22** (1980) 4462.

[4] S. K. Ma, Phys. Rev. Lett. **37**, (1976) 461.

[5] L. P. Kadanoff, Rev. Mod. Phys. **49**, (1977) 267.

[6] K. G. Wilson, in *Progress in Gauge Field Theories*, edited by G. 't Hooft *et al.*, (Plenum, New York 1984).

[7] R. H. Swendsen, in *Real Space Renormalization, Topics in Current Physics,* Vol 30, edited by Th. W. Burkhardt and J. M. J. van Leeuwen (Springer, Berlin, 1982) pg. 57.

[8] J. M. Drouffe and C. Itzykson, Phys. Reports **38** (1978) 133.
J. B. Kogut, Rev. Mod. Phys. **51** (1979) 659 and **55** (1983) 775.
M. Creutz, L. Jacobs and C. Rebbi, Phys. Rep. **95**, (1983) 201.
M. Creutz, *Quarks, Gluons and Lattices* , Cambridge Univ. Press (1984).
J. M. Drouffe and J. B. Zuber, Phys. Reports **102** (1983) 1.

[9] K. Binder, in *Monte Carlo Methods in Statistical Physics*, edited by K. Binder (Springer, Berlin,1979) Vol 7 , and in *Applications of Monte Carlo Methods in Statistical Physics*, (Springer Verlag, Heidelberg, 1983).

[10] I have taken the liberty to use as synonymous the terms action and Hamiltonian since the meaning is clear from the context.

[11] K. G. Wilson and J. Kogut, Phys. Rep. **12C**, (1974) 76.

[12] P. Pfeuty and G. Toulouse, *Introduction to the Renormalization Group and Critical Phenomenon*, (John Wiley & Sons, New York 1978).

[13] D. Amit, *Field Theory, the Renormalization Group and Critical Phenomenon*, (World Scientific, 1984).

[14] N. Metropolis, A. W. Rosenbluth, M. N. Rosenbluth, A. H. Teller and E. Teller, J. Chem. Phys. **21** (1953) 1087.

[15] M. Creutz, Phys. Rev. D **21** (1980) 2308.

[16] D. Callaway and A. Rehman, Phys. Rev. Lett. **49** (1982) 613.
M. Creutz, Phy. Rev. Lett. **50** (1983) 1411.
J. Polonyi and H. W. Wyld, Phys. Rev. Lett. **51** (1983) 2257.

[17] G. Parisi and Wu Yongshi, Sci. Sin. **24** (1981) 483.

[18] G. G. Batrouni, G. R. Katz, A. S. Kornfeld, G. P. Lapage, B. Svetitsky, and K. G. Wilson, Cornell Preprint CLNS-85(65), May 1985.

[19] A. Brandt, in *Multigrid Methods*, Lecture Notes in Math 960, (Springer Verlag 1982) and references therein.

[20] The idea was first discussed by G. Parisi in *Progress in Gauge Field Theories*, edited by G. 't Hooft, *et al.*, (Plenum, New York, 1984).

[21] R. Swendsen, Phys. Rev. Lett. **47** (1981) 1775.

[22] R. Cordery, R. Gupta and M. A. Novotny, Phys. Lett. B **128** (1983) 425.
R. Gupta and A. Patel, Nucl. Phys. B **251** (1985) 789.

[23] D. Callaway and R. Petronzio, Phys. Lett. B (1985) .
This transformation was also known to R. Gupta, B. Svetitsky and K. G. Wilson but not pursued in favor of the first three.

[24] K. C. Bowler, A. Hasenfratz, P. Hasenfratz, U. Heller, F. Karsch, R. D. Kenway, I. Montvay, G. S. Pawley, and D. J. Wallace, Nucl. Phys. **B257** (1985) [FS14] 155, and
A. Hasenfratz, P. Hasenfratz, U. Heller, and F. Karsch, Phys. Lett. **140B** (1984) 76.

[25] A more careful statement is made by M. E. Fischer and M. Randeria, Cornell Note (1985).

[26] R. H. Swendsen, Phys. Rev. Lett. **52** (1984) 2321.

[27] G. Bhanot, Nucl. Phys. **B205** (1982) 168.

[28] H. G. Evertz, J. Jersák, T. Neuhaus and P. M. Zerwas, Nucl. Phys. **B251** (1985) 279.

[29] The presence this phase transition was shown analytically by A. Guth, Phys. Rev. **D21** (1980) 2291; and by J. Frolich and T. Spencer, Commun. Math. Phys. **83** (1982) 411.

[30] T. Banks, R. Mayerson and J. Kogut, Nucl. Phys. **B129** (1977) 493.

[31] T. A. DeGrand and D. Toussaint, Phys. Rev. **D22** (1980) 2478.

[32] R. Gupta, M. A. Novotny and R. Cordery, Northeastern Preprint 2654 (1984). A condensed version to appear in Phys. Lett. B

[33] J. Barber, Phys. Lett. B **147** (1984) 330.

[34] V. Grosch, K. Jansen, J. Jérsak, C. B. Lang, T. Neuhaus and C. Rebbi, CERN preprint 4237/85.

[35] A. N. Burkitt, Liverpool preprint LTH 138 October (1985).

[36] P. de Forcrand and C. Roiesnel, Phys. Lett. **137B** (1984) 213, and **143B** (1984) 453.
P. de Forcrand, Ecole Polytechnique preprint A615.0784 (July 1984).

[37] R. Gupta, G. Guralnik, A. Patel, T. Warnock and C. Zemach, Phys. Rev. Lett. **53** (1984) 1721.
R. Gupta, G. Guralnik, A. Patel, T. Warnock and C. Zemach, Phys. Lett. **161B** (1985) 352.

[38] M. Creutz, Phys. Rev. **D23** (1981) 1815.

[39] F. Gutbrod, DESY preprint 85-092 (1985).

[40] D. Petcher, Private Communication.

[41] P. de Forcrand, G. Schierholz, H. Schneider and M. Taper, Phys. Lett. **143B** (1985) 107.

[42] S. W. Otto and J. Stack, Phys. Rev. Lett. **52** (1984) 2328.
D. Barkai, K. J. M. Moriarty and C. Rebbi, Phys. Rev. **D30** (1984) 1293.

[43] S. Gottlieb, A. D. Kennedy, J. Kuti, S. Meyer, B. J. Pendleton, R. Sugar and D. Toussaint, Phys. Rev. Lett. **55** (1985) 1958.

[44] P. de Forcrand, in the proceedings of this conference.

[45] K. Symansik, in *Proceedings of the Trieste workshop on non-perturbative field theory and QCD (dec. 1982)*, World Scientific (1983) 61.

[46] P. Weisz and R. Wohlert, Nucl. Phys. **B236** (1984) 397.

[47] G. Curci, P. Menotti and G. P. Paffuti, Phys. Lett. **130B** (1983) 205.

[48] Y. Iwasaki, preprint UTHEP-118 (1983).
S. Itoh, Y. Iwasaki and T. Yoshie, preprint UTHEP-134 and UTHEP-146.

[49] K. M. Bitar, S. Gottlieb and C. Zachos, Phys. Rev **D26** (1982) 2853.

[50] K. Bitar, D. Duke and M. Jadid, Phy. Rev. **D31** (1985) 1470.

[51] D. Barkai, K. J. M. Moriarty and C. Rebbi, Phys. Rev. **D30** (1984) 2201.

[52] O. Martin, K. Moriarty and S. Samuel, Nucl. Phys. **B261** (1985) 433.

[53] B. Grossmann and S. Samuel, Phys. Lett. **120B** (1983) 383.

A. Gonzales Arroyo and C. P. Korthals Altes, Nucl. Phys. **B205** [FS5] (1982) 46.

R. K. Ellis and G. Martinelli, Nucl. Phys. **B135** [FS11] (1984) 93.

[54] R. H. Swendsen, Phys. Rev. Lett. **52** (1984) 1165.

[55] H. B. Callen, Phys. Lett. **4B** (1961) 161.

[56] C. Lang in *The Proceedings of the Tallahassee Conference on Advances in Lattice Gauge Theory*, (World Scientific, 1985).

[57] D. Callaway and R. Petronzio, Phys. Lett. **139B** (1984) 189.

[58] K. G. Wilson and C. Umrigar, unpublished.

[59] K. Bitar, FSU preprint SCRI-85-7 (1985).

[60] M. Falcioni, G. Martinelli, M. L. Paciello, G. Parisi, B. Taglienti, Nucl. Phys. **B265** (1986) [FS15] 187.

[61] R. Gupta and A. Patel, Phys. Rev. Lett. **53** (1984) 531 and R. Gupta and A. Patel in *Proceedings of the Argonne Conference on Gauge Theory On a Lattice*, (1984).

[62] M. Creutz, A. Gocksch, M. Ogilvie and M. Okawa, Phys. Rev. Lett. **53** (1984) 875.

[63] P. Stolorz, Caltech Preprint CALT-68-1323 (1986).

[64] K. H. Mutter and K. Schilling, Nucl. Phys. **B230** [FS10] (1984) 275.

[65] A. Konig, K. H. Mutter and K. Schilling, Nucl. Phys. **B259** (1985)33, and in this Conference's Proceedings.

[66] R. Gupta and R. Cordery, Phys. Lett. **A105** (1984) 415.

[67] R. Gupta in *Proceedings of the Tallahassee Conference on Advances in Lattice Gauge Theory*, World Scientific (1985).

[68] R. Shankar, R. Gupta and G. Murthy, Phys. Rev. Lett. **55** (1985) 1812.

[69] R. Gupta, C. Umrigar and K. G. Wilson, under progress.

[70] R. Shankar, Yale preprint YTP 85-25 Nov. (1985).

[71] A. Gocksch and M. Ogilvie, Phys. Rev. Lett. **54** (1985) 1985.

LANGEVIN SIMULATIONS OF QCD, INCLUDING FERMIONS

Andreas S. Kronfeld

Deutsches Elektronen-Synchrotron DESY
Notkestrasse 85, 2000 Hamburg 52
Federal Republic of Germany

INTRODUCTION

Perhaps the first question to ask is, "Why Langevin simulations?"
One way of answering this question is to examine a disease from which most
simulation algorithms suffer. The disease is critical slow down [1], and we
encounter it in updating when $\xi/a \to \infty$ and in matrix inversion (needed to
include fermions) when $m_q a \to 0$. A simulation that purports to solve QCD numerically will encounter these limits, so to face the challenge in the
title of this workshop, we must cure the disease of critical slow down.

One can describe the disease in several ways. Let's focus on updating
first; matrix inversions will be treated in detail below. Figure 1

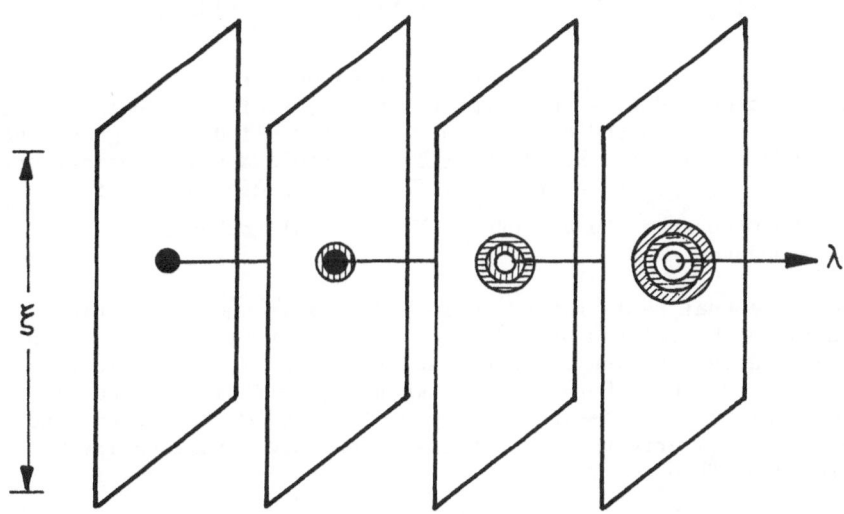

Fig. 1.   Illustration of updating. When ξ is large, many
updates are needed to propagate changes throughout
the lattice.

illustrates the evolution of the system in λ = CPU time during the course of a typical simulation. Consider the effects of changes inside the black disc during the first update. Because the couplings are local, the only portion of the system sensitive to the change is in the circle with vertical stripes. During the second update the circle with the horizontal stripes feels the original changes, but only indirectly, through the changes to the vertically striped circle. As the updating process continues, more of the system feels the changes from the first update of the black circle, yet before one can consider a new configuration to be decorrelated from the original one, many updates are required. Indeed, for typical algorithms the number of updates needed grows as $(\xi/a)^2$. Physically, this critical slow down is due to the reluctance of changes at short distances to propagate to large distances. Numerically, the stability of an algorithm at short wavelengths requires a (moderately) small step size; critical slow down occurs when the effective long wavelength step size becomes tiny.

The remedy for this disease is an algorithm that propagates signals quickly throughout the system; i.e. one whose effective step size is not reduced for the long wavelength conponents of the fields. (Here the effective "step size" is essentially an inverse decorrelation time.) To do so one must resolve various wavelengths of the system and modify the dynamics (in CPU time) of the simulation so that all modes evolve at roughly the same rate. This can be achieved by introducing Fourier transforms. I will show how to implement Fourier acceleration for Langevin updating and for conjugate gradient matrix inversion. The crucial feature of these algorithms that lends them to Fourier acceleration is that they update the lattice globally; hence the Fourier transforms are computed once per sweep rather than once per hit.

SIMPLE LANGEVIN SIMULATIONS

For QCD the simplest Langevin algorithm is given by [2]:

$$u_{x,\mu}^{(\lambda+1)} = e^{-f \cdot T} u_{x,\mu}^{(\lambda)}$$

$$f_i = \sqrt{\varepsilon_{ij}}\, \eta_j + \varepsilon_{ij}[\partial_j S_g - \tfrac{1}{2} Re(\xi^\dagger M^{-1} \partial_j M \xi)] \tag{1}$$

where the T_a are antihermitean adjoint generators of SU(3). S_g is the pure gauge action and M is the fermion hopping matrix. The noise fields, η and ξ have zero mean and dispersion 2. The index i denotes group and position indices: i = (a, x, μ). Finally the field derivatives satisfy the Lie algebra: $[\partial_a, \partial_b] = -f_{abc} \partial_c$. The spatial dependence of ε_{ij} allows us to propagate the effects of $U_j^{(\lambda)}$ to all other links, $U_i^{(\lambda+1)}$, during one update.

The conceptual basis for Langevin updating is the stochastic quantization of Parisi and Wu [3]. When ε_{ij} is infinitesimal one can use the Fokker-Planck equation to show that, for large λ, the field U is distributed as desired. Unfortunately when ε_{ij} is finite the equilibrium action differs from the desired one by $O(\varepsilon)$ terms. On the other hand, the nonlocality of ε_{ij} effects only details of the new terms. Define S = S_g - Tr ln M. Then

$$S_{eq} = (1 + \tfrac{\bar{\varepsilon}}{12} C_A) S + \tfrac{1}{4} \sum_{ij} \varepsilon_{ij} \{2\partial_i \partial_j S - \partial_i S \partial_j S\} \tag{2}$$

$$+ \bar{\varepsilon}(fermionic\ headache)$$

where $\bar{\epsilon}$ = diag(ϵ_{ij}). Later in the talk I will present algorithms whose leading corrections to the equilibrium action are $O(\epsilon^2)$, so please ignore the fermionic headache. Eq.(2) merely emphasizes that the $O(\epsilon)$ terms for nonlocal ϵ_{ij} are not radically different from those that appear when ϵ_{ij} = $\bar{\epsilon}\delta_{ij}$.

FOURIER ACCELERATION OF UPDATING

The idea of Fourier acceleration is to introduce fast Fourier transforms (FFT's) when constructing the drift force f_i in Eq.(1). Then $\epsilon_{ij} \rightarrow \epsilon(p)$, and we pick $\epsilon(p)$ so that the decorrelation time, $N_{dc}(p)$, is the nearly same for all modes of the field. To illustrate how the FFT's can accelerate Langevin updating, consider a free scalar field. In momentum space the update rule reads

$$\phi^{(\lambda+1)}(p) = [1 - \epsilon(p)(p^2 + m^2)]\phi^{(\lambda)}(p) + \sqrt{\epsilon}(p)\eta^{(\lambda)}(p) \tag{4}$$

where p and m are measured in lattice units. For $\phi^{(0)}(p) = 0$ Eq.(4) has a formal solution:

$$\phi^{(\lambda+1)}(p) = \sum_\nu [1 - \epsilon(p)(p^2 + m^2)]^{(\lambda-\nu)} \sqrt{\epsilon}(p)\eta^{(\nu)}(p) \tag{5}$$

Then the correlations in λ are given by

$$\langle \phi^{(N+\lambda)}(p)\phi^{(\lambda)}(p) \rangle = \langle \phi^{(\lambda)}(p)\phi^{(\lambda)}(p) \rangle e^{-N\epsilon(p)(p^2+m^2)} \tag{6}$$

so that the decorrelation time can be defined as

$$N_{dc}(p) = \frac{1}{\epsilon(p)(p^2+m^2)} \tag{7}$$

If ϵ is local in coordinate space than $\epsilon(p)$ is independent of p. Then the decorrelation time for long wavelengths is $N_{dc}(0) \sim m^{-2} \sim (\xi/a)^2$. However, if one chooses $\epsilon(p) = \bar{\epsilon}/(p^2 + m^2)$, then $N_{dc}(p) = \frac{1}{\bar{\epsilon}}$ for all momenta.

For interacting theories one can pursue one of two strategies. If the dynamics is in or near a perturbative regime (as in QCD when $\beta \rightarrow \infty$) one can set $\epsilon(p)$ as above, albeit with a renormalized mass. Otherwise, one can study the correlations of Eq.(6) numerically as the simulation evolves; optimally, this will be done adaptively during the simulation. Both of these strategies have been successful in the XY model. In particular, the numerical determination of the optimal ϵ_{ij} has worked in the phase of the XY model with vortices.

Another illustrative example of how modifications to the Langevin dynamics can accelerate a simulation is given by a roundabout derivation of our fermion formulation. Instead of writing one Langevin equation for the gauge field, consider the equations derived from the action S = S_g + S_f where $S_f = \phi^+ M^{-2}\phi$, which generates the two flavor theory [4]. The Langevin equation for the auxiliary scalar field is then [5]

$$\phi_i^{(\lambda+1)} = (\delta_{ik} - \epsilon_{ij} M_{jk}^{-2})\phi_k^{(\lambda)} + \sqrt{\epsilon_{ij}}\xi_j \tag{8}$$

One wants to pick ϵ_{ij} so that the ϕ field decorrelates as quickly as possible. The obvious choice is $\epsilon_{ij} = \bar{\epsilon} M_{ij}^2$. Then detailed balance can be used to show that the equilibrium action is $S_f = \frac{2}{\sigma} (1 - \frac{1}{2} \bar{\epsilon}) S_f$ exactly. By picking σ, the dispersion of ξ, properly one can also set $\bar{\epsilon} = 1$ so that the update rule becomes $\phi_i^{(\lambda+1)} = M_{ij} \xi_j$: the new ϕ is completely independent of the old, and S_f is simulated exactly! When this expression for ϕ is substituted into the gauge field Langevin equation, Eq.(1) results. As an added bonus one can then adjust the coefficient of the bilinear noise term to give any number of flavors: even, odd, or even fractional.

Gauge fields

Next, let's consider Fourier acceleration of the gauge field update. For several reasons one has to fix the gauge. The most physical reason is that short wavelength gauge artefacts can totally obscure our intuitive notion of momentum, and hence spoil the Fourier decomposition of the field into modes, some of which require acceleration. Another point of some importance is that a nonlocal ϵ without gauge fixing breaks gauge invariance; this could be solved by setting $\epsilon = \epsilon[U]$, but then the equilibrium is determined by the U dependence of ϵ as well as the U dependence of S. Consequently, we fix the gauge completely after each update. Numerical experience has shown that axial gauge gives unsatisfactory performance on an 8^4 lattice in the pure gauge theory; it is better to smooth the axial-gauge-fixed configurations by applying several iterations of a Landau gauge fixing. Details will be published elsewhere.

HIGHER ORDER ALGORITHMS

The problem with the algorithm presented in Eq.(1) is that $\bar{\epsilon}$ must be quite small when the procedure is Fourier accelerated. When ϵ is local, the leading corrections to the equilibrium action do not effect the continuum limit. (See Ref. 1). However, when ϵ is nonlocal this is no longer true, so one neds a higher order differencing scheme for the Langevin equation. Such schemes generally require more memory. For systems without fermions the most efficient methods are similar to the Runge-Kutta algorithms for deterministic differential equations [6]. For simplicity I will leave off the indices needed for Fourier acceleration; they can be found in Ref.[7].

The Runge-Kutta tricks do not quite work for the bilinear noise term in Eq.(1) that introduces fermions, but Batrouni [8] has found a way around the problems. First one calculates a "tentative update" via the simple Euler rule:

$$\tilde{u}_{x,\mu} = e^{-\tilde{\jmath} \cdot T} u_{x,\mu}^{(\lambda)} \qquad \text{. (9a)}$$

$$\tilde{\jmath}_i = \sqrt{\epsilon}\, \eta_i + \epsilon [\partial_i S_g - \tfrac{1}{4} \xi^{\dagger} A_i \xi]$$

followed by the final update, which has drift force

$$\jmath_i = \tfrac{1}{2}\epsilon (1 + \tfrac{g}{6} C_A)[\partial_i S_g + \partial_i \tilde{S}_g - \tfrac{1}{4}\xi^{\dagger} A_i \xi - \tfrac{1}{4}\xi^{\dagger} \tilde{A}_i \xi] + \qquad \text{(9b)}$$

$$+ \sqrt{\varepsilon}\, \eta_j [\delta_{ij} - \tfrac{1}{128} Re(\xi^\dagger A_i \xi\, \xi^\dagger A_j \xi)]$$

In these equations $A_i = M^{-1}\partial_i M^2 M^{-1}$, ζ is an additional fermionic noise, and a tilde implies that S_g or A is evaluated using the tentative update, $\tilde{U}$.

An alternative procedure, which avoids the tentative update, requires a drift force with higher derivative of the action [7]:

$$(10)$$

$$U^{(\lambda)}_{x,\mu} = e^{-f \cdot T}\, U^{(\lambda)}_{x,\mu}$$

$$f_i = \varepsilon (1 + C_A \tfrac{\varepsilon}{24})[\partial_i S_g - \tfrac{1}{2} Re(\xi^\dagger M^{-1}\partial_i M \xi)]$$

$$+ \sqrt{\varepsilon}\, \eta_j \{ \delta_{ij} - \tfrac{1}{4}\partial_i [\partial_j S_g - \tfrac{1}{2} Re(\xi^\dagger M^{-1}\partial_j M \xi)]$$

$$- \tfrac{1}{16} Re[\xi^\dagger (M^{-1}\partial_i M \partial_j M\, M^{-1} + \partial_i M\, M^{-1}\partial_j M\, M^{-1})\xi] \}$$

Although this rule is not simple, it attains $O(\varepsilon^2)$ accuracy for the QCD with fermion loops with only two matrix inversions.

FOURIER ACCELERATION OF MATRIX INVERSION

To compute any of the drift forces in Eqs. (1), (9) or (10) one needs to solve the linear system of equations $M\psi = \xi$, where M is some lattice version of $\gamma_5(\slashed{D} + m_q)$; the γ_5 is included so that M is hermitean. We have used Wilson fermions. The matrix inversion is the most time consuming part of the update, so it is important to consider ways of accelerating routines like the conjugate gradient method [9]. Of course, matrix inversion is also important for quenched hadron spectroscopy.

The conjugate gradient method only works for positive definite matrices, so in practice we consider $M^2\psi = \phi = M\xi$. The convergence is governed by the ratio of the largest and smallest eigenvalues of M^2.

$$(11)$$

$$N_{CG} \approx \frac{|M^2|^{1/2}_{max}}{|M^2|^{1/2}_{min}} \sim (m_q a)^{-1}$$

since $M^2 \sim p^2 + m_q^2$. The gauge interaction produces terms that are off-diagonal in momentum space (the background gauge field exchanges momentum with the fermions), and gauge artefacts can obscure the p^2 dependence of M^2.

We pre-condition M^2 by Fourier transforming (symbolized by $\hat{F}$):

$$(12)$$

$$\varepsilon(p) \hat{F} M^2 \hat{F}^{-1} \hat{F} \psi = A \hat{F} \psi = \varepsilon(p) \hat{F} \phi$$

The pre-conditioner $\varepsilon(p)$ is chosen so that the diagonal elements of A are all unity. In a smooth gauge we expect that

$$\frac{|A|^{1/2}_{max}}{|A|^{1/2}_{min}} \approx 1. \tag{13}$$

As in the acceleration of updating there are two ways to determine the pre-conditioner. In a perturbative regime $\varepsilon(p) \sim (p^2 + m_q^2)^{-1}$, which explains the ε notation. Alternatively, one can compute the components of M^2 diagonal in momentum space and set $\varepsilon(p) = (M^{-2})_{pp}$. In fact, this approach gives a dramatic illustration of the practical necessity of fixing the gauge. Figure 2 shows the numerically determined $\varepsilon(p)$ for a propagator calculation in several gauges. All of the results are for SU(3) on a 8^4 lattice. Curve (a) shows the result for the free theory (all links are set equal to 1). Curve (b) shows $\varepsilon(p)$ for a $\beta = 5.8$ configuration without gauge fixing; evidently a wild gauge artefact has obscured any correlation between momentum and the diagonal elements of M^2. Thus, although all diag(A) = 1, the spread of eigenvalues of A remains enormous. Fixing the same configuration to axial gauge improves this situation, as shown by curve (c), but not dramatically However, Landau gauge, curve (d), produces a pre-conditioner that is qualitatively the same as for the free theory, as desired.

The acceleration provided by this technique is significant on an 8^4 lattice, and will become more significant on larger lattices. Figure 3 shows how many conjugate gradient iterations are needed to obtain convergence to fixed precision with and without Fourier acceleration.

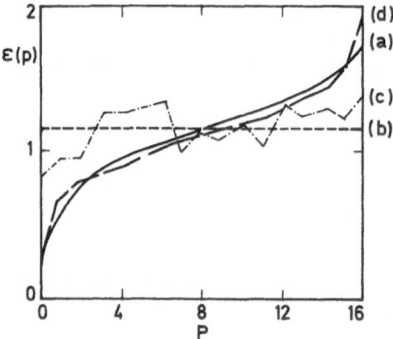

Fig. 2.   Numerical computation of the pre-conditioner $\varepsilon(p)$.
         (a) Free field. (b) No gauge fixing. (c) Axial gauge.
         (d) Landau gauge.

Near physical values of the hopping parameter the accelerated algorithm needs 3 or 4 times fewer iterations. Owing to Eq.(11) this improvement factor should scale with the volume. The overhead for the FFT's is not serious: with everything <u>except</u> the FFT package optimized, a single iteration of the FFT conjugate gradient took only twice the CPU time of a normal iteration. Putting these factors together, we predict that the accelerated algorithm will be about 30 times faster for a 16^4 lattice; in practice, this means that smaller quark masses will be tractible.

CONCLUDING REMARKS

Let me summarize. Fourier acceleration is designed to reduce critical slow down. To make real progress towards the numerical solution of QCD we need algorithms for updating and matrix inversion that attack this problem. The FFT techniques are well suited to the attack, because they exploit our intuition (based on perturbation theory), and because they spread the effects of local changes throughout the system in the course of one iteration. Since the FFT's are inserted before and after each update, it is necessary to have an algorithm where "one update" means "one sweep," rather than "one hit." For matrix inversion this implies an algorithm such as the conjugate gradient, and for updating this implies Langevin or microcanonical [10] simulations. Fourier acceleration has been successful in simulating the XY model and in computing the quark propagator in quenched QCD. A complication when gauge fields are involved is the requirement of gauge fixing.

Finally, let us consider the prospects for a simulation of QCD (including vacuum polarization) on a 8^3x16 lattice. We expect Fourier acceleration to improve the matrix inversion by a factor of 4 and the updating by a factor of 16, so that the overall acceleration will be about

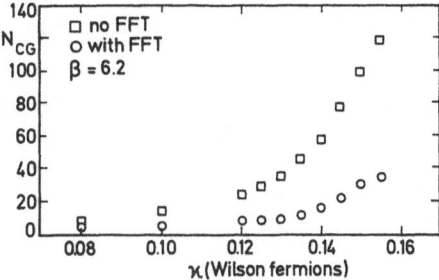

Fig. 3.   Performance of the conjugate gradient with and without FFT's. As K approaches the critical value, the gain with FFT's is impressive.

a factor of 60. More generally, the number of operations needed for a simulation based on Eq.(1), or on the higher order generalizations, will be

$$N_{dc} \left(f_g L^4 + f_f N_{CG} L^4 \right) \qquad (14)$$

where L is the linear size of the system in lattice units. Without Fourier acceleration both N_{dc} and N_{CG} grow dramatically as L increases, as indicated in Eqs.(2) and (11), respectively; this is critical slow down. The aim of our investigations is develop algorithms for which N_{dc} and N_{CG} are at worst slow functions of L. Then the simulation time will grow only as the volume of the system, which is unavoidable.

ACKNOWLEDGEMENTS

This talk summarizes the work of George Batrouni, Christine Davies, Garry Katz, Peter Lepage, Pietro Rossi, Ben Svetitsky, and Ken Wilson. It is a pleasure to recall our many interesting conversations, and I apologize if I have neglected their "pet projects" in preparing this talk. We have used a VAX 750 for program development and an FPS-164 array processor operating from an IBM 3081 host for production computing.

REFERENCES

1. G. Parisi in Progress in Gauge Field Theory, edited by G. 't Hooft, et al, (Plenum, New York, 1984.

2. G.G. Batrouni, G.R. Katz, A.S. Kronfeld, G.P. Lepage, B. Svetitsky, and K.G. Wilson, Phys. Rev. D32 (1985) 2736.

3. G. Parisi and Wu Y.-S., Sci. Sin. 24 (1981) 483

4. D.H. Weingarten and D.N. Petcher, Phys. Lett. 99B (1981) 333.

5. See also A. Ukawa and M. Fukugita, Phys. Rev. Lett. 55 (1985) 1854 for simulations based on this formula.

6. E. Helfand, The Bell System Technical Journal 58 (1979) 2289; H.S. Greenside and E. Helfand, ibid. 60 (1981) 1927.

7. A.S. Kronfeld, DESY report 86 - 006.

8. G.G. Batrouni, Cornell University report CLNS-85/665

9. M.R. Hestenes and E. Stiefel, J. Res. Natl. Bur. Stand. 49 (1952) 409.

10. J.B. Kogut, University of Illinois report ILL-(TH)-85-#75.

THE LOOK-AHEAD FERMION ALGORITHM

Michael Grady

High Energy Physics Division
Argonne National Laboratory
Argonne, IL 60439

ABSTRACT

I describe a fast fermion algorithm which utilizes pseudofermion fields but appears to have little or no systematic error. Test simulations on two-dimensional gauge theories are described. A possible justification for the algorithm being exact is discussed.

INTRODUCTION

I would like to describe a new fermion algorithm with which I have had a great deal of success in two-dimensional applications.[1] It is based on locally updating a small number of pseudofermion fields each time a gauge field is updated. I call it a look-ahead algorithm because the pseudofermion update is tentative. It is made after a trial new gauge field link is selected, but before it is accepted or rejected. I will attempt to show how such a procedure eliminates an important source of systematic error which occurs in the standard pseudofermion technique of Fucito et al.[2,3]

The algorithm has many desirable features. In one and two-dimensional tests, which include the Schwinger Model, I have found no detectable systematic error in the results. This is true even after very high statistics have been achieved through long simulations. A corollary to the lack of systematic error is a lack of dependence on the gauge field hit size. This is the maximum amount that the new trial gauge field link is allowed to differ from its old value in an update. In the standard pseudofermion algorithm, the systematic error increases with increasing hit size, forcing one to adopt rather small hit sizes. This has a detrimental impact on the performance of the algorithm, since it leads to very high correlations between adjacent sweeps. Consequently, one has to perform many sweeps before one has the equivalent of a single uncorrelated measurement; thus the amount of information extracted per sweep is relatively small when compared with the pure gauge theory algorithm, for which the hit size may be optimized for maximum algorithm performance. A limitation to small hit sizes also means that many more sweeps need to be performed to equilibrate lattices before any measurements can take place. It is also more probable that the system will get stuck in a metastable state if the hit size is small. Thus the advantages of an algorithm with complete freedom to set the hit size are great.

On the face, the look-ahead algorithm appears to be an approximation which is exact in several limits which I shall discuss. However, the lack of detectable systematic errors in the numerical examples studied to date lends one to speculate whether the algorithm could possibly be exact, i.e. completely free of systematic error. This could be due to a balancing of errors between the probability of a transition A → A' and that of the transposed process A' → A. I will try to indicate how this could work in what follows. Of course, although it would be very nice to have a fast exact algorithm, one does not need such an algorithm to make progress. If the systematic errors are of sufficiently high order that they affect physical measurements by, say, less than 10% of the total fermionic effect on a system, for reasonable values of parameters, then such an algorithm would be practical for most lattice gauge theory applications. The look-ahead algorithm is fast. It runs only a factor of 3-12 times slower than the corresponding pure gauge theory algorithm in the two-dimensional models studied, depending on the type of interaction. This is with largely unoptimized programs. Realistic four-dimensional problems should not be more than another factor of two slower. Finally, the look-ahead algorithm is simple to implement compared to many of the improved algorithms being suggested, a feature it shares with the standard pseudofermion algorithm.

THE ALGORITHM

The generic action for a system of gauge fields interacting with Fermi fields is

$$S_f(\bar{\psi}, \psi, A) = S_0(A) + \sum_{ij} \bar{\psi}_i O_{ij}(A) \psi_j , \qquad (1)$$

where $S_0(A)$ is the pure gauge action which contains a multiplicative parameter β. $O_{ij}(A)$ is a local operator which depends on A, such as the Dirac operator. In what follows, I shall assume that $O_{ij}(A)$ is positive definite. This is usually not an important restriction, since one can generally use a second order fermion formalism which satisfies the requirement. The partition function is given by

$$Z_f = \int dA \, d\bar{\psi} \, d\psi \, \exp\left(-S_f(\bar{\psi}, \psi, A)\right) . \qquad (2)$$

The Grassmann Fermi fields are then integrated to give

$$Z_f = \int dA \, \exp\left(-S_0(A)\right) \det O(A) . \qquad (3)$$

Eq. (3) is the starting point for numerical simulations, since it is impractical to represent the Grassmann fields themselves on a computer, for all but very small lattices. One could also imagine an analogous system of a complex Bose field interacting with the gauge field through the same interaction matrix $O_{ij}(A)$. This system is formed through the substitution:

$$\bar{\psi}_i \to \phi_i^* , \qquad \psi_i \to \phi_i . \qquad (4)$$

So one has what I will refer to as the boson/gauge system, with action given by

$$S_b(\phi^*, \phi, A) = S_0(A) + \sum_{ij} \phi_i^* O_{ij}(A) \phi_j , \qquad (5)$$

and partition function

$$Z_b = \int dA \ d\phi^* d\phi \ \exp\left(- S_b(\phi^*,\phi,A)\right) . \tag{6}$$

Integration of the ϕ fields yields

$$Z_b = \int dA \ \exp\left(-S_0(A)\right)\left(\det O(A)\right)^{-1} . \tag{7}$$

Thus one sees that the difference between Bose and Fermi statistics is simply the inversion of the determinant factor occurring in (3). Note that the fields ψ_i and ϕ_i carry the same spin, thus the system (5) violates the spin-statistics theorem. This will not be of any consequence, however, as the lattice partition function (6) is still well defined. Unlike the fermion case, for the boson system one can simulate the partition function (6), including the ϕ fields, with Monte Carlo techniques. It is much less practical to use the integrated form (7), due to the non-local determinant factor. However, for the fermion case, one is forced to use the integrated form. This is the primary source of difficulty in incorporating fermions into Monte Carlo simulations. The pseudofermion approach is to somehow use the ϕ fields and the system (6) in an unconventional way so as to simulate the fermion determinant occurring in (3), rather than the inverse determinant which would result from a straightforward simulation of the Gaussian integral.

I restrict consideration to the Metropolis et al. Monte Carlo technique.[4] One produces a chain of configurations for the Fermion/Gauge system in the following way. Choose a gauge field A' which differs from the current gauge configuration, A, on just one link, j. Then accept A' with the probability

$$P(A \rightarrow A') = \mathrm{Min}\left(1,\exp(-S_0(A') + S_0(A))\Delta\right) \tag{8}$$

$$\Delta \equiv \det O(A')/\det O(A) . \tag{9}$$

If rejected the old configuration A is retained. Thus one needs only a fast way of computing the determinant ratio, Δ, of two operators which differ only locally. For the boson/gauge system one needs Δ^{-1} in place of Δ. It can be written

$$\Delta^{-1} = \frac{\int d\phi^* d\phi \ \exp\left(-\phi^* O(A')\phi\right)}{\int d\phi^* d\phi \ \exp\left(-\phi^* O(A)\phi\right)} = \langle\exp\left(-\phi^*\left(O(A')-O(A)\right)\phi\right)\rangle_{\phi(A)} \tag{10}$$

where the notation has been streamlined by replacing the explicit summations in the action with implied matrix-vector multiplication. $\langle \ \rangle_{\phi(A)}$ is defined to be an expectation value in the ensemble with probability distribution $P(\phi,A) \propto \exp\left(-\phi^* O(A)\phi\right)$. This is the distribution which is naturally generated through Monte Carlo simulation of the Gaussian ϕ field. Note that in a standard simulation of the boson system (6), one would not use the full expectation value, Δ^{-1}, for each A field update, but rather the operator inside the expectation value evaluated in a single ϕ configuration. The correctness of both formulations can be verified by detailed balance. They basically differ by an interchange of order of integration. In the latter case the expectation value is evaluated a bit at a time, with an additional determination each time the chain reencounters the transition $A \rightarrow A'$.

For the fermion/gauge system one needs Δ rather than Δ^{-1}. One could simply write

$$\Delta = \langle \exp(-\phi^*(O(A') - O(A))\phi)\rangle^{-1}_{\phi(A)} \ . \tag{11}$$

This is basically the strategy of the standard pseudofermion algorithm. However, this leads to a systematic error when the expectation value is incompletely evaluated at each update. This is simply because the average of the inverse is not the inverse of the average. If one has a number of determinations of some quantity x, then

$$\langle \frac{1}{x} \rangle \neq \frac{1}{\langle x \rangle} \tag{12}$$

if the variance is nonzero; in fact, the left-hand side will always be greater. Another way to represent Δ is to write the same ratio of gaussians as before (10) but interchange A with A', giving

$$\Delta = \langle \exp(\phi^*(O(A')-O(A))\phi)\rangle_{\phi(A')} \ . \tag{13}$$

This differs from the boson expression Δ^{-1} (10) by the sign in the expectation value and in the distribution over which it is taken:

$$P(\phi,A') = \exp(-\phi^* O(A')\phi) \ . \tag{14}$$

Suppose, for the moment, that one had an efficient way of producing this distribution. Then, I claim, in analogy with the Boson algorithm, that instead of using the full expectation value Δ in the Metropolis probability (8), one can instead use the argument of (13) evaluated for a single member of the distribution (14), without introducing a systematic error. The average transition probability $P(A \to A')$ will be the ϕ dependent transition probability $P(A \to A', \phi)$ averaged over the ϕ distribution.

$$P(A \to A') = \frac{\int d\phi \, P(\phi,A')P(A \to A', \phi)}{\int d\phi \, P(\phi,A')}$$

$$= \frac{\int d\phi \, \exp(-\phi^* O(A')\phi)\mathrm{Min}\left(1, \exp(-S_0(A')+S_0(A) + \phi^*(O(A')-O(A))\phi)\right)}{\int d\phi \, \exp(-\phi^* O(A')\phi)}$$

$$\tag{15}$$

$$= \det O(A')\exp(-S_0(A')) \int d\phi \, \mathrm{Min}\left(\exp(S_0(A')-\phi^* O(A')\phi) \ , \right.$$

$$\left. \exp(S_0(A) - \phi^* O(A)\phi)\right)$$

up to a numerical factor. The expression within the integral is symmetric under interchange of $A \hookrightarrow A'$. Now, the equilibrium distribution is given by the detailed balance equation

$$P_{eq}(A)P(A \to A') = P_{eq}(A')P(A' \to A) \ . \tag{16}$$

This can be seen to be solved by

$$P_{eq}(A) = \exp(-S_0(A))\det O(A) \ , \tag{17}$$

which is the desired distribution for the fermion/gauge system. So the systematic error has been eliminated, but at the expense of needing a ϕ distribution equilibrated to the new trial gauge field rather than the old one. How is this distribution to be obtained? One could just equilibrate the entire lattice of ϕ fields to A', each time a gauge field is updated, but this is impractically slow. I suggest, rather, equilibrating only a small neighborhood of ϕ fields about the link, j, being updated. This neighborhood is defined as those ϕ fields which occur with nonzero coefficients in the expression $\phi^*(O(A')-O(A))\phi$, i.e. those which contribute directly to the Metropolis probability. More distant neighbors contribute only indirectly via feedback through the probability distribution. I now define the algorithm explicitly. The steps are as follows:

(1) Choose A' different from A at one site j.

(2) Perform a look-ahead update of ϕ fields within the immediate neighborhood of A_j to $\exp(-\phi^*O(A')\phi)$. Fields must be equilibrated, either with several rounds of updates, or a single multivariate heat bath update.[5]

(3) Accept or reject A according to the Metropolis probability $P(A \rightarrow A',\phi) = \text{Min}(1,\exp(-S_0(A') + S_0(A) + \phi^*(O(A')-O(A))\phi))$, using, for the ϕ fields, those obtained in (2).

(4) It turns out that the ϕ distribution gets distorted by the Metropolis decision in (3), so one must, at this point, update the neighboring ϕ fields again, <u>if</u> A' was accepted. If A' was rejected in (3) then the original ϕ fields are retained.

(5) Go on to the next link.

Note that step (4) is slightly different from the algorithm of Ref. 1. There I also updated the ϕ fields when the A' update was rejected. I eventually found a small systematic error using that algorithm. It also does not satisfy the conditions of the more detailed justification presented later.

What can one say about this algorithm? First, it is exact in zero dimensions (field theory at a point). This is because there is only one ϕ field, and a single heat bath look-ahead update achieves the needed $\exp(-\phi^*O(A')\phi)$ distribution for the ϕ field. This is in contrast to the standard pseudofermion algorithm which already has serious systematic errors in zero dimensions. Second, it may be viewed as the first term in a systematic approximation based on equilibrating larger and larger neighborhoods of the gauge field, A_j, being updated, becoming exact as the neighborhood approaches the full lattice. Third, it is also exact as the gauge field hit size goes to zero, for essentially the same reasons that the standard pseudofermion algorithm is. The hit size is the maximum amount that the new gauge field A_j' is allowed to differ from the old field A_j. This limit is very useful for testing the algorithm. One merely has to run the simulation for several different hit sizes, and look for a hit size dependence in measured quantities. An exact algorithm should be independent of the hit size. If the algorithm has a small systematic error, it can usually be compensated for by extrapolating several runs to zero hit size, or simply by running with a relatively small hit size for which the systematic error is correspondingly small. In the systems studied so far, I have found no evidence for a hit size dependence using the look-ahead algorithm. Even a completely open hit, one in which the new field A_j' is completely unrestricted, seems to produce correct results. For instance, in the spinless Schwinger Model (Table I), internal energies for runs with an open hit ($-\pi < A_j' < \pi$), and runs with a fairly restricted hit ($A_j - 0.5 < A_j' < A_j + 0.5$), as well as runs with an exact algorithm, agree within statistical errors of about 5% of the fermionic effect on the system. By this I mean the difference between the measured quantity with the fermions present, and the same

Table 1. Spinless Schwinger Model

Algorithm	Internal Energy	$\langle\bar{\psi}\psi\rangle$	$\langle\bar{\psi}UU\psi\rangle$
look-ahead, $\eta=1$	0.5597(10)	---	---
look-ahead, $\eta=2$	0.5595(5)	---	---
look-ahead, $\eta=2\pi$	0.5591(3)	0.4035(6)	.0594(6)
exact	0.5589(7)	0.4036(5)	.0590(4)
quenched	0.5450(14)	0.4066(8)	.0627(6)
no look-ahead, $\eta=1$	0.5713(15)	---	---
no look-ahead, $\eta=2$	0.5847(18)	---	---
no look-ahead, $\eta=2\pi$	0.6600(7)	0.3948(6)	.0536(4)

Internal energy, $\langle\bar{\psi}\psi\rangle$, and $\langle\bar{\psi}UU\psi\rangle$ (a two lattice spacing correlation function) for the look-ahead algorithm with three different hit sizes (η). $\eta = 2\pi$ corresponds to a completely open hit. Values are also given for the exact determinant calculating algorithm, and for the quenched or pure gauge system. The last three entries are for an algorithm which had no look-ahead update; the ϕ fields in a neighborhood were updated only after each gauge field update. This last algorithm is similar to the standard pseudofermion algorithm with only one ϕ configuration used for averages. $\langle\bar{\psi}\psi\rangle$ and $\langle\bar{\psi}UU\psi\rangle$ were measured from a sample of gauge configurations using a separate pseudofermion Monte Carlo. Gauge configurations were not stored for every run, so this measurement was not performed in every case. Simulations are for $\beta = 1.02$, $m = 0.02$ on a 4×4 lattice. Errors are from binned correlations, with a bin size of 1000 sweeps.

quantity in the quenched approximation, where the fermion determinant is set to unity. The spinless Schwinger model is an unphysical two-dimensional model of a U(1) gauge field, A, interacting with a ψ field which has Fermi statistics but the interactions of a scalar field. Because of its relative simplicity, this model is a good testing ground for lattice fermion algorithms, since the main problem of such algorithms is getting the Fermi statistics right. The precise nature of the interaction matrix O_{ij} is of secondary importance. Of course, one should then go on to test more realistic models, if trials with the simplest models are successful.

The excellent accuracy of the algorithm, for quantities which depend on the gauge fields, is all the more remarkable if one observes the behavior of the running ϕ fields of the simulation. I do not propose using these pseudofermion fields for anything but performing the gauge field simulations; however, it is interesting to monitor $\langle\phi^*\phi\rangle$ from these running fields, and compare it to the same quantity one gets if one takes a sample of gauge fields from the simulation and equilibrates the pseudofermion fields to each gauge configuration using a separate Monte Carlo. This is a standard matrix inversion technique used to measure $\langle\bar{\psi}\psi\rangle$. (For the spinless Schwinger model $\langle\bar{\psi}\psi\rangle = \langle\phi^*\phi\rangle$; for the Schwinger Model with Kogut-Susskind fermions discussed later $\langle\bar{\psi}\psi\rangle = 2m\langle\phi^*\phi\rangle$.[6]) There is a very noticeable lag in the value of $\langle\phi^*\phi\rangle_r$, from the running fields of the simulation which increases in an apparently linear fashion with the hit size. For example a run with an open hit gave $\langle\phi^*\phi\rangle_r =$

0.3721 ± 0.0003 and one with the restricted hit of ±0.5 gave $\langle \phi^* \phi \rangle_r$ = 0.3999 ± 0.0003. On the other hand, the relaxing pseudofermion run on the sample of fixed gauge fields from the open hit run gave $\langle \phi^* \phi \rangle$ = 0.4035 ± 0.0006 and the same procedure applied to a sample of gauge fields from a run with an exact algorithm which calculates the full determinant analytically at each update gave $\langle \phi^* \phi \rangle$ = 0.4036 ± 0.0005. The agreement of these last figures is another indication that the algorithm is producing a correct, or nearly correct, gauge field distribution. The lag in $\langle \phi^* \phi \rangle_r$ for an open hit is a large effect, 8% in absolute terms and 1000% in terms of the fermionic effect as earlier defined. However, this appears to affect the gauge fields produced by the simulation very little, if at all (i.e. within the ~5% statistical error). There would be no lag in $\langle \phi^* \phi \rangle_r$ if the whole lattice were equilibrated at each update, so it is clearly an effect of not equilibrating the more distant neighbors. This would seem to indicate a very high degree of cancellation of errors as far as the gauge fields are concerned. It should be noted that there certainly is room for such cancellations in a Monte Carlo algorithm. One would expect that not equilibrating more distant neighbors would result in a reduction of the Metropolis transition probability $P(A \to A')$. This is because the more distant neighbors will retain memory of A, and likely weight the system toward keeping the old configuration. However, the same effect will also reduce $P(A' \to A)$. If the amount of reduction is the same, i.e. if each probability $P(A \to A')$ is augmented by an extra factor $f(A,A')$, satisfying $f(A,A') = f(A',A)$, then the equilibrium distribution will be unaffected, since the detailed balance equation (16) is unchanged upon inclusion of such a factor.

Details of simulations of the spinless Schwinger model are given in Ref. 1. Runs have since been extended to approximately 500,000 sweeps on a 4 × 4 lattice for the data presented in Table I. Each run consumed 15-20 hours of CPU time on a VAX 780. I have also studied the Schwinger model with Kogut-Susskind fermions,[7] which have the correct spin interactions, although there is still the doubling problem to contend with. Here lattices up to 70 × 70 have been studied, with modest amounts of CPU time on a CRAY XMP. The Schwinger Model is complicated by a symmetry breaking effect. In the continuum, there are an infinity of vacuums distinguished by a parameter θ which does not appear in the Lagrangian.[8] Although, in the massive model, they are not degenerate, they are nevertheless stable. This means that on an infinite lattice there will be no tunneling between different θ-vacua. However, on a finite lattice, there is tunneling among a band of θ's centered around θ = 0, which is, perturbatively, the minimum energy vacuum. Not surprisingly, the tunneling rate increases with increasing hit size, resulting in a hit size dependence of results. I find that these effects can be eliminated by, for each hit size, extrapolating results to θ = 0. On a lattice, θ can be obtained by measuring the expectation value of the imaginary part of the plaquette variable, which, to lowest order, can be equated with the continuum field strength. This is true since θ can be interpreted as an overall background electric field. Larger values of θ can be induced in the simulations by adding a term α·Im (plaquette) to the action, and varying the parameter α. Several runs for each hit size were performed, some with α ≠ 0. Physical measurements such as internal energy are then plotted vs. <Im(plaquette)>, and linearly extrapolated to θ = 0 as in Fig. 1. Although the raw data for widely different hit sizes does not appear to agree, extrapolations to θ = 0 do agree within statistical errors which, in this case, are again around 5% of the fermionic effect on the system. Thus, once the symmetry breaking effect is understood and controlled, possible systematic errors of the algorithm are again below an acceptably small limit, even for an open hit size. A detailed report on the Schwinger model with Kogut-Susskind fermions will appear soon.

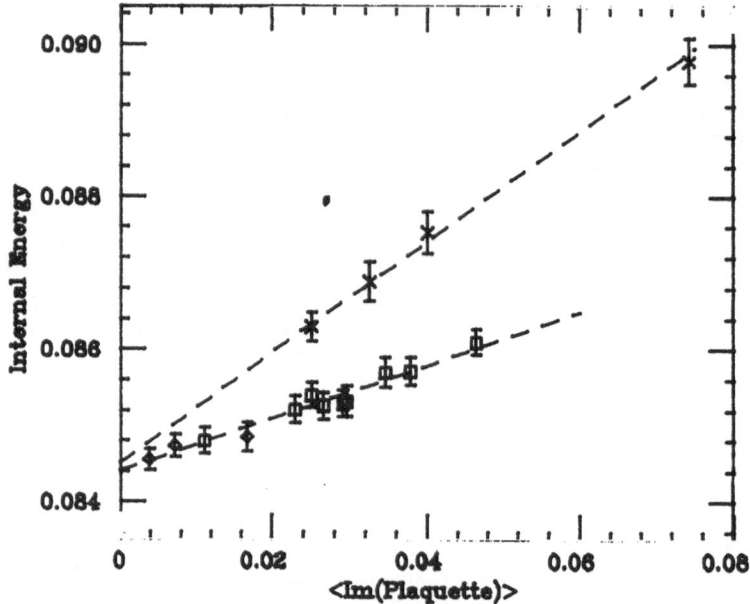

Fig. 1. Extrapolation to θ = 0 for the Schwinger Model with Kogut-Susskind fermions. Data are for β = 6, M = 0.07 on 32 × 32 (x, □) and 70 × 70 (◇) lattices. Data plotted as x are for an open hit size, η = 2π, and as □ and ◇ for the restricted hit size η = 1. Although the data as taken show a hit size dependence, their extrapolations to θ = 0 do not. This suggests that the observed hit size dependence is due to increased vacuum tunneling, and not a systematic error in the algorithm.

The look-ahead algorithm shows promise in that possible systematic errors are apparently well under control, at least for the two-dimensional gauge theories studied. Of course, it must be checked that the same is true for non-abelian four-dimensional theories, which will be done soon. On the surface there does not appear to be any feature of the algorithm which is inherently low-dimensional. One may therefore hope with some confidence that it will work well enough to be of practical value in four dimensions, especially since the performance in two-dimensions is far above the minimum requirements of practicability. Finally, I should mention the speed of the algorithm which is, of course, of primary importance. As an example, for the Schwinger model with Kogut-Susskind fermions, one must update four pseudo-fermion fields for each gauge field update, and then, about half of the time (if the gauge field was accepted) these four fields must be updated again. So one has an average of six pseudofermion updates per gauge field update. In four

dimensions one will have twelve pseudofermion updates per gauge field
update. Thus, one can guess roughly, that the program with fermions will
run about one order of magnitude slower than the corresponding pure gauge
theory algorithm. Considering the high quality of gauge configurations
obtained, due to the freely adjustable hit size, this is extremely fast
for a fermion algorithm. In actual programs I have achieved about one-
half of this guessed performance; however the current programs have some
inefficiencies which, when corrected, should nearly double the speed.
For non-abelian theories this ratio may actually go down, since, for an
SU(N) theory, the number of color components of the gauge field increases
faster with N than do those of the pseudofermion field.

A POSSIBLE JUSTIFICATION

I have attempted to make the previous sections self-contained in
that they describe a fast fermion algorithm which is an approximation to
a much slower exact algorithm, and which works well in two-dimensional
examples. One needn't really analyze the situation further than this.
The method can simply be tested in four dimensions, either by varying the
hit size and looking for consistency or by comparing results to those
obtained using other established methods. However, the complete lack of
a detectable systematic error in the examples tested so far, even for a
completely open hit size, is intriguing. Could it be that the
cancellation of errors which appears to be occurring is exact, making
what appears to be an approximation into an exact algorithm? I wish to
spend some time pursuing this question. The analysis will be along
unconventional lines, as the usual sort of detailed balance proofs do not
seem to shed much light on this question. The reason for this is that
the ϕ distribution in the fermion/gauge system is dependent on the
history of the gauge field updates, as the decision whether to update ϕ
fields is contingent upon acceptance of the previous A field update.
Also, unlike the algorithm for the boson/gauge system, which uses a
single joint probability distribution for both A and ϕ updates, the
pseudofermion fields are updated to a different probability function than
are the gauge fields. Thus the ϕ fields are constantly chasing the A
fields, and don't attain their true equilibrium values. They do as the
hit size, and thus the speed of changes in A, is reduced. On the other
hand, they reach a sort of equilibrium distribution in that operators
which depend on ϕ achieve stable values. Apparently, for a given hit
size, the ϕ distribution achieves a stable, uniform, lag behind the true
equilibrium distribution. As mentioned before, a symmetric bias will not
upset the balance between the transitions A $\rightarrow$ A' and A' $\rightarrow$ A. One needs
to know to what extent bias in the transition probabilities due to the
lag in the ϕ distribution is symmetric.

My approach follows from the observation that, neglecting for a
moment the pure gauge piece of the action, the probability ratios
$P(A \rightarrow A')/P(A' \rightarrow A)$ for the fermion/gauge system and for the analogous
boson/gauge system are simply the inverses of each other

$$\frac{P_f(A \rightarrow A')}{P_f(A' \rightarrow A)} = \frac{P_b(A' \rightarrow A)}{P_b(A \rightarrow A')} = \frac{\det O(A')}{\det O(A)} . \tag{18}$$

The probabilities themselves differ by interchanging the identity of the
old and new gauge configurations, A and A', i.e.

$$P_f(A \rightarrow A') = P_b(A' \rightarrow A) . \tag{19}$$

This is true both for the integrated form of the algorithms, using explicit determinants as above, and for the ϕ-dependent probabilities used in the look-ahead algorithm and the standard boson algorithm.

One can view the production of a chain of gauge configurations from P_f as being closely related to the backward construction of chains for the bosonic system from P_b, i.e. the determination of probable past states of the bosonic chain, given the present state. Past and present, of course, refer to Monte Carlo time. The basic idea of what follows is to interpret the look-ahead algorithm as a procedure which produces a representative chain of configurations for the <u>bosonic</u> system, out of an ensemble of possible chains for a specific correct algorithm. This chain <u>ends</u> with a specified final configuration and can, through applying the look-ahead procedure, be extended indefinitely into the past. The ϕ distribution is viewed as a dependent distribution, produced by the specified boson algorithm which is specially chosen so that it is possible to reconstruct past updates. If one can really reconstruct the ϕ distributions that go with past gauge configurations according to probabilities that correspond to a correct boson algorithm, then the usual detailed balance arguments will apply, implying that the ϕ fields are simulating the determinant in the gauge field transition probabilities without bias. Thus the overall philosophy is not to directly analyze the fermion algorithm. Rather, one first establishes a relationship between it and a correct algorithm for the analogous boson system. Then one can make use of the extensive knowledge one has of the boson algorithm, including full detailed balance.

It is interesting to first consider the idea of these "backward" Monte Carlos in general, without the complications of the ϕ fields. The backward Monte Carlo is defined to be the procedure one gets upon interchanging the roles of the new and old gauge fields in the Metropolis transition probability. For a system with a desired equilibrium distribution $P_{eq}(A) = \exp(-\beta H(A))$, the normal Metropolis probability would be

$$P(A \rightarrow A') = \text{Min}\left(1, \exp(-\beta H(A'))/\exp(-\beta H(A))\right) . \qquad (20)$$

One can easily check that the given distribution solves the detailed balance condition (16). For the backward Monte Carlo one uses a transition probability

$$P'(A \rightarrow A') \equiv P(A' \rightarrow A) . \qquad (21)$$

Interestingly, this does not yield a nonsensical or totally random equilibrium distribution, but rather

$$P'_{eq}(A) = \exp(\beta H(A)) = P_{eq}(A)^{-1} . \qquad (22)$$

The equilibrium distribution produced by the backward Monte Carlo is the negative temperature distribution for the original system, i.e. a distribution with all of the relative equilibrium probabilities inverted. Actually, one wants the pure gauge piece of the probability distribution to be the same for the two systems; however, it too will be inverted under $A \leftrightarrow A'$. This can be corrected by simultaneously taking $\beta \rightarrow -\beta$ in S_0. Thus the actual relationship between the fermion/gauge and boson/gauge transition probabilities is

$$P_f(A \rightarrow A', \beta) = P_b(A' \rightarrow A, -\beta) . \qquad (23)$$

If the transition probabilities are the same, then why are the forward and backward chains, and equilibrium distributions so different? The reason is two-fold. First, there is a selection effect, in that, for the forward constructed chain an initial state is given and the Monte Carlo produces a distribution of final states. In the backward case, a final state is given and an initial state distribution is sought. The second, related, reason is that the backward transition probabilities are not exactly the transpose of the forward ones; they differ in the diagonal elements, i.e. the probability of rejection. It is useful to consider $P(A \rightarrow A')$ as a matrix operating on probability distributions, with indices A and A'. For the boson system

$$M_{bA'A} \equiv P_b(A \rightarrow A') . \tag{24}$$

A particular state A_0 is represented by the distribution $\delta(A - A_0)$. What is of interest is the eigenvector of M with eigenvalue 1, which is the equilibrium distribution. It turns out that this is the largest eigenvalue, so that $P_{eq}(A)$ can be obtained by raising M to a high power and operating on an arbitrary initial distribution

$$P_{eq}(A) \propto \lim_{N \rightarrow \infty} M_{AA'}^N P_0(A') . \tag{25}$$

This is simply a restatement of the Monte Carlo procedure. Now, although it is true that

$$M_{fA'A} = M_{bAA'} = M_{bA'A}^T , \qquad A' \neq A , \tag{26}$$

where T represents the transpose, the diagonal elements are not the same. $M_{fA'A}$ is defined so that it has columns which sum to unity

$$M_{fAA} = 1 - \sum_{A' \neq A} M_{fA'A} , \tag{27}$$

whereas

$$M_{bAA}^T = M_{bAA} = 1 - \sum_{A' \neq A} M_{bA'A} = 1 - \sum_{A' \neq A} M_{fAA'} . \tag{28}$$

In other words, M_f is column normalized whereas M_b^T has the same off-diagonal elements but has diagonal elements chosen so that it is row normalized (normalized in the sense of an arithmetic sum to unity). So

$$M_f = M_b^T + D , \tag{29}$$

where D is a diagonal matrix. The extra term D is really due to the same selection effect mentioned above, but occurring at each updating step. The rejection probability is different if the initial state is fixed and the final state is to be determined, or vice versa. Therefore it is this selection effect, occurring at each stage in the process, which makes for the difference in the forward and backward chains.

The next step is to introduce ϕ updates in between the A updates, and to use the ϕ-dependent probabilities for the A updates, $P(A \rightarrow A', \phi)$. The ϕ updates will be heat bath updates for complete ϕ neighborhoods of a given link, as defined earlier. The idea is to choose the ϕ updates so that both the forward and backward (i.e. look-ahead) algorithms end up with the same ϕ distributions along the chain, for fixed sequences of A configurations. Then, since it is known that the ϕ fields in the forward algorithm correctly simulate the determinant factor, the same will be true of the backward algorithm. There are two requirements for the ϕ updates. First, the ϕ distribution must be insensitive to $P(A \rightarrow A)$ since this differs between the two chains. This can be done by forbidding an update of a ϕ neighborhood if the previous A field update was rejected. Consider, for example, an algorithm depicted as follows:

$$-A_1 - \phi_1 - A_2 - \phi_2 - A_3 - \phi_3- \ . \tag{30}$$

"A_1" means to update the field A_1; "ϕ_1" to update the ϕ neighborhood of A_1, etc. The states of the chain are represented by the dashes and can be labelled by the previous update. The above requirement can be met by considering the $A_1 - \phi_1$ update as a unit, rather than as two independent updates, with probability $P(A, \phi \rightarrow A', \phi')$. The rule can be implemented by augmenting this probability with a factor

$$\left(1 + \delta_{AA'} \cdot (\delta_{\phi\phi'} - 1) \right) \ . \tag{31}$$

This prevents ϕ from being changed if A is not changed. It does not upset detailed balance since it is symmetric under interchange of primed and unprimed configurations. The second requirement is that the ϕ updates are invertible in the sense that later updates can be undone, uncovering earlier states of the chain, and reproducing the past ϕ configurations. One wants forward and backward produced chains that contain no gauge update rejections to have identical probabilities except for end selection effects. It turns out that this is all that is needed to make use of the known ϕ distribution for the forward algorithm in demonstrating the lack of systematic error for the backward (fermion) algorithm.

On the face, the ϕ updates in (30) are not invertible since, e.g. the ϕ_2 update only partially overlaps the ϕ_1 update. Thus the last time some of the fields in the ϕ_2 neighborhood were updated was much further back on the chain. It would be impossible to recreate the conditions of such an "old" update. However, it is not necessary to exactly recreate these conditions. This can be seen as follows. There is a very large amount of freedom in choosing a boson algorithm, in that the order of ϕ and A updates is immaterial, as far as the equilibrium distribution is concerned. The ensembles of Monte Carlo chains produced by different algorithms will, however, differ in detectable ways, such as correlations along the chain or in average rejection probability. Thus some of the information in a particular ensemble of chains is irrelevant. The nature and extent of this irrelevant information can be characterized by considering artificial combinations of pieces of chains produced by one algorithm, which are constructed so that they belong to the ensemble of a different algorithm. Consider the algorithm where the ϕ neighborhood of a gauge field link is updated both immediately before and immediately after the gauge field is updated.

$$-\phi_{1b} \diagdown A_1 - \phi_{1a} \diagdown \phi_{2b} - A_2 - \phi_{2a} - \phi_{3b} - A_3 - \phi_{3a}-$$

$$-\phi'_{1b} - A'_1 - \phi'_{1a} - \phi'_{2b} \diagdown A'_2 - \phi'_{2a} \diagdown \phi'_{3b} - A'_3 - \phi'_{3a}- \tag{32}$$

$$-\phi''_{1b} - A''_1 - \phi''_{1a} - \phi''_{2b} - A''_2 - \phi''_{2a} - \phi''_{3b} \diagdown A''_3 - \phi''_{3a}- \diagdown \ .$$

The subscript "b" refers to a "before" hit and "a" to an "after" hit. One constructs an artificial chain as follows. Choose among the ensemble a chain (primed) which happens to be in the same state after the ϕ'_{2a} update as the double-primed chain is after the ϕ''_{3b} update. Then find another (unprimed) which is in the same state after ϕ_{1a} as the primed chain is after ϕ'_{2b}. This procedure can be extended indefinitely. The artificial chain is made by making the indicated jumps between chains, i.e. of the states

$$-A_1 - \phi_{1a} - A'_2 - \phi'_{2a} - A''_3 - \phi''_{3a} - . \tag{33}$$

Does this artificial chain have the correct equilibrium distribution? Surprisingly, the answer is yes, because the transition probabilities and the order of updates in the artificial chain are the same as in an algorithm where the ϕ neighborhood is updated only after the corresponding gauge field update (30). Thus the artificial chain (33) does not belong to the ensemble of chains from which it was created (32) but it nevertheless does belong to an ensemble associated with a correct algorithm.

The concept of the artificial chain is the final ingredient needed to complete the argument. Although it does not appear possible to recreate ϕ updates for past positions on chains of types (30) or (32), it is possible for the artificial chain. Suppose one is given the state of the chain ϕ''_{3a}. One wishes to have the ϕ distribution at position ϕ''_{3b}, for use in the gauge field update probability

$$P(A_3''^{old} \to A_3'', \phi) . \tag{34}$$

One can achieve this ϕ distribution by inserting the trial old value of the gauge field into the gauge configuration and simply performing the update ϕ''_{3b}. This is just the look-ahead update procedure described earlier. If one wanted to stay on the double-primed chain, then one would have difficulty in undoing the ϕ''_{3b} hit, since the last time some of those fields were updated was long ago. Use of the artificial chain obviates this problem by allowing a skip to a place on a different chain which is reversible. The artificial chain argument is just making explicit the extent to which information carried by the ϕ fields is irrelevant to producing the correct equilibrium distribution. It is telling us that the ϕ_3 update in (30) can be erased, uncovering the ϕ distribution that exists at the point ϕ_2 simply by re-updating the ϕ_3 fields to the conditions that existed before the A_3 update.

Recapitulating, one considers the ensemble of all Monte Carlo chains produced by the algorithm (30) for the boson/gauge system. For this system, detailed balance in the ϕ fields can be used to show that if the ϕ fields are ignored, which is equivalent to integrating over the ϕ distribution, the effective binary gauge field transition probability for $A \neq A'$ is, from the bosonic analog of (15),

$$P_b(A \to A') = \det O(A) \exp\big(S_0(A)\big) f(A,A') ; \qquad f(A,A') = f(A',A) . \tag{35}$$

The look-ahead algorithm reproduces almost the same ensemble of chains, but works backward from a given configuration rather then the usual forward direction. The forward and backward produced chains differ only in the probability of retaining the same gauge configuration, i.e. of rejecting an update, because the normalization conditions of forward and backward transition probabilities are necessarily different. However, since neither algorithm updates a ϕ field if the corresponding gauge field update is rejected, the dependent ϕ distributions are not sensitive to this difference. The artificial chain concept was used to prove that the results of individual ϕ updates used by the two algorithms are the same. Since the look-ahead algorithm, therefore, produces the same ϕ distribution as the normal boson algorithm, the average binary transition probabilities (35) must also be the same. Thus, after integrating out the ϕ distribution one obtains an average fermion transition probability

$$P_f(A \to A',\beta) = P_b(A' \to A,-\beta) = \det O(A')\exp\left(-S_0(A')\right)f(A,A') \qquad (36)$$

which, using (16), gives

$$P_{feq}(A) = \exp\left(-S_0(A)\right)\det O(A) \qquad (37)$$

the desired fermion equilibrium distribution. Also, it is now clear why the running ϕ distribution does not appear to be in equilibrium with the fermion gauge distribution. Rather it is in equilibrium with the corresponding gauge distribution for the negative β boson/gauge system, which has a different weight for a given gauge configuration because of the different probability of retention compared to the fermion case.

The above justification is relatively complex and admittedly unconventional. I do not consider it at the level of a rigorous proof, although I think it contains most of the necessary conceptual ingredients for one. I am still searching for possible flaws in the logic and a clearer and more precise notation in which to express the relevant ideas. The look-ahead algorithm is in a new class of algorithms in which an auxiliary field is updated according to a different function than is the main field one wishes to simulate. For this reason, detailed balance in the auxiliary field is not satisfied. Relaxing the full detailed balance condition allows for a greater range of algorithms, some of which might be able to simulate distributions which cannot be simulated by local algorithms which are fully balanced. In the absence of analytical proofs, such algorithms may still be used by testing them empirically, either against themselves, such as looking for hit size dependences, or against slower, but more easily justifiable, algorithms. I hope to have convinced the reader, however, that analysis of such algorithms, although difficult, is not impossible. It is hoped that further progress in this area will result in a completely rigorous justification of the look-ahead algorithm, or a variant of it.

REFERENCES

1. M. Grady, Phys. Rev. D32 (1985) 1496, and Argonne Report No. ANL-HEP-PR-84-69 (1984) unpublished.
2. F. Fucito, E. Marinari, G. Parisi, and C. Rebbi, Nucl. Phys. B180 (1981) 369.
3. A variation of the pseudofermion algorithm which is more similar in spirit to the one discussed here is given in G. Bhanot, U. M. Heller, and I. O. Stamatescu, Phys. Lett. 129B (1983) 440.
4. N. Metropolis, A. W. Rosenbluth, M. N. Rosenbluth, A. H. Teller, and E. Teller, J. Chem Phys. 21 (1953) 1087.
5. I thank B. Bunk and M. Blairon for suggesting that a multivariate heat bath be used here. The heat bath method is discussed in M. Creutz, Phys. Rev. D21 (1980) 2308.
6. E. Marinari, G. Parisi, and C. Rebbi, Nucl. Phys. B190 [FS3] (1981) 734.
7. T. Banks, S. Raby, L. Susskind, J. Kogut, D. R. T. Jones, P. N. Scharbach, and D. K. Sinclair, Phys. Rev. D15 (1977) 1111; L. Susskind, Phys. Rev. D16 (1977) 3031.
8. S. Coleman, R. Jackiw, and L. Susskind, Ann. of Phys. 93 (1975) 267; S. Coleman, Ann. of Phys. 101 (1976) 239.

DYNAMICAL FERMIONS USING LANCZOS

Ian M. Barbour

Department of Natural Philosophy

University of Glasgow, Glasgow G 2 8QQ

ABSTRACT

Preliminary results are presented for QCD simulated on relatively small lattices with quark loops included. The method is based upon the metropolis updating procedure with the ratio of fermion determinants calculated using a modified version of the Lanczos algorithm for tri-diagonalizing a matrix.

1) INTRODUCTION

Several schemes (1) have been proposed recently to allow studies of lattice Quantum Chromodynamics beyond the quenched approximations, i.e. with the inclusion of quark loops. The pseudofermion and the micro-canonical methods have been used with considerable success in such studies and have shown that ambitious measurements are possible with present day supercomputing facilities. However, studies of lattice QCD in the quenched approximation have shown that it is essential that the non-perturbative contributions, which are dominant in controlling the small eigenvalues of the fermion matrix, must be calculated correctly. It is these contributions which provide the underlying mechanisms which mani-fest themselves in the low energy hadron spectrum.

We describe here a numerically exact metropolis update scheme (2) which ensures that all fermion loops are calculated correctly and which is tractable enough to enable simulations to be performed on reasonably sized lattices.

2) THE UPDATE SCHEME IN SU(3)

We use Susskind fermions on a hypercube lattice. Integration over the fermion fields gives an effective action of the form

$$S_{eff} = S_G(U) - \frac{n_f}{4} \text{ Tr } \ln(D(U) + m)$$

We take the standard Wilson form on the lattice for the gluonic action. The non-local fermionic contribution is given by the discretized form of the Dirac operator describing n_f flavours of mass m. The method can be generalized to deal with alternative forms of the fermionic lattice action and non-degeneracy in the masses.

Implementation of a metropolis update procedure to bring the gauge fields into thermal equilibrium requires knowledge of the ratio of the determinants

$$\frac{|D(U) + m + \Delta|^{n_f/4}}{|D(U) + m|^{n_f/4}} = |1 + (D(U) + m)^{-1}\Delta|^{n_f/4}$$

where Δ is the change in the fermion matrix due to a random change of one link. This ratio is determined if the 6x6 block of the inverse of $D(U)+m$ labelled by the link coordinates is known. Similarly to update at random any one of the 32 links in a given hypercube one requires knowledge of the corresponding 48x48 block of the inverse.

A fast, efficient and accurate method for obtaining the block of the inverse is via an extension of the Lanczos algorithm (3) which basically tridiagonalizes a matrix, i.e.

$$X^+ D X = T \text{ with } X^+ X = 1$$

if D is hermitian. In its block form we write

$$X = \left\{ X_1, X_2, X_3, \ldots \right\} \qquad \text{where } X_i \text{ are } (3n_s^3 n_t \times 24) \text{ block vectors}$$

and T is a block tridiagonal form with off-diagonal elements B_i (24x24) Note that it is more efficient to invert separately for even and for odd labelled columns of the block.

Then, for a given X_1, the tridiagonal form is obtained by the iterative scheme

$$D\, X_1 = X_2 B_1$$
$$\qquad\qquad\qquad\qquad\qquad\qquad (*)$$
$$D\, X_2 = X_1 B_1^+ + X_3 B_2 \quad \text{etc}$$

with diagonal elements m (for Susskind fermions). The B_i can be chosen to be triangular matrices.(4) Then after N steps

$$(D + m)^{-1} X_1 = \sum_{n=1}^{N} X_n A_n + R_N$$

where the A_n are determined by elementary recursion relations. The convergence of the procedure (i.e. $R_N \rightarrow 0$) is determined by the vanishing of a 24x24 matrix. Numerical accuracy is not lost even although the Lanczos vectors X_i lose their orthogonality after several iterations. The block version of the algorithm is 5 to 10 times faster than single row inversion depending upon the coupling β and the quark mass m. A possible reason for this gain is that the blocked form of eqn. (*) is less constrained than non-blocked tridiagonalization. The method is valid for all quark masses $m \gtrsim 0$ and can be generalized to the non-hermitian form found in finite density lattice QCD.

The change of one link on the hypercube will change the 48x48 block of the inverse. The new block is found using rank annihilation. Given a change to one element of the block, say

$$\Delta M = a U V^+$$

with U and V unit column vectors, then

$$(M + a \, UV^+)^{-1} = M^{-1} - \frac{a(M^{-1}u)(V^+M^{-1})}{1 + a \, V^+M^{-1}u}$$

The new block is obtained by applying the above 18 times.

The procedure followed in performing one 'sweep' of the lattice is

1) Invert on a hypercube to obtain the 48x48 block of the inverse.

2) Select a link on the hypercube and update with say 10 hits.

3) With the final accepted link perform rank annihilation to calculate the new block of the inverse due to the change in that link.

4) Select another link on the hypercube and repeat from step 2.

With this procedure one can go round a hypercube several times, say 6 laps, and bring it into local equilibrium with the rest of the lattice. One then picks another hypercube and repeats. One 'sweep' of the lattice requires inverting on $n_s^3 n_t/8$ hypercubes.

The method has several advantages. Implementation of block Lanczos is efficient and accurate. Rank annihilation is fast and, by dealing with a hypercube, essentially reduces the number of degrees of freedom by a factor of 32 by enabling the links on the hypercube to be updated many times. Block Lanczos is valid for all quark masses and ensures that the large fermion loops are correctly handled. The accuracy of the procedure can also be checked by varying the criterion for the vanishing of the 24x24 matrices for convergence and checking that after 1 or more sweeps from a given initial configuration the final configuration is unaltered. The method is minimal in storage requirements and can be extended to objects larger than an elementary hypercube.

However, the scale of the calculation does grow as the space-time volume squared, contrary to other methods. At its present stage of development lattices of size $8^3 16$ in SU(2) and 8^4 in SU(3) are feasible.

3) PRELIMINARY RESULTS

Initial measurements with SU(3) fields were made on a 4^4 lattice for the average plaquette and the chiral condensate at $\beta = 5.2$ for 1 flavour at quark mass of .03 in lattice units. Runs were performed on an IBM(4361) and CRAY XMP. The results, for two different quenched starting configurations are shown in Figs. 1 and 2. Several points emerge. With a sensible quenched configuration initially, convergence in both observables is rapid taking place after 15 'sweeps' and agrees with the IBM run which started from the other side of the phase transition. Fluctuations are relatively large over a small number of sweeps (and apparently correlated in both quantities) indicating that updating hypercube by hypercube is effectively reducing the number of degrees of freedom. Relative 'sweep' times between the IBM and Cray XMP were 900 min and 8 min. These times have subsequently been reduced.

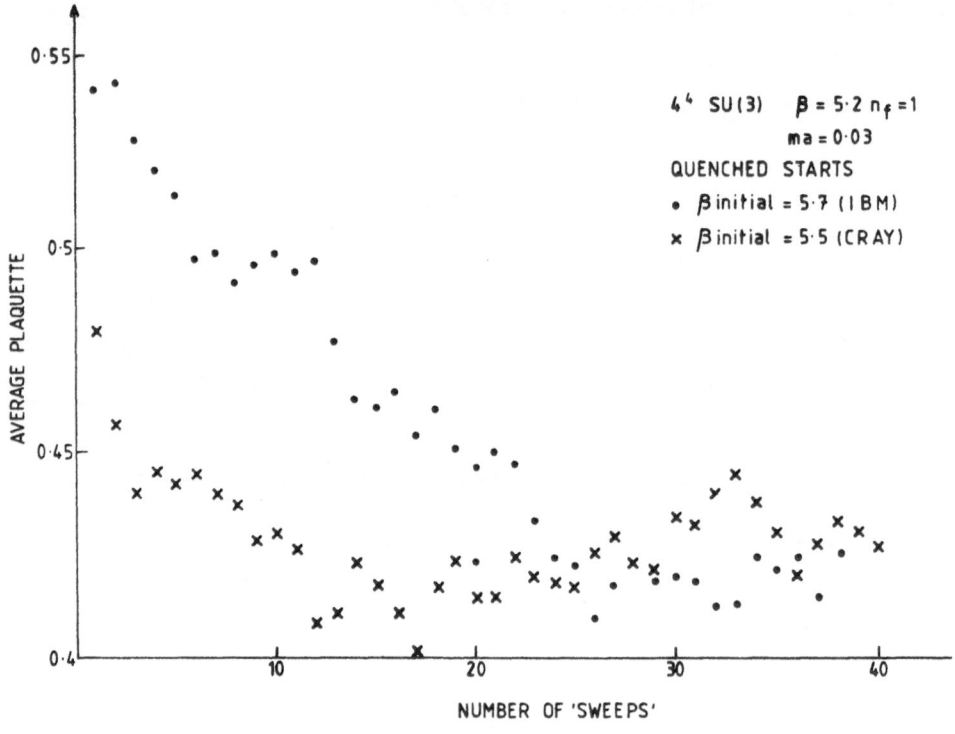

Figure 1

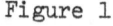

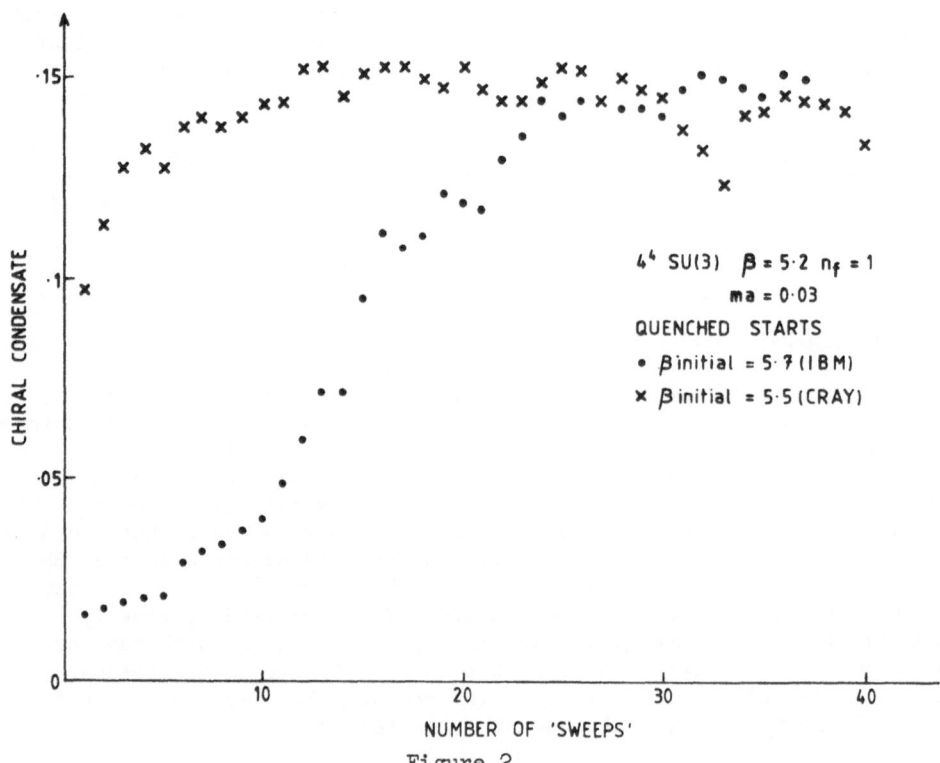

Figure 2

In SU(2), over and above a gain in time from the smaller dimensionality, there is an underlying symmetry structure in the inverse of the fermion matrix. Hence larger scale calculations are possible. The Table summarizes preliminary results (albeit for a small number of sweeps) for the average plaquette and condensate on 6^4 and 8^4 lattices for varying quark mass and coupling and 2 flavours. There is a clear signal at $\beta = 2.2$ on a 6^4 lattice for quark mass $ma = .03$ and $.01$ that the chiral symmetry is restored: the spectral density is essentially zero in both cases for small eigenvalues. The time per sweep is based upon single hypercube updating as described above. Again, these times have been substantially reduced. The above mentioned symmetry of the inverse will allow for larger objects to be updated with a further reduction in sweep and equilibration time.

TABLE

$$SU(2) \quad n_f = 2$$

Lattice	β	m a	'sweeps'	ave.plaq.	$<\bar{\psi}\psi>$	Time(hours)
6^4	2.2	.01	9	.570	.039	1.1
6^4	2.2	.03	12	.576	.053	.9
6^4	2.1	.03	13	.535	.106	1.4
6^4	2.0	.03	25	.500	.136	1.5
8^4	2.1	.035	18	.530±.002	.118±.002	13
8^4	2.1	.05	40	.529±.002	.125±.002	12

4) CONCLUSIONS

Preliminary calculations using the updating procedure and algorithms described above have been encouraging. Calculations can be performed on lattices big enough that will hopefully confirm or refute measurements via other methods such as microcanonical or pseudofermion methods especially at the phase transitions for deconfinement and chiral symmetry restoration. These measurements are underway at present.

ACKNOWLEDGEMENTS

I would like to thank my colleagues who have contributed to the program described above, in particular, Phillip Gibbs and Gerrit Schierholz and also Nasr Behilil, Mohammed Rafiq and Conrad Burden.

REFERENCES

(1) Weingarten, D. and Petcher, D., Phys. Lett. B99, 33 (1981)
Fucito, F., Marinari, E., Parisi, G. and Rebbi, G., Nucl. Phys. B B180[FS2], 369 (1981)
Callaway, D. and Rahman, A., Phys. Rev. Lett. 49, 613 (1982)
Polonyi, J. and Wyld, H.W., Phys. Rev. Lett. 51, 2257 (1983)
Polonyi, J., Wyld, H.W., Kogut, J.B. Shigemitsu, J. and Sinclair,D.K. Phys. Rev. Lett. 53, 644 (1984)
Guha, A. and Lee, S.-C., Phys. Rev. D27, 2412 (1983)

(2) Barbour, I.M., Behilil, N.E., Gibbs, P.E., Rafiq, M., Moriarty,K.J.M. and Schierholz, G. DESY preprint, Jan., 1986

(3)   Barbour, I.M., Behilil, N.E., Gibbs, P.E., Schierholz, G., Teper, M.
      DESY preprint 84-087 (1984), in Lecture Notes in Physics, "The
      Recursion Method and Its Applications" (Springer, Berline, Heidelberg,
      New York, Tokyo, 1985)

(4)   Scott, D.S.: in Sparse Matrices and Their Uses, ed. Duff, I.S.
      (Academic Press, London, New York, Toronto, Sydney, San Francisco,
      (1981)

QCD WITH WILSON FERMIONS ON INTERMEDIATE SIZE LATTICES*

Ion Olimpiu Stamatescu

Institut für Theorie der Elementar-
teilchen, F.U. Berlin, West-Germany

ABSTRACT

Fast fermionic algorithms with local updating, developed in continuation of earlier work are presented, discussed and used to investigate the effect of dynamical Wilson fermions in the SU(3) Yang Mills theory on 4x4x4x4 and 8x8x8x4 laticces. A rapid change in the ρ-behaviour is observed for light quarks. We discuss here also other algorithms and optimization procedures.

1.INTRODUCTION

The present increase in computer power (beyond the characteristic number of 10 Giga: flops, bits) and the developement of new, improved algorithms may provide the factor of 10-100 making the difference between possible and impossible in realistic calculations with dynamical quarks.

The present situation in QCD calculations seems to be the following: due to the strong increase with the lattice volume of the number of operations required by exact algorithms, in order to obtain fast, local algorithms working on large lattices (beyond 6x6x6x6 with Wilson, still larger with Susskind fermions) we must introduce approximations. These may be related to the assumed factorization of the fermionic determinant in a finite step algorithm (like Metropolis, etc.) or the discretization of an infinitesimal algorithm (Random Walk, Langevin Equation), etc.

Between waiting for the better times of a perfect algorithm to come, or keep working with approximative methods without woring, I think that a realistic attitude should be the following: we stay aware of the fact that our "instrument" is not perfect and we try to have a tight control on the various errors which may be introduced by the method.

More precisely, this means that:

(i)   - we must try to avoid biased algorithms, which systematically underestimate (or overestimate) the transition probabilities (the Boltzmann factors ratios), such that even a small error in the local updating adds up over a whole sweep to acquire a volume factor;

---

*) Elaborated version of the talk at the Wuppertal Lattice Meeting, 1985.

(ii) - we must estimate and monitor any error introduced by our approximations, to have an estimation of the uncertainty in the results;

(iii) - we must be able to tune our algorithms such as to decrease any imprecision below a previously set level (at the expence of more work, of course, e.g., by taking farther appart the links over which factorization is assumed, or by reducing the discretization step, etc.);

(iv) - we must try to eliminate sources of metastability and unknown correlations.

These requirements seem to me necessary if we want to obtain results characterizing the model and not be bothered by artefacts of the method.

The situation may be better with global updating algorithms, which in principle can be made exact while still needing only few calculations of the determinant to perform a Yang Mills sweep. In practice we can at least hope to turn some of the systematic errors characteristic to the fast local algorithms into statistical errors. An example of such an algorithm was given in /1,2/ and its analysis is under way /3/.

In an earlier work we presented local algorithms based on the pseudo-fermionic integration /4,5/ used to obtain the determinant ratios /6/ needed in a Yang Mills Monte Carlo with the effective action:

$$S_{eff} = S(YM) - f \; \ell n \; Det(w) \qquad (1.1)$$

Here S(YM) is the pure Yang Mills (plaquette) action and Det(w) is the fermionic determinant. These algorithms allowed us to analyse the problem of the finite temperature transition on a $4^3 \times 2$ lattice. As it was reported in /1/, a strong effect of the dynamical quarks was measured, in spite of their apparently small contribution to the updating probability. The behaviour with β of various observables (Polyakov loops, physical gluonic energy and plaquette) was found to be very flat, indicating smooth or no transition between a "hadronic" and a "plasma" regime for time size of 2 and the fermionic parameters used there (f=3 flavours and hopping parameter k=.12, i.e. an intermediate mass). See Fig.1 below.

On a 4x4x4x4 lattice these algorithms run 4' per Yang Mills sweep. Most of the time is spend in the pseudo-fermionic Monte Carlo producing the determinant ratios. To increase the speed we can try to reduce the number of pseudo-fermionic sweeps, N(PF), and thus the quality of the estimate of the determinant ratios. The uncertainty introduced in this way can be partially removed by using the updating algorithm of Kuti /7/.

In section 2 we shall discuss such an algorithm. The results in section 3 are obtained with a program running 10"/sweep on the 4x4x4x4 lattice and 1' on the $8^3 \times 4$. Other algorithms are discussed in section 4.

The work described here is primarily done in collaboration with Philippe de Forcrand from CRAY Research, Chippewa Falls. Part of the work involves collaborations with G.Feuer, C.Hege, V.Linke and A.Nakamura at the Freie Universität Berlin. I am very indebted to all the above mentioned and to M.Creutz, J.Kuti and other people for very instructive discussions. I want to thank the organizers of the Wuppertal meeting for inviting me and for making this contribution possible.

I would also like to thank G.Spix and the CRAY Research and J.Guralnik and the LANL for supporting my visit to CRAY Research and Los Alamos during which part of the work described here was done. This work is supported in part by the Deutsche Forschung Gemeinschaft.

2. AN OPTIMIZED LOCAL ALGORITHM BASED ON PSEUDO-FERMIONIC INTEGRATION

We shall discuss here a local algorithm optimized in the frame of Kuti's updating method /7/. Various improvements of the pseudo-fermionic Monte Carlo are discussed in Appendix A.

With W the coupling matrix for Wilson fermions:

$$W_{m\,m} = 1\!\!1 - k \sum_{\mu=1}^{4} \left(\Gamma_+^{\mu}(r)\, U_{m\mu}\, \delta_{m\,m+\mu} + \Gamma_-^{\mu}(r)\, U_{m\mu}^+\, \delta_{m\,m-\mu} \right) \; ; \quad \Gamma_{\pm}^{\mu}(r) = r \pm \gamma^{\mu} \qquad (2.1)$$

the quantity needed for the simultaneous updating of N links is:

$$\rho_N\left(\{u_\ell\} \to \{u_\ell'\}\right) = \left(\frac{Det\,(W(\{u'\}))}{Det\,(W(\{u\}))} \right)^{f} \; ; \quad \ell = (m,\mu) \qquad (2.2)$$

with f the number of flavours, k the hopping parameter, r the Wilson parameter, γ^{μ} euclidean Dirac matrices and $U(n,\mu)$ SU(3) matrices.

Using a pseudo-fermionic Monte Carlo /4,5/ we can obtain ρ_N as /6/:

$$\rho_N\left(\{u\} \to \{u'\}\right) = \left\langle exp(-\Delta S(PF)) \right\rangle_{S(PF)}^{-f/2} =$$

$$\left\langle exp(\Delta S(PF)) \right\rangle_{S'(PF)}^{f/2} = \left(\frac{\left\langle exp(-\frac{1}{2}\Delta S(PF)) \right\rangle_{\frac{1}{2}(S(PF)+S'(PF))}}{\left\langle exp(\frac{1}{2}\Delta S(PF)) \right\rangle_{\frac{1}{2}(S(PF)+S'(PF))}} \right)^{f/2} = \ldots \qquad (2.3)$$

where the pseudo-fermionic action $S(PF;\{U\})$ is:

$$S(PF;\{U\}) = S(PF) = \sum_{m,m} \phi_m^* \left(W^+ W(u)\right)_{m\,m} \phi_m \; ; \quad \Delta S = S' - S \qquad (2.4)$$

If we want to perform the pseudo-fermionic calculation of ρ only once for the updating of the N links, we must either use global updating, or find some approximative factorization of ρ_N allowing us to write down a local algorithm, or use the discretize form of an infinitesimal algorithm, which is again an approximation.

The approximative factorization given by the direct linearization:

$$\rho_N \simeq \rho_N' \equiv exp\left(\frac{f}{2} \left\langle \Delta S(PF) \right\rangle_{S(PF)} \right) \qquad (2.5)$$

which corresponds to the "pseudo-fermionic method" /4/ is biased due to Jensen's inequality (see Appendix B):

$$e^{\langle x \rangle} \geqslant \langle e^{-x} \rangle^{-1}$$

more precisely:

$$\rho_N' - \rho_N \simeq \frac{f}{4} \sum_{\ell=1}^{N} \left\langle \left(\Delta_\ell S(PF) - \langle \Delta_\ell S(PF) \rangle \right)^2 \right\rangle_{S(PF)} + \ldots \qquad (2.6)$$

As was illustrated in /1,2/ - see also Fig.1a,b - the error accumulates over each sweep leading to a substantial systematic underestimation of the fermionic effects. This error does not depend on the convergence of the pseudo-fermionic Monte Carlo and growths with the number of flavours. There are a number of results obtained with the pseudo-fermionic method, mainly for staggered (or Susskind) fermions - see, e.g., /8,9/. Because of the systematic underestimation indicated above these results can be considered as representing a lower bound for the fermionic effects.

In the spirit of the requirement (i) in section 1 we prefer a less drastic (while computationally similar) approximation, namely /1,2/:

$$\rho_N \simeq \tilde{\rho}_N \equiv \prod_{\ell=1}^{N} \left\langle exp(-\Delta_\ell S(PF)) \right\rangle_{S(PF)}^{-f/2} \qquad (2.7)$$

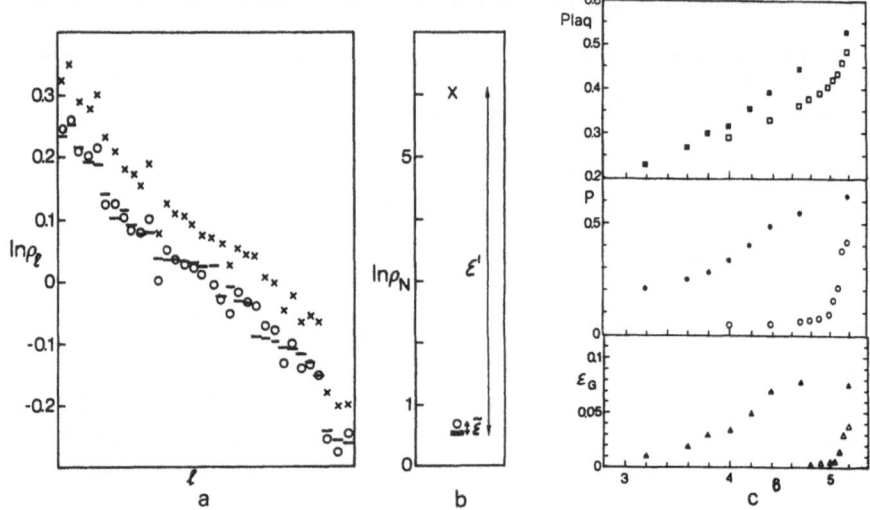

Fig. 1. Factorization approximations. (a) The determinant ratios for each link being changed independently by exp(0.3iX), X SU(3) generators, on a 2x2x2x2 hot (β=4.7) lattice: exact ratios (-), $\langle exp(-\Delta_\ell S)\rangle^{-f/2}$ (o) and $exp\left(\frac{f}{2}\langle\Delta_\ell S\rangle\right)$(x), both with N(PF)=400. (b) Accumulated sweep error: exact ratio (-), $\tilde{g}$ approximation eq.(2.7) (o) and g' approximation eq.(2.5) (x); $\tilde{\varepsilon}$ is the error eq.(2.8), ε' the error eq.(2.6). (c) Plaquette, Polyakov loop and gluonic energy vs β on a $4^3 2$ lattice (f=3, k=.12).

At variance with the direct linearization eq.(2.5), the error introduced by eq.(2.7) has no definite sign and is significantly smaller as it involves only the nondiagonal part of the correlation matrix:

$$\tilde{g}_N - g_N \simeq \frac{f}{2} \sum_{\ell \neq \ell'} \left(\langle \Delta_\ell S(\text{PF}) \Delta_{\ell'} S(\text{PF})\rangle - \langle \Delta_\ell S(\text{PF})\rangle \langle \Delta_{\ell'} S(\text{PF})\rangle \right) + .. \quad (2.8)$$

This error does not need to accumulate. It can be easily monitored using eq.(2.8) and its magnitude can be controlled by thinning out the bushes*, such that the change Δ U(1) needs not be strongly restricted, as in algorithms based on eq.(2.5). Thus we can comply with requirements (ii), (iii), section 1, without contradicting requirement (iv). See Fig.1a,b.

With the factorized form eq.(2.7) we can do a Metropolis algorithm and the results in /1,2/ - see Fig.1c - were obtained in this way.

However, calculating the determinant ratios by a statistical procedure (the pseudo-fermionic Monte Carlo) suggests using Kuti's algorithm /7/ which needs a Boltzmann factor determined only up to some statistical error. The transition probability which we shall then use is:

$$p(u \rightarrow u') = p(\text{YM}) . \alpha \left(\theta(u - u') + \hat{g}(u \rightarrow u') . \theta(u' - u) \right) \quad (2.9)$$

where we use as estimate $\hat{g}$ of the true determinant ratio g the average $\langle exp(-\Delta S(\text{PF}))\rangle^{-f/2}$ obtained over few pseudo-fermionic configurations. Here α is a parameter which should ensure $\alpha\hat{g} \leq 1$ and θ some ordering relation. The a priori probability p(YM) takes into account the Yang Mills action. Of course, the parameter α influences the acceptance and therefore it cannot be chosen too small, as this increases the relaxation time.

---

*) Notice that each class of loops in the determinant expansion /10/ does not couple links beyond a certain distance, although it appears in all k-orders larger than its geometric perimeter. The quality of the factorization approximation is determined by the correlation length, which is only indirectly related to k, but directly controls eq.(2.8).

The updating rule eq.(2.9) fulfils detailed balance in an average sense if $\hat{\varrho}$ is an unbiased estimate of ϱ , i.e. "on the average" $\hat{\varrho} \simeq \varrho$ /7/. Symbolizing this average by $[\ ]$, here we only have, however:

$$[\hat{\varrho}_N{}^{-2/\mathfrak{f}}]^{-\mathfrak{f}/2} = \varrho_N \qquad (2.10)$$

and as:

$$[\hat{\varrho}_N] = \varrho_N + \mathcal{O}\ (1/\sqrt{N(PF)}) \xrightarrow[N(PF)\to\infty]{} \varrho_N \qquad (2.11a)$$

we obtain:

$$[\hat{\varrho}_N] - \varrho_N \simeq \frac{1}{8}\ \mathfrak{f}\ (\mathfrak{f}+2) \sum_{\ell=1}^{N}\ \left(\ [\langle\Delta_\ell S(PF)\rangle^2_{S(PF),\,N(PF)}]\ -\right.$$
$$\left.[\langle\Delta_\ell S(PF)\rangle_{S(PF),\,N(PF)}]^2\right)\ \simeq\ \mathcal{O}\ (1/\sqrt{N(PF)}) \qquad (2.11b)$$

Thus the statistical error introduced by using an imprecise $\hat{\varrho}$ obtained with a small number N(PF) of pseudo-fermionic sweeps is taken care of by Kuti's algorithm. The residual error reflected by eq.(2.11)* has to be monitored and controlled by increasing N(PF) if necessary, or by correction factors to eq.(2.3), using e.g.:

$$\langle\ exp\ (-\Delta_\ell S(PF))\ \cdot\ exp\ (-\tfrac{\mathfrak{f}(\mathfrak{f}+2)}{8}\ \Delta_\ell S(PF)\big(\Delta_\ell S(PF) - \langle\Delta_\ell S(PF)\rangle\big))\rangle\ \rangle \quad (2.12)$$

Notice that the error described by eq.(2.11) is not of the same kind as the one in eq.(2.6): in particular, the error given by eq.(2.11) goes to zero with increasing N(PF) while the one in eq.(2.6) is independent on N(PF) since it affects the averages themselves. How large we should take N(PF) is matter of optimization balancing speed vs precision.

In the following we use this algorithm with the $\hat{\varrho}_\ell$'s calculated once for a whole bush of N links to be changed $\{U(1)\} \to \{U'(1)\}$, as:

$$\hat{\varrho}_\ell = \left(\frac{1}{N(PF)} \sum_{n=1}^{N(PF)} exp\ (-\bar{c}\ (\ell,n)\ \Delta_\ell u)\right)^{-\mathfrak{f}/2} \qquad (2.13)$$

where "n" represents a pseudo-fermionic configuration thermalized with $S(PF;\{U\})$ (before any change in this bush). This corresponds to the approximation eq.(2.7) and thus introduces the error eq.(2.8). For the coefficients $\bar{c}(1,n)$ we use the improved formulae /1,2/ whereby part of the pseudo-fermionic integration is already included analytically (see also Appendix A). This increases the efficiency by a factor 10-100.

3.CALCULATIONS ON 4x4x4x4 AND 8x8x8x4 LATTICES

We shall present here results obtained on 4x4x4x4 and 8x8x8x4 lattices using the algorithm defined by eqs.(2.7,9,13). We continuously monitor the three sources of error as given by eq.(2.8) - the factorization error - and by eq.(2.11) and the "occasional overflow" - which are errors we introduce when using Kuti's algorithm**. Maximal

---

*) The possibility of such an error was pointed out to us by J.Kuti.
**)We expect from the occasional overflow an error roughly of the order of the overflow frequency divided by the average $\ln\hat{\varrho}$. It was pointed out to us by Ulli Wolff and by J.Kuti that eq.(2.3) has infinite variance. This is a quantitative effect, showing up here in very large while seldom deviations of $\hat{\varrho}$ from 1. As an overflow chance allways exists in the practical implementations of Kuti's algorithm, unless $\hat{\varrho}$ is bounded by some reasonably small number, and we continuously monitor this overflow, no qualitatively new effects are introduced by the large variance. Note that using the improved formulae /1,2/ strongly reduces this variance.

values observed were 10%±10% for eqs.(2.8) and (2.11) and 1% overflow, while typical values are much smaller. In general, accounting also for statistical uncertainties, we conclude that we can trust our results up to an overall 20% error concerning the fermionic effects. We use throughout antiperiodic boundary conditions for the fermions and f=3.

The quantities which we measure all the time are the plaquette and the Polyakov loop (thermal Wilson string) averages:

$$A = \langle \tfrac{1}{3} \, Re \, Tr \, (U(Plaq.)) \rangle \qquad (3.1)$$

$$P = \langle \tfrac{1}{3} \, Re \, Tr \, (U(Polyakov)) \rangle \qquad (3.2)$$

and the "fermionic condensate":

$$S = \tfrac{1}{12} \langle \bar{\Psi} \Psi \rangle \qquad (3.3)$$

On the $8^3 \times 4$ lattice we also measure:

$$\varepsilon_G = 3\beta \, (A(time-space) - A(space-space)) \qquad (3.4)$$

and other quantities like Wilson loops, etc. (the latter results are not yet complete and will be published elsewhere). At the level of our precision we put together the two sets of data considered as representing a finite temperature model. Remember that on lattices of time-size 4 pure Yang Mills theory has a deconfining transition at β=5.7.

In Fig.2 we show some results obtained using Kuti's algorithm with various N(PF). As it can be seen, the spread of the points remains inside 20% of the distance to the pure Yang Mills values and no definite trend is seen. Therefore even working at the lowest N(PF) values we should not depass the general level of 20% uncertainty in the fermionic effects. The overall convergence is however slower for small N(PF), roughly by a factor 1/α (see eq.(2.9)), since small N(PF) implies larger fluctuations and requires therefore smaller α to limit the overflow. On the other hand it seems that the number of heating sweeps can safely be reduced to 1-4.

Fig.3 gives A, P and S-S(free) vs k, showing the installation of the "light" quarks regime. The transition (whether a real one or only a cross over is difficult to say) goes from k~.15 at β=5.2 down to k~.12 at β=5.6.

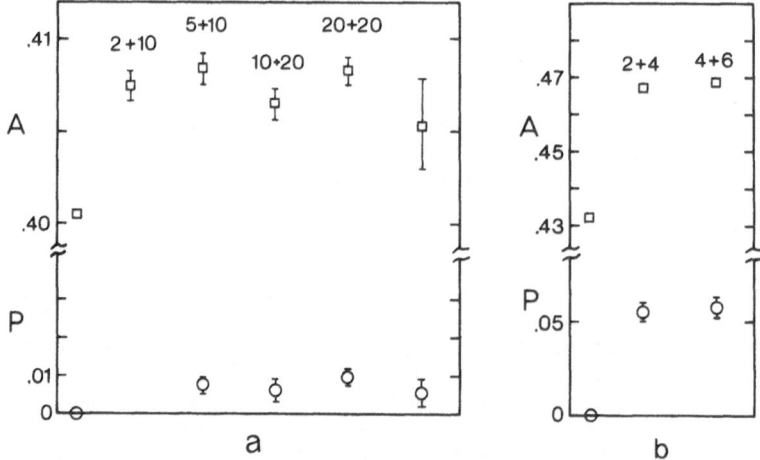

Fig. 2. Plaquette and Polyakov loops using various numbers of heating sweeps (1st number) and N(PF) (2nd number). (a) β=5, k=.12 on the 4x4x4x4 lattice; last point represents a Metropolis run. (b) β=5.2, k=.15 on the $8^3 \times 4$ lattice. First point is pure YM.

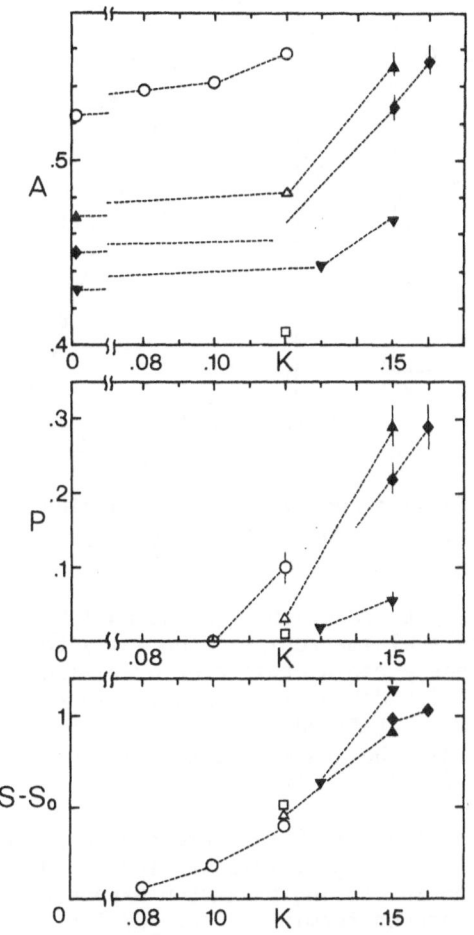

Fig. 3. k-dependence of A, P and S-S(free). Empty symbols are 4x4x4x4 data, full symbols are $8^3 4$ data, for f=3 and β = 5(□), 5.2(▼), 5.3(♦), 5.4(△,▲) and 5.6(○). The dashed lines are only eye-guides.

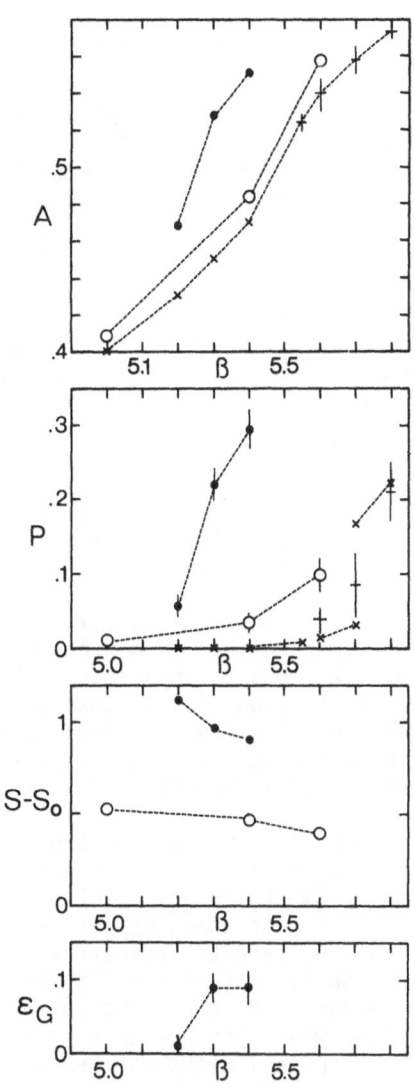

Fig. 4. β-dependence of A,P, S-S(free) and ε_G at fixed k = .15(●, $8^3 4$ lattice), .12(○, 4x4x4x4) and O.(pure Yang Mills, +:4x4x4x4, ✗:$8^3 4$). The dashed lines are only a guide for the eye.

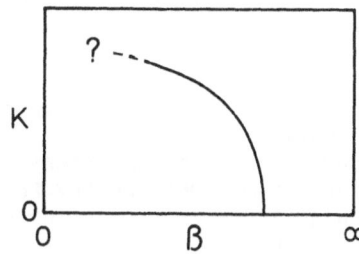

Fig. 5. Expected appearence of the finite temperature transition in the β, k plane.

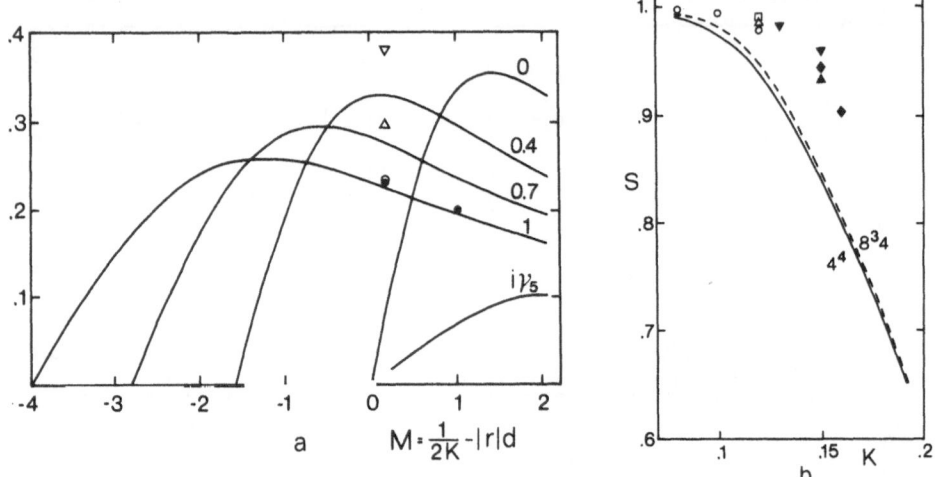

Fig. 6. (a) 4x4x4x4: S at β=5.2, S(free) curves vs M=.5/k-4, various r.
(b) S, S(free) vs k for r=1 (same symbols as in Fig.3).

Fig.4 shows A, P, S-S(free) and $\mathcal{E}_c$ as function of β indicating the change in behaviour from a hadronic to a plasma regime. Especially the data at k=.15 are compatible with either a rapid cross-over or a smooth phase transition. Recall however that with decreasing β k(crit) increases and so does the "renormalized" quark mass at fixed k, roughly ~k-k(crit). Fixing k and decreasing β means thus that the quarks become heavier, which by itself implies a drop in the fermionic effects.

In fact Figs.3,4 represent different views of the same region of the phase diagram and we expect a picture as in Fig.5. This was indeed found in a fundamental Higgs model /2,12/ and in hopping parameter expansion /13/; a similar trend appears with staggered fermions - see /8/ for a review. Our results indicate a similar behaviour here, but the precision is not sufficient to clarify the differences. In particular we must warn the reader that we have found indication that the points at large k, small β (where the pseudo-fermionic fluctuations are large and small α must be chosen) might not be converged and the Polyakov loop might grow.

In Fig.6 we present S, which should be compared with S(free). Most of our data are taken at r=1, but we can show some points also for r<1. The significance of S-S(free) for the chiral transition is however not clear to us, therefore we attempt no interpretation. See /15/ for r=$i\gamma_5$.

Summarizing, we see a clear effect driven by the small quark masses at high k-values which decrease with increasing β. The β-behaviour at these large k-values is consistent with either a relatively rapid cross over or a smooth phase transition, moving to strong coupling with increasing k. The algorithm can ensure an acceptable level of precision.

4.DISCUSSION

The simulation of dynamical fermions is a challenge in lattice theory calculations, and as long as no exact and fast algorithm offers itself straightforwardly we need to consider various methods and try to optimize their performances. As the promising algorithms are still in the build up phase no definitive cross check can be done, yet some comparison between them can help estimate the developement possibilities.

Recall that algorithms with pseudo-fermionic integration using local finite step updating are exact only with sequential updating. That is, one needs to calculate ρ after each updating, either on the basis of the present configuration ("forward"), or on the basis of the proposed new configuration ("backward"), or on some intermediate configuration - see eq.(2.3) /6/. The number of pseudo-fermionic sweeps N(PF) can be reduced, in the extreme case down to 1, using Kuti's (forward) algorithm /7/ or Grady's (backward) algorithm /16/. Except for Grady's method in the special case f=2, a systematic error of the type eq.(2.11) is thereby introduced, maximal for N(PF)=1 (for backward algorithms $f \rightarrow -f$ here)*. Moreover, as in Kuti's or Grady's algorithms detailed balance holds only on the average, reducing N(PF) worsens the overall convergence.

In a sense one has here two diffusion processes with different diffusion constants and N(PF) should be compatible with the ratio of the latters /17/. Since the relaxation time is dominated by the slowest of the two processes we effectively have a lower bound on   N(PF) x Nr.of YM sweeps   and it does not help to try to reduce N(PF) too much.

Notice incidentally that using, as we do here, the improved observables $\bar{O}$ /1,2/ with Kuti's or Grady's algorithms means that even for N(PF)=1 part of the pseudo-fermionic integration is already done, implying a quite large "effective" standard N(PF). In particular, the following algorithm (which can be developed from Grady's):
- propose a change $U \rightarrow U'$ using the Yang Mills action,
- produce pseudo-fermions thermalized with exp(-S(PF;U')),
- update with Metropolis probability:

$$p(U \rightarrow U') = p(YM;U \rightarrow U') \min(1,\bar{O}(U-U'))$$

$$\bar{O}(U-U') = \frac{\int [d\phi_1 d\phi_2] \, exp(-S(PF;U))}{\int [d\phi_1 d\phi_2] \, exp(-S(PF;U'))} \; ; \; \phi_1 \bullet\!\!\!-\!\!\!-\!\!\!\overset{u,u'}{-\!\!\!-\!\!\!-\!\!\!\blacktriangleright}\!\!\!\bullet \phi_2 \quad (4.1)$$

also fulfils detailed balance on the average**. Here the xplicite formulae for the improved observables $\bar{O}$ can be found in /1,2/ ($U \leftrightarrow U'$). We are currently trying to implement this algorithm on bushes of V/16 or V/12 links on which $\bar{O}$ factorizes (see Appendix A) /18/. Of course one can write backward algorithms using eq.(2.3) with every finite N(PF), interpolating between the algorithms in /6/ and /16/ - similar to Kuti's.

Simultaneous local updating of many links treats the reactions of the fermionic determinant to changes in different links as independent. The quality of this approximation can be optimized and should only depend on the correlations between links. Here again one has the trade between speed and precision, when the links must be taken more and more far appart to reduce the correlations.

In our calculations as described above we found possible to reduce N(PF) to 2-10 improved pseudo-fermionic sweeps, without generating unacceptable errors via eq.(2.11) and the occasional overflow. Then we need $O(10^3)$ Y.M. sweeps to get convergence at $k \gtrsim .12$. On the bushes used here, of V/2 links which have no sites or plaquettes in common, the nondiagonal correlation eq.(2.8) builds up to maximally $2\% \pm 2\%$ for large ρ and k. To stay in the frame of < 20% total error in the fermionic effects, we shall need both to increase N(PF) and to thin out the bushes when going to higher ρ, k and on larger lattices (see also /19/). But as by now we only need $O(10^3)$ improved pseudo-fermionic sweeps per Y.M. sweep.

For comparison, the global heat bath algorithm introduced in /1,2/ seems to need an N(PF) of some hundreds /3/. As it uses 64 bushes on

---

*) Notice that larger f can also be simulated by more pseudo-fermions.
**)This result was derived incited by a discussion with M.Creutz.

lattices of sides multiples of 4 (48 bushes for multiples of 6), for which factorized improved formulae $\bar{O}(\{U\} \to \{U'\})$ are given - see App.A -, it needs some 10000 improved pseudo-fermionic sweeps per Yang Mills sweep - again independently on the lattice size (and again this is true only in part, here because for larger bushes the pseudo-fermionic averages fluctuate more and larger N(PF) will be needed). The overall convergence however is very good, characteristic to the heat bath updating. Therefore we hope to lose only a factor of 10-30 against the present algorithm, but by which factor we buy the elimination of any systematic approximation.

Of course, one can envisage also a Kuti or Grady version of the global algorithm, by which we mean in general forward or backward algorithms using eq.(2.3) with restricted N(PF), applied to update a whole bush of links at a time. The transition probability would be of the type eqs.(2.9) or (4.1), respectively, rewritten for the whole bush with help of $\bar{O}(\{U\} \to \{U'\})$ and $\bar{O}(\{U'\} \to \{U\}) = 1/\bar{O}(\{U\} \to \{U'\})$.

As already mentioned, another possibility is to use an algorithm based on the diffusion equation. Such algorithms show an error introduced by the discretization employed for the Fokker-Planck or Langevin Equation, which can be dealt with either through an extrapolation /1,2/ or by using refined versions of the discretization /20/. A Random Walk algorithm was applied in /1,2/ to QCD on a $4^3 2$ lattice, and it has been shown that for vanishing step size the results extrapolate nicely to the Metropolis values. Methods for reducing the discretization error and speed up the convergence of the Langevin Equation are given in /20,21/.

A Langevin Equation algorithm based on the pseudo-fermionic Monte Carlo can be defined via the following updating rule (compare /20,21/):

$$U_\ell \to U_\ell' = U_\ell \, exp \left(i \sum_s X_s \left(\varepsilon \, K_{\ell,s}(YM) + \varepsilon \, K_{\ell,s}(PF) + \sqrt{\varepsilon} \, \omega_{\ell,s} \right) \right) \quad (4.2)$$

with:

$$K_{\ell,s}(YM) = \frac{-1}{2\eta} \left(S(YM; U_\ell e^{i\eta X_s}) - S(YM; U_\ell e^{-i\eta X_s}) \right) \simeq -S(YM; U_\ell; X_s) \quad (4.3)$$

while for the force $K(PF; U \to U')$ we use an estimation obtained with few pseudo-fermionic sweeps (in the extreme case just one):

$$K_{\ell,s}(PF) \simeq \hat{K}_{\ell,s}(PF) = \frac{1}{4\eta} \, \ell n \frac{\langle \bar{O}(U_\ell(e^{i\eta X_s}-1)) \rangle_{N(PF)}}{\langle \bar{O}(U_\ell(e^{-i\eta X_s}-1)) \rangle_{N(PF)}} \simeq \frac{-f}{2N(PF)} \sum_{n=1}^{N(PF)} Re Tr \left(\bar{c}(\ell,n) U_\ell; X_s \right) \quad (4.4)$$

over the configuration $\{U\}$ in place. Here $\varepsilon = \eta^2$ is the discretization step and X_s are SU(3) generators (see Appendix C for some general formulae /22/). The Langevin noise

$$\langle \omega_{\ell,s}(\tau) \, \omega_{\ell',s'}(\tau') \rangle = 2 \, \delta_{\ell\ell'} \, \delta_{ss'} \, \delta_{\tau\tau'} \quad (4.5)$$

will be renormalized by the noise introduced via the approximation eq.(4.4), but only in the next order in the step size. Also here applies the remark made above, about the inefficiency of reducing N(PF) too much. Therefore it is advantageous to use the improved formulae for the $\bar{c}(1,n)$.

The question we didn't touch yet is: what kind of physics can be made on the intermediate size lattices on which these algorithms are practicable? We know from pure Yang Mills analyses that to study scaling properties and see continuum theory we probably need very large lattices. I think, however, that there are interesting physical questions which can be studied on lattices of intermediate size (by which I mean volumes of 1-10 $\times 10^3$ for Wilson, larger for staggered fermions). For instance:
 - Thermodynamik properties and ordering effects: we have seen in /1,2/ and here that we can obtain nontrivial information on the finite temperature properties already on $4^3 2$ to $8^3 4$ lattices.

- Qualitative effects: hadronization and distorsion of the heavy quark potential can already be observed on an $8^3 4$ lattice /12/.
- Corrections to other approximations: for instance, spectrum calculations with dynamical quarks can be compared with quenched results on the same lattice, to see the trend and understand the corrections.

Therefore the developement of fast, reliable algorithms which are practicable on intermediate size lattices is in my opinion both important for the algorithms building program and interesting for selected physical questions.

APPENDIX A

We shall discuss here shortly some optimization techniques for the pseudo-fermionic Monte Carlo.

Following /23,24/ the improved observable is defined as /1,2/:

$$\bar{O}(u \to u') = \frac{\int d\phi_1^* d\phi_1 \, d\phi_2^* d\phi_2 \, \exp(-S(PF;u'))}{\int d\phi_1^* d\phi_1 \, d\phi_2^* d\phi_2 \, \exp(-S(PF;u))} \quad ; \quad \partial u = \partial u' = \phi_2 - \phi_1 \quad (A.1)$$

and we have:

$$\langle \bar{O}(u \to u') \rangle_{S(PF)} = \langle O(u \to u') \rangle_{S(PF)} = \langle \exp(-\Delta S(PF)) \rangle_{S(PF)} \quad (A.2)$$

The analytic formulae can be written down for simultaneous changes of the links in a bush such that the end-sites of the links do not communicate via W*W (see eq.(2.1)), with the coefficients $\bar{c}(1,n)$ given in /1,2/:

$$\bar{O}_N(\{u\} \to \{u'\}) = \exp \left(- \sum_{\ell=1}^{N} Re \, Tr \, (\bar{c}(\ell, \{\phi\}) \Delta u_\ell) \right) \quad (A.3)$$

There are 64 such bushes on lattices of sides multiples of 4, and 48 for multiples of 6. A small routine to construct these bushes can be obtained any time as file or list from the author. The efficiency of the pseudo-fermionic Monte Carlo increases by a factor of 10-100 - see Fig.7.

Other optimization possibilities run under the general name of pre-conditioning. As a first example we consider a renormalization of the pseudo-fermionic fields, to wit:

$$\varrho^{-2/5} = \frac{\int [d\phi] \exp(-\phi^* M' \phi)}{\int [d\phi] \exp(-\phi^* M \phi)} = \lambda^{\alpha r} \langle e^{-\phi^*(\lambda M' \lambda - M)\phi} \rangle_M \quad (A.4)$$

and we look for λ such as to minimize the variance of eq.(A.4). For a diagonal coupling matrix M=W*W the exact solution is:

$$\lambda^{-2} = \frac{1}{N} \langle \phi^* M' \phi \rangle_M = 1 + \frac{1}{N} \langle \phi^* \Delta M \phi \rangle_M \quad (A.5)$$

leading to zero variance, as this is already the exact result. In the realistic case we can take as trial eq.(A.5) evaluated on a subensamble.

As a second example, we can envisage extracting known terms of the hopping parameter expansion from the determinant. Writting (M=W*W):

$$I(R;\{u\}) = \int [d\phi] e^{-\phi^* R^* M R \phi} = (Det(R) \, Det(M))^{-2} \quad (A.6)$$

we chose R such that Det(R) and 1/R are trivial, and so that R*MR lacks the lower terms in the k-expansion - to be overtaken in the prechoice, by adding the corresponding loops /10/ to the Yang Mills action. Thus, e.g.:

$$R_{m,m} = \delta_{m,m} \cdot \exp \left(-4k^4 \sum_{\mu < \nu} Plaq(m,\mu\nu) - \frac{6}{N_t} f \, (2k)^{N_t} Pol.(m) \right) \quad (A.7)$$

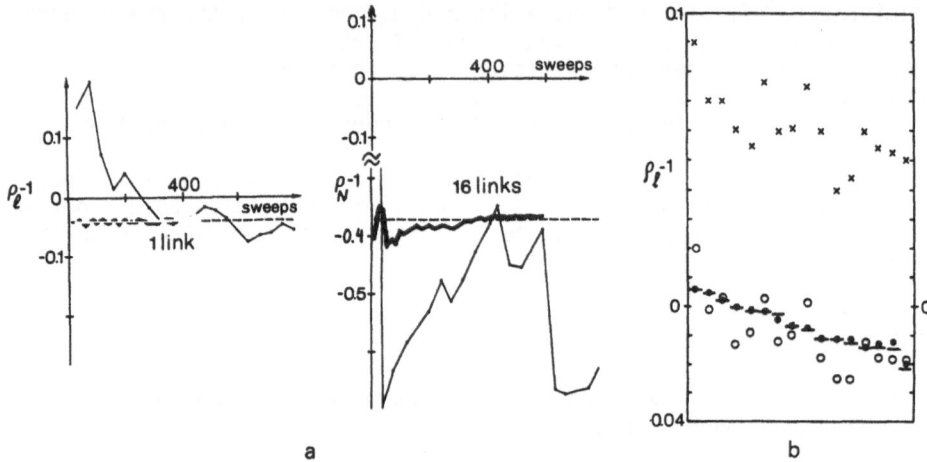

Fig. 7. Improved vs standard pseudo-fermionic averages on a hot
(β=4) $2^3 4$ lattice (U'=U exp(0.4iX), k=.12). (a) Convergence
of the standard (thin) and improved (fat) pseudo-fermionic
Monte Carlo (f=2). (b) Exact (-), standard (o), improved (•)
and linear approximation eq.(2.5) (x) (f=3).

where the coefficients has been arranged such that $\ln I(R;\{U\})$ has no
plaquettes and Polyakov loops in its k-expansion , is used in a
pseudo-fermionic Monte Carlo to calculate

$$\rho(R) = \left(I(R,\{u'\}) / I(R,\{u\}) \right)^{-f/2} = \left\langle e^{-\Delta S(PF,R)} \right\rangle_{S(PF,R)}^{-f/2} \qquad (A.8)$$

which will be used in the updating with the prechoice taken according to
the action $S(YM) - (f/2) \ln Det(R)$.

APPENDIX B

Let us denote
$$\langle f(x) \rangle = \int_a^b dx \; f(x) \; \rho(x) \qquad , \; \rho(x) > 0 \qquad (B.1)$$

Generalized Jensen inequalities can easily be obtained if the convexity
of the functions f(x) can be simply compared. More precisely, we have:

$$f^{-1}(\langle f(x) \rangle) \geqslant g^{-1}(\langle g(x) \rangle) \quad \text{if} \quad f''/f' \geqslant g''/g' \geqslant 0, \ldots \text{(B.2)}$$

in particular:

$$\exp(\text{arccosh} \langle \cosh x \rangle) \geqslant \langle e^x \rangle \geqslant e^{\langle x \rangle} ; \; \langle e^{px} \rangle \geqslant \langle e^x \rangle^p \text{(B.3)}$$

APPENDIX C

For SU(n) groups a simple parametrization of the generators is

$$B_+^{(ab)} = \tfrac{1}{2}(A^{(ab)} + A^{(ba)}); \; B_-^{(ab)} = \tfrac{i}{2}(A^{(ab)} - A^{(ba)}); \; a < b; \; C^{(a)} = A^{(aa)}/\sqrt{2 - \tfrac{2}{n}} \quad (C.1)$$

where
$$A^{(ab)}_{i\,j} = \delta_{ia}\delta_{jb} \quad ; \quad P^{(ab)}_{i\,j} = (\delta_{ia} + \delta_{ib})\delta_{ij} \qquad (C.2)$$

Then (η is the step):
$$\exp(i\eta B_\pm^{(ab)}) = 1 + (\cos\tfrac{\eta}{2} - 1) P^{(ab)} + i \sin\tfrac{\eta}{2} . B_\pm^{(ab)}$$
$$\exp(i\eta C^{(a)}) = \delta_{ij}(1 + (e^{i\eta} - 1)\delta_{ai}) e^{-i\tfrac{\eta}{n}\frac{1}{\sqrt{2-2/n}}} \qquad (C.3)$$

and the change $U \to U' = U \exp(i\eta X)$ can simply be written as a set of routines involving only a fraction of the number of operations implied by matrix multiplication /22/. This can be advantageous for large n.

REFERENCES

1   Ph.de Forcrand and I.O.Stamatescu, Nucl.Phys. B261 (1985) 613

2   I.O.Stamatescu, in "Advances in Lattice Gauge Theory", D.W.Duke and J.F.Owens eds., World Scientific 1985

3   A.Nakamura and I.O.Stamatescu, work in progress

4   F.Fucito, E.Marinari, G.Parisi and C.Rebbi, Nucl.Phys. B180 /FS2/ (1981) 360

5   D.H.Weingarten and D.N.Petcher, Phys.Lett. 99B (1981) 333

6   G.Bhanot, U.Heller and I.O.Stamatescu, Phys.Lett. 128B (1983) 440

7   A.Kennedy and J.Kuti, Phys.Rev.Lett. 54 (1985) 2473

8   F.Karsch, talk at the Meeting "Lattice Gauge Theory - a Challenge in Large Scale Computing", Wuppertal, Nov. 1985

9   C.Rebbi, talk at the Meeting "Lattice Gauge Theory - a Challenge in Large Scale Computing", Wuppertal, Nov.1985

10 I.O.Stamatescu, Phys.Rev. D25 (1982) 1130

11 Ph.de Forcrand and I.O.Stamatescu, preprint Berlin, Jan.1986, to be published

12 Ph.de Forcrand and I.O.Stamatescu, in preparation

13 F.Karsch, E.Seiler and I.O.Stamatescu, Phys.Lett. 157B (1985) 60

14 P.Hasenfratz, F.Karsch and I.O.Stamatescu, Phys.Lett. 133B (1983) 221

15 K.Osterwalder and E.Seiler, Ann.Phys.(N.Y.) 110 (1978) 440

16 M.Grady, Phys.Rev. D32 (1985) 1496

17 D.Zwanziger, Phys.Rev.Lett. 50 (1983) 1886
    P.Rossi and D.Zwanziger, Nucl.Phys. B243 (1984) 261

18 Ph.de Forcrand and I.O.Stamatescu, work in progress

19 D.Weingarten, IBM Watson Research Center Report (1985)

20 G.Batrouni et al., Cornell preprint CLNS-85(65), May 1985

21 J.B.Kogut, Illinois preprint, ILL-(TH)-85-#75, Oct.1985

22 A.El'Kadra, thesis Berlin 1986

23 G.Parisi, R.Petronzio and F.Rapuano, Phys.Lett. 128B (1983) 418

24 Ph.de Forcrand and C.Roiesnel, Phys.Lett. 151B (1985) 77

THE POTENTIAL BETWEEN STATIC QUARKS IN SU(2) LATTICE GAUGE THEORY WITH DYNAMICAL FERMIONS*

E. Laermann, F. Langhammer and P.M. Zerwas

Inst. f. Theoret. Physik
RWTH Aachen
D-5100 Aachen, West-Germany

and

I. Schmitt
Physics Department
University of Wuppertal
Gaussstr. 20, D-5600 Wuppertal 1

ABSTRACT

We investigate the interquark potential on a 16 x 8^3 lattice, including dynamical fermions in the Monte Carlo simulation of the SU(2) lattice gauge theory. At large distances we find indications for a deviation from the linear rise of the potential, expected from a break-up of the flux tube between the heavy quarks through spontaneous creation of light quark pairs. The results are consistent with general physical expectations and show reasonable agreement with results obtained for the quark condensate and meson masses.

1. INTRODUCTION

In pure non-Abelian gauge theories of the strong interactions quarks are expected to be confined. Evidence has been accumulated in recent years that the potential between static quarks in fact rises linearly with distance [1]. The presence of light dynamical quarks, however, should alter this picture: spontaneous quark-pair creation in the field stretched between the static quarks screens the color charge of the heavy fermions at a scale 0 (1 fm) and thus turns the linearly rising potential at large distances into the (asymptotically constant) short range potential between bound states of heavy and light quarks. At small distance, the decreasing vacuum polarization charge should in turn lead to a stronger Coulomb part of the force than in a pure gauge theory.

The lattice formulation provides, at present, the only calculational scheme in which quantum chromodynamics can be solved at large distances. Tackling the problem of color screening needs to go beyond the quenched

---

*presented by I. Schmitt

approximation. Through the fermion determinant highly non-local terms are introduced into the effective gauge field action requiring a formidable computational effort in the simulation of lattice gauge theory with dynamical fermions. Employing the pseudofermion method |2|, we have approached this problem for four-fould degenerate Kogut-Susskind fermions in color SU(2) on a 16 x 8³ lattice at ß-values between 1.85 and 2.5 with quark masses generally chosen between .05 and .2 in units of the inverse lattice spacing |3|. We expect to approach approximate asymptotic scaling in this ß range, allowing us to map the values of the potential measured at various couplings ß onto one curve. The scale is fixed by assigning the string tension the (experimental) value $\sqrt{\sigma}$ = 400 MeV. Because of the somewhat modest lattice size, the necessity to extrapolate to small quark masses and and possible systematic errors of the pseudofermion method we don't expect our results to be correct at a few percent level. Our primary target is rather of qualitative nature: to see that at small to medium distances the Coulomb part of the force is strengthened and to look for indications of color screening at large distances when light dynamical quarks are included.

We have carried out cross checks by determining the π, ρ masses and the chiral condensate |4|. The results are mutually consistent. We furthermore compare our results with data obtained by means of the microcanonical method, wherever such data are available |5,6|. The agreement we encounter in this comparison makes us hope that the systematic errors inherent to the pseudofermion approach are well under control.

2. WILSON LOOPS AND POTENTIALS

In this section we will report on the results we have gotten for the interquark potential. Measurements of the quark condensate and the π,ρ masses will be discussed subsequently. The average plaquette does not change dramatically if dynamical quarks are attached to the gauge boson system, Fig. 1. At ß = 2.1, for example, it decreases by 10% for m=.1 compared with the quenched value |7|. This was theoretically anticipated in a weak coupling expansion of the unquenched plaquette |8|. The data slowly approach the perturbative value but still deviate from it even for the largest ß's considered here. For large quark masses the average plaquette converges to its quenched value. Choosing the quark mass m = 10, we observe that the fermions have frozen out. We want to mention that the measurements of the plaquette in ref. |6| and the data presented here nicely agree with one another.

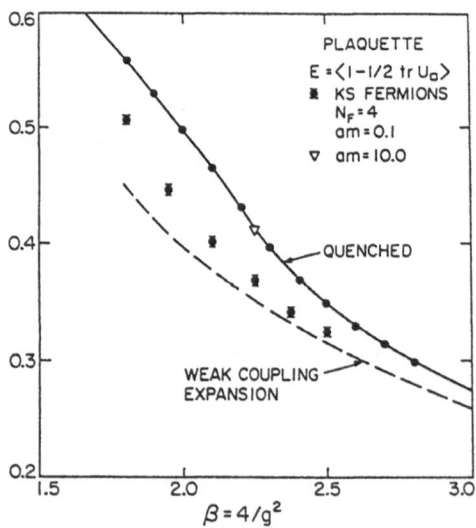

Fig.1: The average plaquette with and without Kogut Susskind fermions.

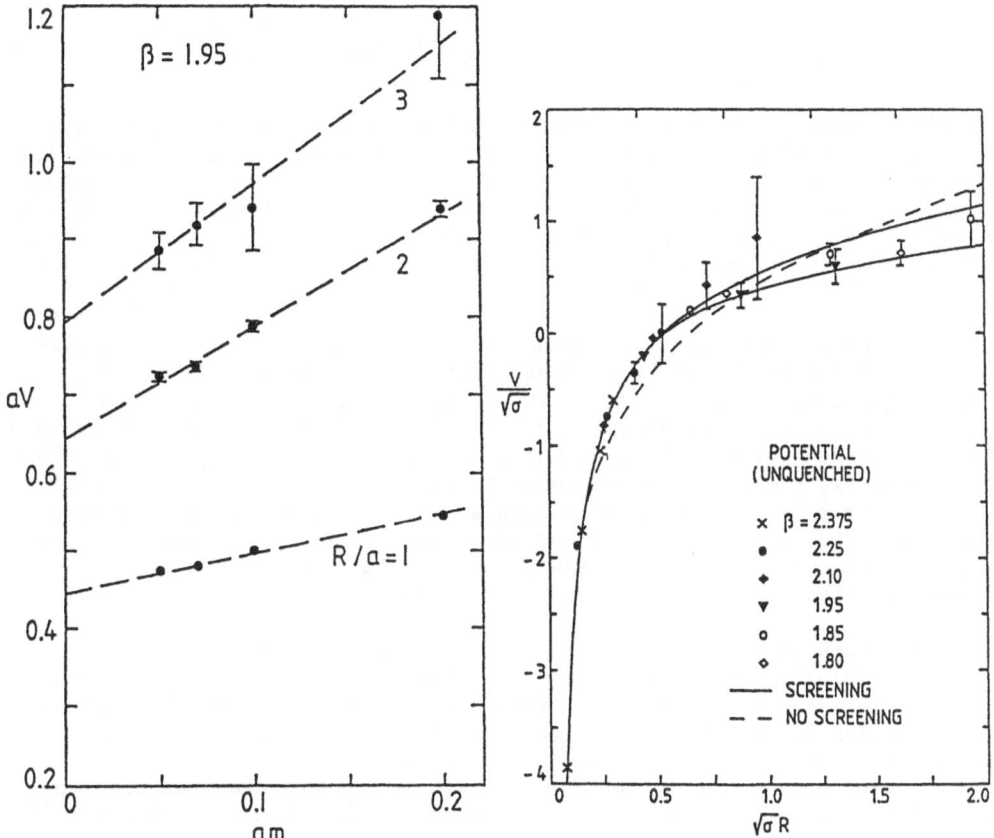

Fig. 2: The potential vs. the quark mass at ß=1.95 including the linear extrapolation to zero quark mass.

Fig. 3: The potential of the complete theory.

Large Wilson loops are more affected by the presence of fermions, raising their values by up to an order of magnitude at a given ß. We extract the potential from the Wilson loops $W(R,T)$ by standard procedures

$$V(R) = - \lim_{T \to \infty} \frac{1}{T} \log W(R,T).$$ (1)

The unquenched potential aV depends on the fermion mass am. This dependence can be approximated by a linear function to extrapolate aV down to zero quark mass, as is shown in Fig. 2 for ß = 1.95. In order to map the values of the potential at am = 0, measured for different couplings ß onto one curve, we assume that asymptotic scaling holds approximately in the ß range covered by our analysis. This allows us to express the potential as a function of the distance in one common, yet to be determined physical unit. The quark self-energy is subtracted off by adjusting one point of the potential values in the region where they overlap.

The result of this analysis is shown in Fig. 3. We paramatrize the potential by an ansatz that corresponds to a cut-off confinement form with screening length μ^{-1},

$$V(R) = const. + \left(-\frac{\alpha}{R} + \sigma R\right)\frac{1 - e^{-\mu R}}{\mu R} \qquad (2)$$

At small to medium distances R this parametrization maintains a Coulombic plus linear behavior whereas the R-dependent part asymptotically approaches a constant $\delta = \sigma/\mu$, the splitting energy of the heavy quark pair. Setting the string tension to $\sigma = (400\ MeV)^2$, the fit returns a Coulomb coefficient $\alpha = 0.28 \pm 0.02$, a lattice spacing $a(\beta=1.95) = 0.24 \pm 0.03$ fm and a screening length $\mu^{-1} = 0.7 \pm 0.2$ fm. The full lines in Fig. 3 correspond to the two extreme values $\mu^{-1} = .9$ (upper curve) and $.5$ (lower curve) respectively.

In order to test the significance of the screening effect we parametrized the unquenched data by a superposition of a linear and Coulombic term without a screening factor in the potential. The result of the fit, corresponding to $\alpha = 0.28 \pm 0.02$ and $a(1.95) = 0.16 \pm 0.02$ fm is represented by the dashed curve in Fig. 3. While such an explanation cannot be ruled out completely, the dashed curve tends to intersect the unquenched data points, lying below at medium distances and overshooting at large distances. This feature is even more pronounced when one attempts to describe the unquenched data only by a renormalization of the lattice spacing a while leaving the other quenched parameters unchanged.

The screening length $\mu^{-1} = .7$ fm obtained from the fit seems reasonable. The Coulomb coefficient α is larger than the analogous quenched value for which we got $\alpha = .21$. This is plausible since α is an effective parameter accounting for string as well as short distance effects in which the leading term of the coupling constant is proportional to $1/(22-2N_F)$. Also the splitting energy $\delta = \sigma/\mu \sim 600$ MeV falls into the range expected from quarkonium spectroscopy |9|. The $\Lambda_{\overline{MS}}$ value extracted from the potential is lower in the unquenched case by approximately a factor of 2 compared with the quenched approximation, where we get $\Lambda_{\overline{MS}}(N_F = 0) = 110 \pm 10$ MeV. Even though it cannot be ruled out that the absolute normalization might change with increasing lattice size, we want to mention that similar observations have been made in other analyses |10| and by ourselves |4| in the investigation of hadron masses.

3. CROSS CHECKS

To check the concistency of the results obtained in the preceding section we want to mention a few points, which will be discussed in more detail in ref. |4|.

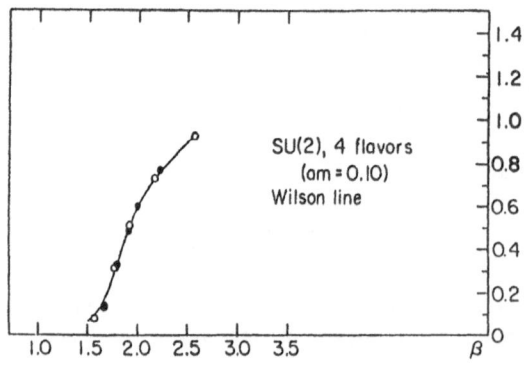

Fig. 4: The Wilson line on a 4×8^3 lattice.

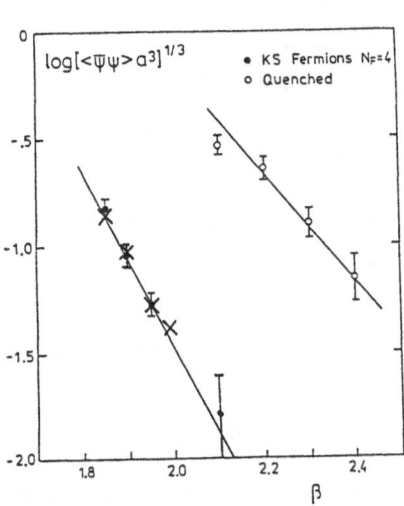
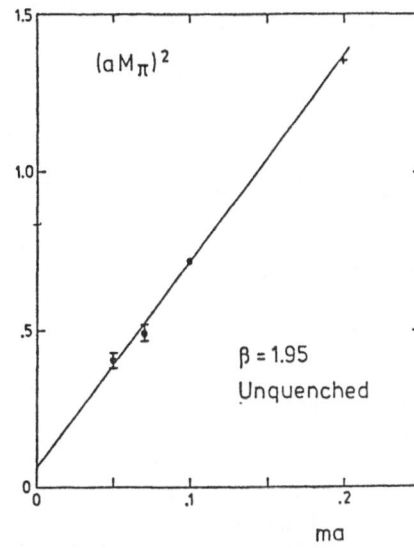

Fig. 5: Scaling properties of the un-Fig. 6: The pion mass squared vs. the
quenched and quenched quark   quark mass at ß = 1.95.
condensate.

I. Wilson line

In pure gauge theories a non-zero value of the Wilson line W, the product of link variables taken along the temperature axis, characterizes the
deconfinement phase of a system. This appears to remain so even if fermions
are incorporated which destroy the center symmetry of the action |5|. For
our 16 x 8³ lattice we found very small values of the Wilson line (.02−.01),
so that our analysis is not affected by any obvious deconfinement effects.
This changes for a typical thermodynamical lattice of small temporal size.
In Fig. 4 we present the measurement of the Wilson line as a function of
ß on a 4 x 8³ lattice. Our data (full circles) are in good agreement with
those of Kogut et al. |5| (open circles) obtained by means of the micro-
canonical method.

II. Quark condensate

A non-zero value of the quark condensate $< \bar{\Psi} \Psi >$ for vanishing quark
masses signals spontaneous breaking of chiral symmetry. We have measured
the condensate at ß = 1.85, 1.9, 1.95 and 2.1 for various quark masses. Extrapolating these values down to zero mass quarks, we find non-vanishing
values for $< \bar{\Psi} \Psi >$ in accord with spontaneous breaking of chiral symmetry.
In Fig. 5 we compare   1/3 log $< \bar{\Psi} \Psi >$ with the quenched results for the
same quantity |11|. We also draw the respective scaling curves and 4 points
of the analysis presented in ref. |6| (crosses). Note that the numbers are
compatible with asymptotic scaling. Taking the current algebra result of
190 ± 20 MeV |12| for the renormalization group invariant value of the
SU(3)-quark condensate (for one flavour) as input, we find a lattice spacing a (1.95) = .2 fm, a little lower than but not inconsistent with the
.24 fm obtained from the analysis of the potential.

III. π, ρ masses

We have measured the π and ρ masses for various values of the quark
mass m at ß = 1.85 and 1.95. The pion mass squared extrapolated down to zero
at vanishing quark masses within 2 standard deviations (see Fig. 6). The
relationship m_π^2 = const. x m, supported by current algebra yields a result
consistent with the expectation that a Goldstone particle exists in non-

Abelian gauge theories with spontaneous breaking of chiral symmetry.

The ρ-mass value extrapolated to zero quark mass is given as $am_\rho = .74 \pm .05$ at $\beta = 1.95$. Given the physical ρ-mass of 770 MeV, this corresponds to a lattice spacing a (1.95) = 0.19 $\pm$ 0.02 fm in satisfactory agreement with the values obtained from the potential and the chiral condensate. At $\beta = 1.85$ we find $am_\rho = 1.07$, compatible with approximate asymptotic scaling which gives a (1.95) / a (1.85) = .66. It should be noted that the values quoted here depend within an error margin of 15% on the way we extrapolate to zero quark mass (see ref. |4| for more details).

4. SUMMARY

We have shown in this note that results for the potential between heavy quarks in lattice gauge theories with dynamical fermions are compatible with general physical expectations. The Coulombic force is stronger than in the quenched approximation. At large distances we find indications for a break-up of the color flux tube between the quarks as expected from spontaneous creation of light quark pairs, though a purely Coulombic + linear behavior of the potential cannot be completely ruled out.

A finite quark condensate together with a vanishing pion mass at zero quark mass hint to spontaneous breaking of chiral symmetry. The value for the lattice spacing a obtained from the potential, the quark condensate and the ρ mass are consistent with one another.

Given the intricacies of strong interaction physics, this note could illuminate only a few points. The results seem encouraging though. A realistic determination of hadron properties at large distances certainly needs the employment of much more powerful tools than those available to us at present.

ACKNOWLEDGEMENT

We thank the Computer Centers at RWTH, Aachen and Bochum University for continuous support.

REFERENCES

1. J.D. Stack, Phys. Rev. D27 (1983) 412;
   A. Hasenfratz, P. Hasenfratz, U. Heller and F. Karsch, Z. Phys. C25 (1984) 191;
   D. Barkai, K.J.M. Moriarty and C. Rebbi, Phys. Rev. D 30 (1984) 1293;
   J.D. Stack, Phys. Rev. D29 (1984) 1213;
   S. Otto and J.D. Stack, Phys. Rev. Lett. 52 (1984) 2328;
   (Erratum: Phys. Rev. Lett. 53 (1984) 1028);
   R. Sommer and K. Schilling, Wuppertal preprint WU B 85-6;
   F. Gutbrod and I. Montvay, Phys. Lett. 136B (1984) 411;
   M. Flensburg and C. Peterson, Phys. Lett. 153B (1985) 412.
2. F. Fucito, E. Marinari, G. Parisi and C. Rebbi, Nucl. Phys.B180 (1981) 369;
   H.W. Hamber, E. Marinari, G. Parisi and C. Rebbi, Phys. Lett. 124B (1983), 95.
3. E. Laermann, F. Langhammer, I. Schmitt and P.M. Zerwas: The Interquark Potential: SU(2) Color Gauge Theory with Fermions, to be published.
4. E. Laermann, F. Langhammer, I. Schmitt and P.M. Zerwas: ρ, π Masses and Chiral Symmetry Breaking:SU(2) Color Gauge Theory with Dynamical Fermions, to be published.
5. J. Kogut, J. Polonyi, H.W. Wyld and D.K. Sinclair, Phys. Rev. D31 (1985) 3307.
6. J. Kogut, J. Polonyi, H.W. Wyld and D.K. Sinclair, Ill-(TH)-85-43.

7. B. Berg and J. Stehr, Z. Physik C9(1981) 333 and references quoted therein.
8. U. Heller and F. Karsch, Nucl. Phys. B258 (1985) 29;
   H.W. Hamber and C.M. Wu, Phys. Lett. 127B (1983) 119.
9. S. Güsken et al, Nucl. Phys. B269 (1985) 393.
10. I. Montvay, Desy 85-072;
    F. Fucito and S. Solomon, Proceedings, Nuclear QCD Workshop, Santa Barbara 1985.
11. A. Billoire, R. Lacaze, E. Marinari and A. Morel, Nucl. Phys. B251 (1985) 581.
12. J. Gasser and H. Leutwyler, Phys. Rep. C87 (1982) 77.

THE SU(2) CHIRAL MODEL IN AN EXTERNAL FIELD:

A COMPLEX STOCHASTIC PROCESS ON A NON-ABELIAN GROUP [†]

Jan Ambjørn and Sung-Kil Yang[*]

The Niels Bohr Institute, University of Copenhagen
Blegdamsvej 17, DK-2100 Copenhagen Ø, Denmark

1. INTRODUCTION

The path integral with a complex valued distribution can be evaluated by solving an equivalent stochastic process under some conditions [1,2]. The stochastic evolution of dynamical variables is described by the complex Langevin equation. This opens the new possibility to perform the numerical simulation of the complex action system. The method was tested for a number of toy models [3-6]. The results can be described as encouraging, although the problem of convergence to an equilibrium distribution is non-trivial, both from a theoretical and a practical point of view [7]. In this report we present a result of the complex Langevin simulation of the two-dimensional lattice SU(2) chiral model in an external field. This is an interesting test of the complex Langevin equation in a highly non-trivial situation where the solution for the model is known (to some degree of confidence) by the Bethe-ansatz method [8]. A full detail of the following can be found in ref. [9].

2. COMPLEX NON-ABELIAN STOCHASTIC PROCESS

In stochastic quantization, one introduces a fictitious time t and a probability distribution $P(U,t)$ satisfying the Fokker-Planck (FP) equation [10]

$$\partial_t P(U,t) = - \nabla^a [\nabla^a + (\nabla^a S(U))] P(U,t) , \qquad (1)$$

where the action S is a functional of a matrix U belonging to the group G, the right Lie derivative $\nabla^a = (UT^a)_{ij} \partial / \partial U_{ij}$ and T^a the generator of G.

To make an importance sampling that realizes the distribution $P(U,t)$ it is convenient to use the corresponding Langevin equation

$$-i U^{-1}(t) \partial_t U(t) = [-i \nabla^a S(U) + \eta^a] T^a , \qquad (2)$$

---
[*] Supported by a Nishina Memorial Foundation fellowship.

[†] Presented by S.-K. Yang

where η^a is a Gaussian noise vector normalized as $\langle \eta^a(t) \eta^b(t')\rangle_\eta$ $= 2 \delta^{ab} \delta(t-t')$ and $\langle \cdots \rangle_\eta$ stands for the Gaussian stochastic average. One can now show that [10]

$$\langle O(U(t))\rangle_\eta = \int dU\, O(U)\, P(U,t) \tag{3}$$

which connects the Langevin equation and the FP equation. Therefore the fact that the FP equation has the stationary distribution $P(U,t)|_{t\to\infty} \propto \exp[-S(U)]$ establishes the equivalence between stochastic quantization and the ordinary path integral formulation.

Even if the action $S(U)$ is complex the FP equation (1) still makes sense. However the problem of convergence of $P(U,t)$ to $\exp[-S(U)]$ is more subtle. For the ordinary real action one proves the convergence by looking at the corresponding FP Hamiltonian which is a Hermitian positive semi-definite operator. If $S(U)$ is complex the FP Hamiltonian is no longer Hermitian and not much is known about the spectrum. Of course $\exp[-S(U)]$ is realized by the ground state, but it is quite non-trivial to prove the convergence to $\exp[-S(U)]$ [3] .

The connection between the Langevin equation (2) and the FP equation (1) in the complex case is obtained in the following way [2,9]. If the group is $SU(N)$ a complex action $S(U)$ will immediately move the changes in $U(t)$ from the tangent space of $SU(N)$ to that of the complex extension $SL(N,\mathbb{C})$ during the stochastic process. It is then usual to introduce a positive definite distribution on $SL(N,\mathbb{C})$, $P_{SL}(M,M^\dagger,t)$ with $M \in SL(N,\mathbb{C})$, which is defined by

$$P_{SL}(M,M^\dagger,t) = \langle \delta(M,M(t))\rangle_\eta \tag{4}$$

or equivalently

$$\langle O(M(t))\rangle_\eta = \int dM\, O(M)\, P_{SL}(M,M^\dagger,t) . \tag{5}$$

This distribution is not to be confused with the complex one $P(U,t)$ defined on $SU(N)$. The relation between the two is taken to be

$$\int dU\, O(U)\, P(U,t) = \int dM\, O(M)\, P(M,M^\dagger,t) . \tag{6}$$

Knowing $P_{SL}(M,M^\dagger,t)$ allows us in principle to construct $P(U,t)$ provided the class of observables for which the RHS of (6) is convergent is large enough. Under this main assumption we arrive at the relation (3). Using the Langevin equation and (3) it is shown that the complex valued distribution $P(U,t)$ satisfies the FP equation (1) [9][*].

We must stress that we have not derived eq. (3), only shown that if there is a connection between a stochastic process on $SL(N,\mathbb{C})$ and the FP equation on $SU(N)$, it must be of the way stated.

---

[*] The FP equation for $P_{SL}(M,M^\dagger,t)$ is derived in ref. [9].

3. THE MODEL

We test the complex Langevin equation by simulating the SU(2) principal chiral model in an external field. The model has been solved in the continuum by Polyakov and Wiegmann [8]. If the system is put in a constant external field H conjugate to the conserved Noether charge, a second order phase transition occurs in the infrared weak field regime. The value of the critical external field is given by $H_c = 2m$ with m

being the mass gap of the theory. The other interesting phenomenon is a cross-over to the ultraviolet strong field regime, where the perturbative method is applicable by virtue of the asymptotic freedom.

Numerical evidence of the phase transition was found in ref. [11] by measuring the free energy by the conventional Monte Carlo simulation of the complex action [12]. In this method, however, it is quite hard to obtain reliable data for either strong fields or weak coupling (i.e., larger lattice size). We want to improve these parts of the calculation in ref. [11] by means of the complex Langevin equation.

The model is described by the action

$$S = -\frac{\beta}{4} \sum_{n,\mu} \left(\text{tr } U_n e^{h_\mu} U_{n+\hat{\mu}}^{-1} + \text{tr } U_{n+\hat{\mu}} e^{-h_\mu} U_n^{-1} \right), \qquad (7)$$

where U_n is an SU(2) matrix on the lattice site n and $\hat{\mu} \in \{\hat{x}, \hat{\tau}\}$ is a unit vector in the positive μ -direction. A constant external field is chosen to be

$$h_\mu = \delta_{\mu\tau} h \frac{\sigma^z}{2} \quad ; \quad h \in \mathbb{R}, \quad \sigma^z = \begin{pmatrix} 1 & 0 \\ 0 & -1 \end{pmatrix}. \qquad (8)$$

Since $e^{\pm h_\mu}$ do not belong to the SU(2) group the action becomes complex for non-zero external field h. The lattice external field is related to the continuum one H through $h = Ha$ with a being the lattice spacing. It is shown that the lattice action (7) corresponds to the ordinary quantum Hamiltonian plus external field term, where the diagonal Noether charge is coupled to H linearly [11].

The lattice charge density operator is given by

$$q(n) = \frac{\beta}{4} \left[\sinh\left(\frac{h}{2}\right) \text{tr} U_n U_{n+\hat{\tau}}^{-1} + i \cosh\left(\frac{h}{2}\right) \text{tr} U_n i\sigma^z U_{n+\hat{\tau}}^{-1} \right] \qquad (9)$$

and its average value is

$$\langle q(n) \rangle = \frac{\partial}{\partial h} \left\{ \lim_{vol \to \infty} \frac{1}{vol} \ln\left[\int dU e^{-S(U)} \right] \right\}. \qquad (10)$$

If we are in the scaling region, it is predicted that [8]

$$\langle q(n) \rangle (\beta, h) \begin{cases} = 0 \quad ; \quad 0 \le h < h_c, \\ \sim (h/h_c - 1)^{1/2} \quad ; \quad h > h_c, \end{cases} \quad \text{for } 0 < h - h_c \ll h_c$$

$$(11)$$

where $h = h(\beta) = Ha(\beta)$ and $h_c = h_c(\beta) = 2ma(\beta)$. For $h \gg h_c$ a one-loop perturbative computation gives [11]

$$\langle q(n) \rangle (\beta, h) = \frac{h}{4\pi} \left\{ \ln \left[\frac{h}{C f(\beta)} \right] + \frac{1}{2} \right\},$$

$$C = \sqrt{128} \, e^{(\frac{\pi}{2} - 1)/2},$$

(12)

where the two-loop scaling function is [13]

$$f(\beta) = a(\beta) \Lambda = (\pi \beta)^{1/2} \, e^{-\pi \beta}$$

(13)

with Λ being the lattice Λ-scale. The scaling law for the charge density $\langle q(n) \rangle (\beta, h) = q^z(H)a$ reads

$$\langle q(n) \rangle (\beta, h) \Big|_{h = h(H/m, \beta)} = \frac{q^z(H)}{\Lambda} f(\beta),$$

(14)

where H/m is fixed and $h(H/m, \beta) = (H/m)(m/\Lambda)f(\beta)$ with the mass gap $m/\Lambda = 22.1 \pm 1.9$ [14].

4. LANGEVIN SIMULATION

Our algorithm is the ordinary first order discretized Langevin equation with a step size Δt:

$$U(t_n + \Delta t) = U(t_n) \exp \left\{ i \left[-i \, \nabla^a S(U(t_n)) \Delta t + \sqrt{2 \Delta t} \, \eta_n^a \right] \right\}$$

(15)

and $\langle \eta_n^a \eta_{n'}^b \rangle = \delta^{ab} \delta_{nn'}$, but with multihit on each site in such a way that the total Langevin time step forward for that site is equal to a prefixed value Δt. The individual $(\Delta t)_i$ was adjusted such that the numerical value of the drift term times $(\Delta t)_i$ was always bounded by a fixed number. In this way we could avoid the numerical instabilities which are essentially caused by the fact that the complex action is not bounded from below (see refs. [3,9] for details). We worked with a quite small average value of Δt (around 0.01 - 0.02). The length of each run corresponded (for the large lattices) to a total Langevin time $T \sim 300$. Thermalization of $\langle q \rangle$ was reached for T around 10-20.

We have measured the charge density and the nearest neighbor spin correlation. Whenever possible we have also measured the same quantities by the indirect method (using heat bath algorithm for updating) employed in ref. [11]. For the nearest neighbor correlation we have always found nice agreement between the two methods, both for the correlation in x-and τ-directions.

For the charge density the situation is more subtle. Let us compare the results of both methods at $\beta = 2.0$, where a correlation length is approximately 10 lattice spacings [14]. For small lattices of the order of a correlation length we found again nice agreement between the two methods, see Fig. 1. The almost linear dependence of $\langle q(n) \rangle$ on h is ascribed to the finite size effect.

In order to examine whether $\langle q \rangle$ goes to zero for $h < h_c$ ($h_c \simeq 0.2$ at $\beta = 2.0$) the lattice size has to be at least that of correlation lengths. In Fig. 2 we present a comparison of data obtained

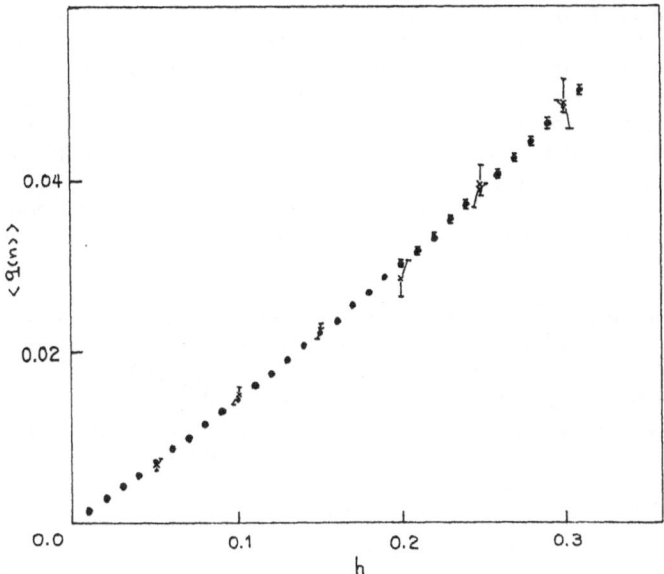

Fig. 1. Charge density $\langle q(n) \rangle$ versus h. Crosses are Langevin results
and dots are Monte Carlo results at $\beta = 2.0$ on a 10x10 lattice.

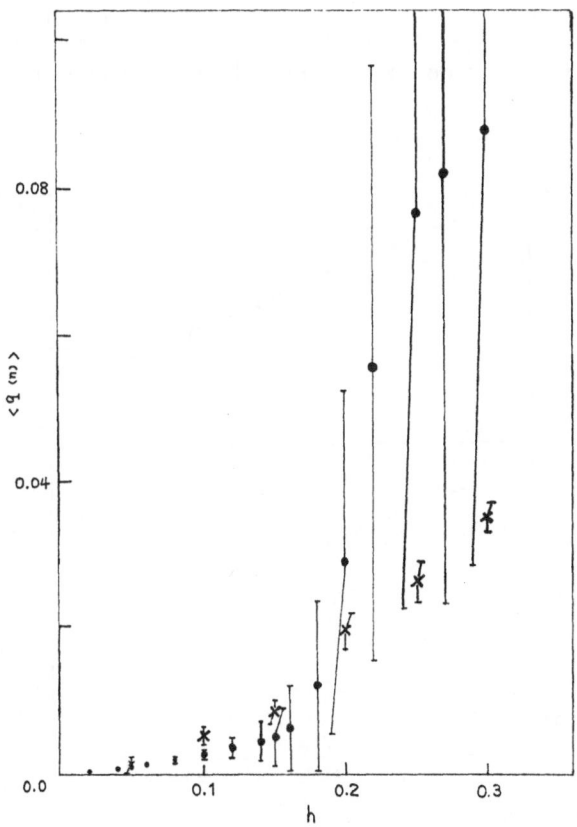

Fig. 2. Charge density $\langle q(n) \rangle$ versus h. Crosses are Langevin results
and dots are Monte Carlo results at $\beta = 2.0$ on a 30x30 lattice.

by the measurements on a 30x30 lattice. It is seen that the use of the complex Langevin equation superficially looks as a success: we can go to large fields and large volume without problems and the results seem not incompatible with the heat bath results. However, a closer study performed for small h, where the heat bath results can be extracted with good precision reveals that the measured <q> is higher when using the Langevin equation for lattice sizes larger than two correlation lengths. We would also remark that the convergence of <q> does not converge in a very clear way as $\sqrt{\text{Volume}}$. This is in sharp contrast to the behavior of the nearest neighbor correlation.

Finally we have checked various aspects of scaling for large external fields, a region inaccessible by using the conventional indirect method. In Fig. 3 is shown the scaling of <q> (β,h) for two values of H/m, where we have used the central value m/Λ = 22.1 to convert h into H/m. Fig. 4 shows the behavior of <q> as a function of h at β = 2.0, and we examine the approach to the perturbative strong field regime. The curve drawn is the one-loop result (12). Both Figs. 3 and 4 show nice agreement with the scaling law.

5. DISCUSSION

We have tested the non-abelian complex Langevin equation for the principal chiral model in an external field. In the weak field regime (0 < h < h$_c$) the data obtained are compatible with the results obtained by conventional methods in the range where the conventional method can be applied. The agreement is perfect for small lattices of the order of a correlation length. For large lattices (three or more correlation lengths) the agreement is no longer perfect when we compare the measured

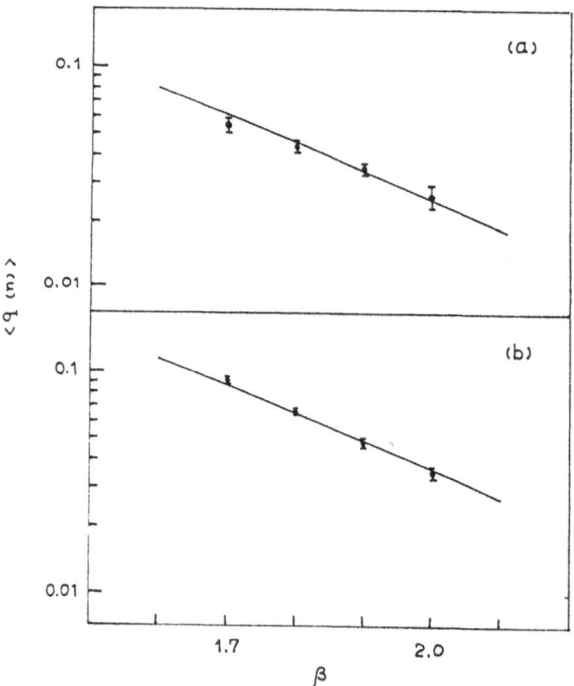

Fig. 3. Charge density <q(n)> versus β for a) H/m = 2.4 and b) H/m = 2.9. The solid line represents the scaling curve (14) with
a) $q^z(H)/\Lambda$ = 5.53 and b) $q^z(H)/\Lambda$ = 7.81.

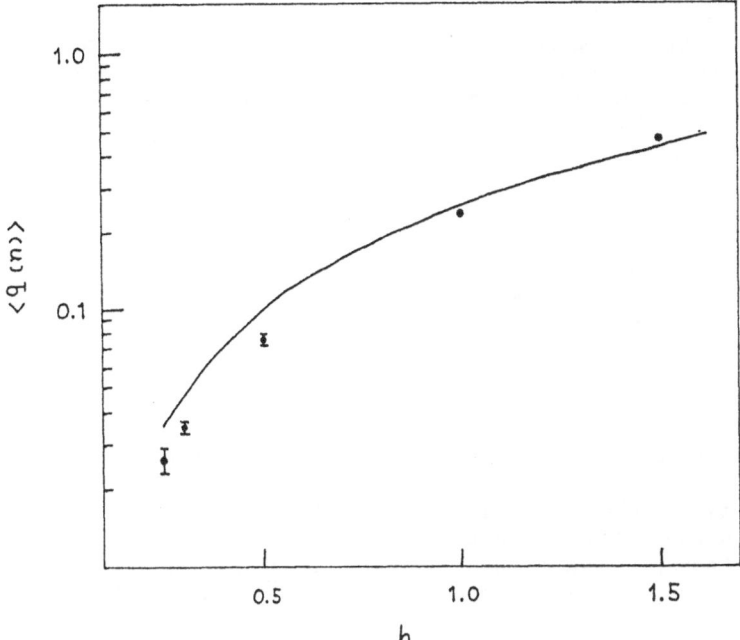

Fig. 4. Charge density $\langle q(n) \rangle$ versus h at $\beta = 2.0$. The solid line represents the one-loop result (12).

charge densities $\langle q \rangle$. The charge measured by the complex Langevin equation has, for $h < h_c$ very long time fluctutations and its expectation value does not scale with the volume.

In order to understand these observations let us first remember that the physics which governs the behavior of $\langle q \rangle$ in the weak field regime (see eq. (11)) is non-perturbative threshold effects associated with the excitation of massive particles in the ground state of the full Hamiltonian [8]. On the other hand we have checked that the stationary point of Langevin equation is still the perturbative vacuum U = const. (though the importance sampling is of course different from the h=0 case). This implies that the fluctuations shall describe not only the non-perturbative mass gap, but also the threshold effect. In particular the low-momentum components of the fluctuating fields are important. The evolution of the low-momentum parts has a very large correlation time

$\sim$ (correlation length)^2 as clearly explained in ref. [15]. Therefore we can argue that the appearance of the long time correlation (and hence the poor convergence of the data) in the weak field regime is due to the fact that the long-wavelength structure of the fluctuations is responsible for the threshold effect. If one keeps these remarks in mind we feel that the complex Langevin equation did its best to simulate the relevant physics.

The situation changes quite remarkably as we enter somewhat into the perturbative strong field regime $h \gg h_c$. The fluctuations become small and no long time correlation is observed. In this region the stationary point of the Langevin equation is still U=const. which is now a correct vacuum configuration by virtue of asymptotic freedom. Further the high-momentum components of the field, which evolve with a very short correlation time $\sim$ 1 [15], are relevant in the strong field regime, thanks again to asymptotic freedom. This explains the fast convergence of $\langle q \rangle$ we have observed. Consequently we have found nice agreement with the scaling law in this strong field regime, which is totally inaccessible by conventional methods. In this regime the complex Langevin equation has been very successful.

ACKNOWLEDGEMENT

One of us (S.-K.Y.) would like to thank the organizers of the Wuppertal workshop for the opportunity to present this material in such a stimulating environment.

REFERENCES

[1] - J.R. Klauder, Acta Phys.Austriaca, Suppl. XXV (1983) 251, (Springer Verlag).
[2] - G. Parisi, Phys.Lett. 131B (1983) 393.
[3] - J.R. Klauder and W.P. Petersen, J.Stat.Phys. 39 (1985) 53.
[4] - H.W. Hamber and H.-C. Ren, Phys.Lett. 159B (1985) 330.
[5] - J. Ambjørn, M. Flensburg and C. Peterson, Phys.Lett. 159B (1985) 335.
[6] - F. Karsch and H.W. Wyld, "Complex Langevin Simulation of the SU(3) Spin System with Non-Zero Chemical Potential", ILL-(TH)-85#60.
[7] - J. Ambjørn and S.-K. Yang, Phys.Lett. 165B (1985) 140.
[8] - A. Polyakov and P.B. Wiegmann, Phys.Lett. 131B (1983) 121;
    - P.B. Wiegmann, Phys.Lett. 141B (1984) 217.
[9] - J. Ambjørn and S.-K. Yang, "The SU(2) Chiral Model in an External Field: a complex stochastic process on a non-Abelian group", NBI-HE-86-03.
[10] - I.T. Drummond, S. Duane and R.R. Horgan, Nucl.Phys. B220 [FS8] (1983) 119;
     - M.B. Halpern, Nucl.Phys. B228 (1983) 173;
     - A. Guha and S.C. Lee, Phys.Rev. D27 (1983) 2412.
[11] - S.-K. Yang, "Universality of the Bethe-Ansatz Solution for the Chiral Non-Linear o-Model", NBI-HE-85-30 (to appear in Nucl.Phys. B).
[12] - G. Bhanot, E. Rabinovici, N. Seiberg and P. Woit, Nucl.Phys. B230 [FS10] (1984) 291.
[13] - E. Brézin and J. Zinn-Justin, Phys.Rev. B14 (1976) 3110.
[14] - M. Fukugita and Y. Oyanagi, Phys.Lett. 123B (1983) 71.
[15] - G.G. Batronni, G.R. Katz, A.S. Kronfeld, G.P. Lepage, B. Svetitsky and K.G. Wilson, Phys.Rev. D32 (1985) 2736.

THE t-EXPANSION AND LATTICE QCD[*]

David Horn

School of Physics and Astronomy
Tel Aviv University
Tel Aviv 69978, Israel

The method that will be reviewed here was developed as a systematic nonperturbative scheme[1] which provides an analytic tool for the study of quantum Hamiltonian systems on the lattice. We will start by following the definitions of ref. 1 and one of the solvable models on which it was tested, and then review the results of its application to the SU(2) lattice gauge theory in 3+1 dimensions[2] and the SU(3) theory[3] in the pure gauge sector.

The simple idea behind the method is that, given any variational ground-state $|\psi_0>$, one can improve on it by applying to it the operator $e^{-tH/2}$ since, for positive t, it will enhance the relative importance of the true vacuum component in $|\psi_0>$. Moreover, taking the $t \to \infty$ limit of

$$(1) \quad |\psi_t> = Z^{-\frac{1}{2}} e^{-tH/2} |\psi_0> \qquad Z = <\psi_0|e^{-tH}|\psi_0>$$

one contracts the initial trial state onto the vacuum wave-functional. All the physical observables which are evaluated in $|\psi_0>$ will tend toward their vacuum expectation values. In particular, the energy function

$$(2) \quad E(t) = <\psi_t| H |\psi_t> = -\frac{\partial}{\partial t} \ln Z(t)$$

will converge to the vacuum energy as $t \to \infty$.

For practical calculations we have to expand the exponential in a power series of the Hamiltonian. The calculation calls therefore for an evaluation of a series of the moments of H. Note however that every moment $<\psi_0| H^n |\psi_0>$ has a volume dependence like (volume)^n hence a series for the norm-function Z(t) is very difficult to handle. On the other hand E(t) is an extensive quantity. This guarantees that its Taylor expansion will have terms proportional to the volume only. These are the connected matrix elements $<H^n>^c$:

$$(3) \quad E(t) = \sum_{n=0}^{\infty} \frac{(-t)^n}{n!} <H^{n+1}>^c$$

They can be defined recursively by the relation:

───────────
[*] Work supported in part by the U.S.-Israel Binational Science Foundation.

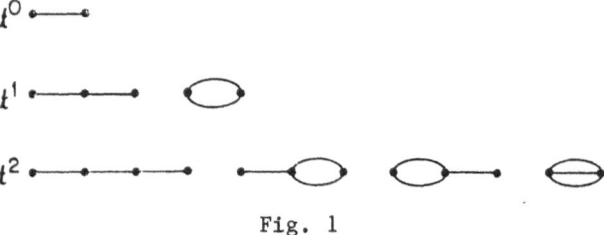

t^0

t^1

t^2

Fig. 1

(4) $\quad <H^{n+1}>^c = <\psi_0|H^{n+1}|\psi_0> - \sum_{\gamma=0}^{n-1} \binom{n}{\gamma} <H^{\gamma+1}>^c <\psi_0|H^{n-\gamma}|\psi_0>$

Eq. (4) is a cluster expansion. By using a mean field trial-state $|\psi_0>$ one can obtain a simple diagrammatic representation in which only configurations which are connected on the lattice lead to non-vanishing contributions to this expansion.

Let us illustrate the method on the solvable problem of the Heisenberg antiferromagnet in 1+1 dimensions, defined by

(5) $\quad H = \frac{1}{4} \sum_i \vec{\sigma}(i) \cdot \vec{\sigma}(i+1)$

on an ordered chain with periodic boundary conditions. If we represent the expression $\frac{1}{4} \vec{\sigma}(i) \cdot \vec{\sigma}(i+1)$ by a line on the link connecting the point i to i+1, we find that the connected diagrams for the first few terms in the expansion are those shown in fig. 1. Choosing the antiferromagnetic trial-function

(6) $\quad |\psi_0> = \prod_i |\sigma_z(i) = (-1)^i>$

we obtain the following t-expansion for the energy density:

(7) $\quad \mathcal{E}(t) = \frac{E(t)}{V} = -\frac{1}{4} - t + 2t^2 + \frac{4}{3} t^3 - 16 t^4 + \frac{112}{5} t^5 + \frac{224}{3} t^6 + \cdots$

The first term, $-\frac{1}{4}$, is the mean-field result. The exact value of the vacuum energy density is $0.25 - \ln(2) = -.4431$. This should result from the infinite series of eq. (7) in the limit $t \to \infty$. In every calculation we are however faced with the fact that we can determine only a finite number of terms in this series. Hence we have to devise a method which will allow us to extract from the finite series an estimate of the infinite one. Fig. 2 shows some of the methods that we have tried. First note that the Taylor expansion does show, for low t, an improvement in the right direction but it blows up inevitably as t increases. Replacing it with diagonal Padé approximants

$$\mathcal{E}_{(L,M)}(t) \simeq \frac{P_L(t)}{Q_M(t)}$$

we obtain successive improvements as L=M increases from 1 to 3. However the best results are obtained by using a D-Padé method in which we fit a non-diagonal Padé, with $M \geq L+2$, to $d\mathcal{E}/dt$ and present the result of the integration of this approximant. In this example it turns out that four such D-Padé approximants converged beautifully to within 1% of the exact value. This is therefore the method of analysis of the energy-density that we decided to apply to the lattice gauge theories in 3+1 dimensions.

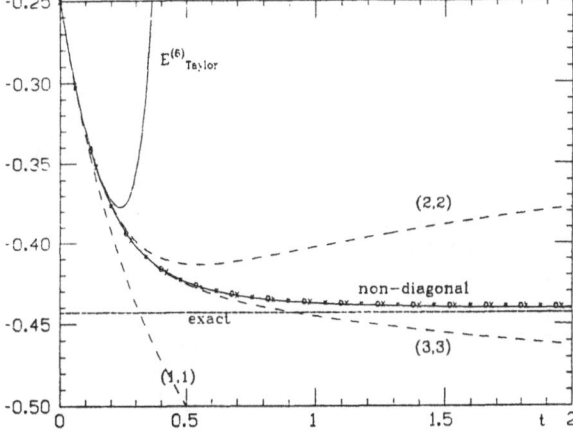

Fig. 2. Energy density of
the Heisenberg antifer-
romagnet.

The Kogut-Susskind Hamiltonian of the SU(2) lattice gauge theory
has the form

$$(8) \quad H = \frac{g^2}{2} \left(\sum_\ell \vec{E}_\ell^2 + x \sum_p (2 - \mathrm{tr} U_p) \right)$$

where $x = 4/g^4$, U_ℓ are the 2x2 matrices which are the group elements of
SU(2), U_p are their plaquette products and $\vec{E}_\ell$ are the conjugate quantum
variables

$$(9) \quad [E_\ell^a , U_{\ell'}] = \frac{\sigma^a}{2} U_\ell \delta_{\ell \ell'}$$

which carry the color electric flux. For our analysis we choose as $|\psi_0\rangle$ the
strong-coupling vacuum

$$(10) \quad \vec{E}_\ell |\psi_0\rangle = 0.$$

This is the only state which is both gauge-invariant and mean-field in
nature. In addition to the vacuum energy we calculate also the string-
tension by starting with a trial state which has an infinite string of
$\prod_{\ell = -\infty}^{\infty} U_\ell$ attached on to the strong-coupling vacuum and evaluating the
resulting increase in energy per unit string-length.

In our analysis we found it convenient to work with

$$(11) \quad \bar{H} = \sum_\ell \vec{E}_\ell^2 - x \sum_p \mathrm{tr} U_p$$

which is related to eq. (8) by an overall multiplicative and additive
constant. The vacuum energy-function is defined by

$$(12) \quad E(t,g) = \frac{\langle \psi_0 | H e^{-t\bar{H}} | \psi_0 \rangle}{\langle \psi_0 | e^{-t\bar{H}} | \psi_0 \rangle}$$

and can be expressed as a double series in t and $y = \frac{2}{g^2}$.

In fig. 3 we see the results of performing a D-Padé procedure on
$\partial \mathcal{E}/\partial y$. This is a quantity which should be positive definite. The fact that
our approximants cross zero at $y \simeq 2$ means that they should not be trusted
from there on. Integrating the $\partial \mathcal{E}/\partial y$ curve we obtain the results for the

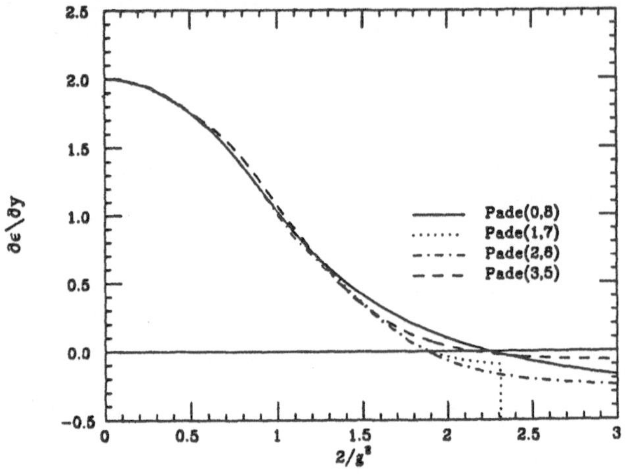

Fig. 3. The leading
D-Padé approximants
for the y-derivative
of the SU(2) energy
density.

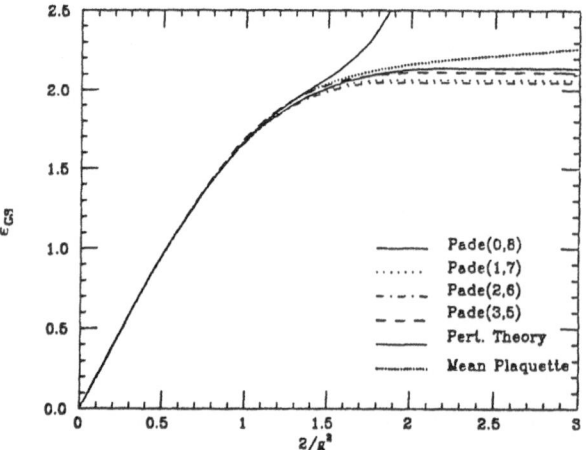

Fig. 4. The energy
density of SU(2) obtained
by integration of the
curves in fig. 3.
Comparison is made with
perturbation theory and a
mean-plaquette upper
limit.

Fig. 5. Specific heat
curves for SU(2)
obtained by differentiation
of fig. 3 (the full curves).
Dashed curves represent
results of the (0,5) and
(1,4) approximants.

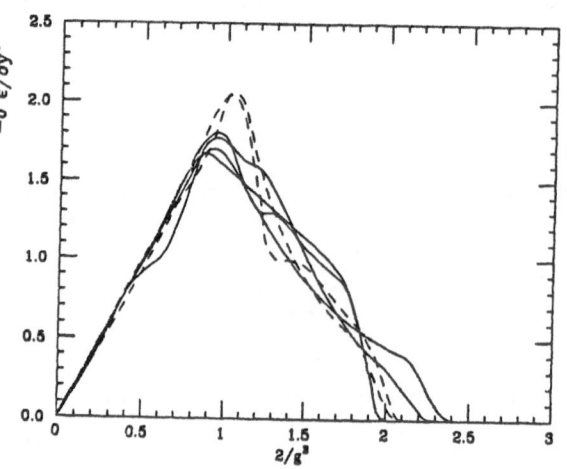

128

energy density displayed in fig. 4. They are compared with a strong-coupling curve (which changes trend at $y \simeq 1.3$ and runs off to infinity) and a mean plaquette upper limit from a variational lattice calculation[4]. It is clear from these curves that they can describe successfully the transition region from the strong- to the weak-coupling regime. This conclusion follows also from the curves of fig. 5 which are obtained by a further derivative of fig. 3 with respect to y. We observe a peak in the specific-heat which signifies the location of the cross-over around $y \simeq 1$. The region from $y \simeq 1$ to $y \simeq 2$ is therefore the place where we may look for scaling results.

The vacuum energy is not a quantity which is of direct physical relevance to continuum QCD. However, from our energy function we can deduce a physical quantity: the mass of the 0^{++} glueball. One finds[2,5] that this can be represented by

$$(13) \quad M(t) = -\frac{g^2}{2} \frac{\partial}{\partial t} \ln \left(-\frac{\partial E}{\partial t}\right)$$

Having obtained in this way a t-expansion for the mass, we combine it with our t-expansion for the string-tension (σ) and obtain a double series in t and y for their ration $R = \frac{m^2}{\sigma}$. This is a quantity which should scale in the weak-coupling regime. The curves in fig. 6, which are the results of different approximants applied to the t^6 expansion of this ratio, indicate a region of stability around $R \simeq 13$ for the range $1 < g^2 < 2$, which should be our window to scaling. We conclude therefore that our calculation indicates that the 0^{++} glueball mass is approximately $3.5\sqrt{\sigma}$.

The generalization of the model and its analysis to SU(3) is straightforward. The Hamiltonian takes the form

$$(14) \quad H = \frac{g^2}{2} \left[\sum_{\ell} \vec{E}_{\ell} + x \sum_{p} (6 - \text{tr } U_p - \text{tr } U_p^+) \right]$$

where U_p are now 3x3 matrices which are unitary but no longer hermitian. The color electric flux operators generate SU(3) transformations according to

$$(15) \quad [E_{\ell}^a , U_{\ell'}] = \frac{\lambda^{\alpha}}{2} U_{\ell} \delta_{\ell \ell'}$$

and x is now chosen as $x = 2/g^4$. The evaluation of the connected diagrams for SU(3) is much more complicated than the one for SU(2). We have calculated the first few moments of H by hand and then turned to REDUCE for a partial automation of the procedure. This was particularly useful for calculating a simplified problem - the one-plaquette Hamiltonian - on which we could test our numerical methods. An example is provided in fig. 7. Here we compare the first excitation energy obtained from the diagonalization of the one-plaquette Hamiltonian with the results of several diagonal Padé approximants to the mass, calculated from the t-expansion of the energy via eq. (13). Only the average of the Padé approximants (evaluated at t=1000) lies close to the exact value as one moves into the weak-coupling regime.

The connected matrix elements calculated in the one-plaquette problem can be used in the cubic lattice problem. One has to add connected diagrams which lie on several neighbouring plaquettes. Such diagrams up to order x^6 are shown in fig. 8. Each square represents the operation of $\text{tr}U_p + \text{tr}U_p$ on a plaquette and the number shows how many times the same operator

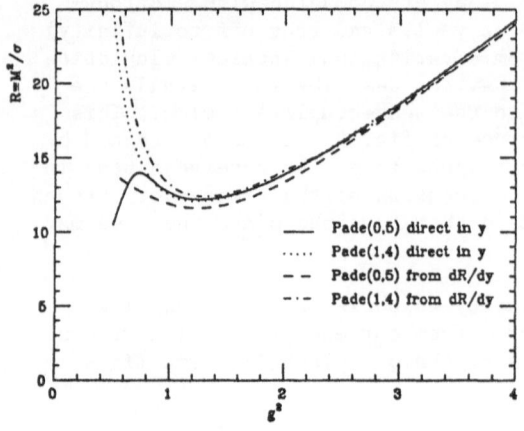

Fig. 6. Ratio M^2/σ displays scaling properties for $1 < g^2 < 2$.

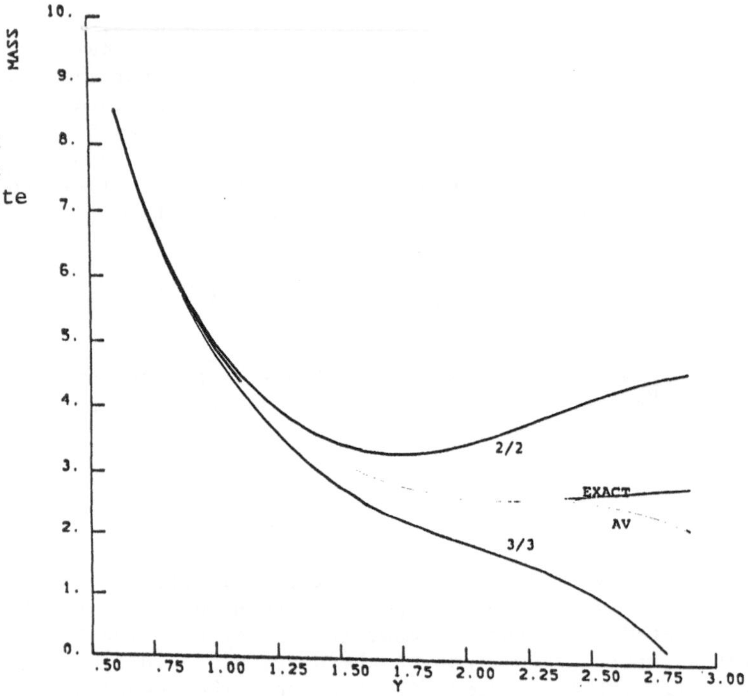

Fig. 7. The first excitation in the SU(3) one-plaquette problem.

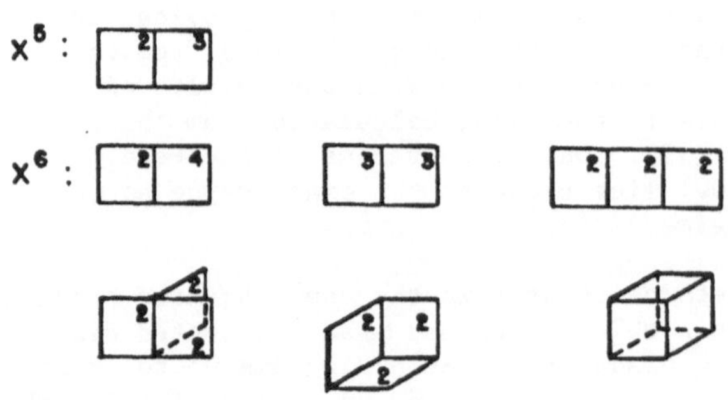

X^4 :

X^5 :

X^6 :

Fig. 8. Lowest connected diagrams in the SU(3) problem.

130

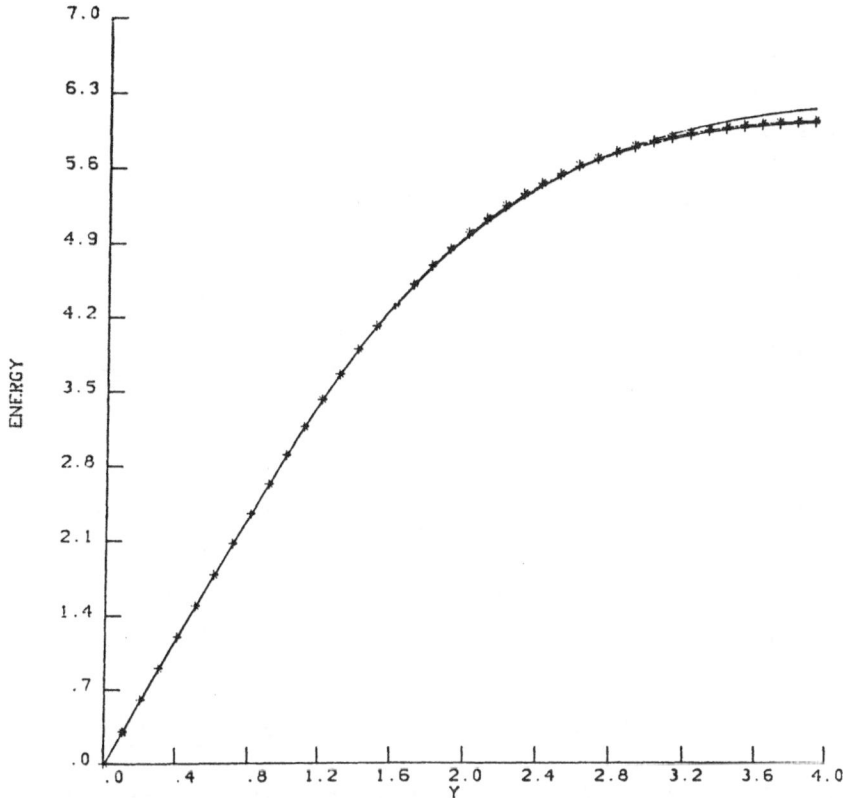

Fig. 9. The energy density in the SU(3) problem.

.was applied. In addition to these magnetic terms one has to take into account the electric ones (i.e. Casimir operators). By applying them to the various links we obtain contributions of higher powers of t.

Using a t-expansion to order t^8 (i.e. H^9) we constructed[3] D-Padé approximants to the energy-density. The three approximants (0,6) (0,7) and (1,8) are shown in fig. 9. Their second-derivatives peak near $y \simeq 1.6$ which is therefore presumed to be the cross-over point..

The interesting quantity remains, as before, the ratio $R=M^2/\sigma$. Some approximants for this ratio which have a common region of stability are presented in fig. 10. They include a (2,2) Padé for the ratio as well as the result of a y-integration of a (2,2) Padé to dR/dy. In both cases the Padé approximant is evaluated at t=1000. On the same figure we find also the average of the D-Padé (0,4) and (1,3) (denoted AV5 because it includes information of the t-expansions out to the t^5 term) and the average of (0,5) and (1,4) denoted AV6. All the Padés are approximants of dR/dy integrated in y. In the region around $y \simeq 2$ they lie around 10, therefore the 0^{++} glueball is estimated to be near $3\sqrt{\sigma} \simeq 1.3$ GeV.

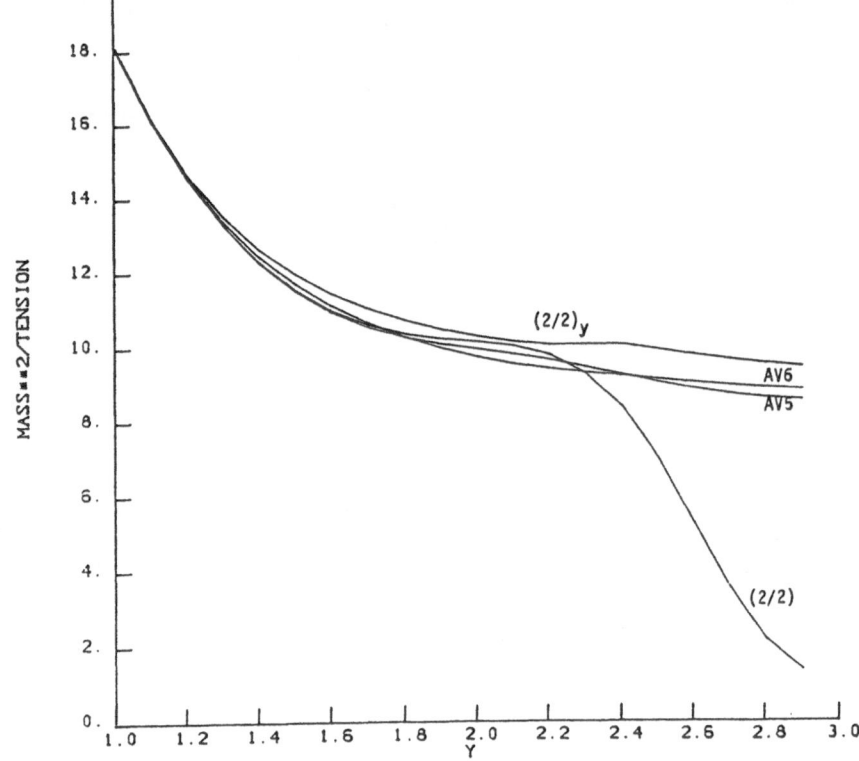

Fig. 10. The ratio M^2/σ in SU(3) displays a plateau around 10.

Let us emphasize that all Padé tricks are applied in t and not in
the coupling variable y. We allow ourselves however the freedom to apply
the analysis to the y-derivative of the series and then integrate the
result along y. Not all approximants agree with one another. Some are
singular or show non physical features. We trust a result only if many
different approximants agree on it. Clearly the longer our series gets
the easier it will be to get meaningful and trustworthy results.

Presently we employ a new computer code, developed by van den Doel
and Roskies[6] , which performs the complete calculation of the connected
part of any vacuum diagram up to order H^{10} . We are using the same
procedure to calculate also other states, in particular the axial glueball
for which we hope to have meaningful results in the near future. The
inclusion of fermions in this approach is straightforward and we plan to
carry out similar calculations for lattice QCD with quarks. The success
of our approach so far indicates that it may develop into a viable
algebraic method for QCD calculations. It remains to be seen how far we
can extend our t-series. Hopefully it will be sufficient to obtain
reliable results for the structure of hadrons.

REFERENCES

1. D. Horn and M. Weinstein, Phys. Rev. D30 (1984) 1256.
2. D. Horn, M. Karliner and M. Weinstein, Phys. Rev. D31 (1985) 2589.
3. C.P. van den Doel and D. Horn, Phys. Rev. D, to be published.
4. D. Horn and M. Karliner, Nucl. Phys. B235 (1984) 135.
5. D. Horn in "From SU(3) to Gravity" (Ed. E. Gotsman and G. Tauber)
    Cambridge Univ. Press 1985, p. 123.
6. C.P. van den Doel and R. Roskies, preprint in preparation.

" It seems to me that the problem No. 1
of high energy physics are scalar particles."

L.B. Okun '81 [1]

LATTICE HIGGS MODELS

J. Jersák

Institute of Theoretical Physics, E
Technische Hochschule Aachen
Physikzentrum
D-5100 Aachen, West Germany

I. INTRODUCTION

This year has brought a sudden interest in lattice Higgs models. After five years of only modest activity we now have many new results obtained both by analytic and Monte Carlo methods. This talk is a review of the present state of lattice Higgs models with particular emphasis on the recent development.

1. Why Higgs models should be studied on the lattice

Gauge fields coupled to the scalar matter fields (Higgs models) [2,3] are of crucial importance in several areas of contemporary physics:
- In the SU(2)*U(1) electroweak theory of Glashow, Salam and Weinberg, and in its generalizations, the masses of the vector bosons are usually assumed to arise through the Higgs mechanism.
- The inflationary models of early Universe are based on the Coleman-Weinberg model [4] of the first order Higgs phase transition (PT).
- The phenomenological theory of superconductivity by Ginzburg and Landau is closely analogous to abelian Higgs models in three dimensions.

Higgs models are also extensively used in attempts to formulate a theory of confinement in gauge theory in the presence of matter fields. Last but not least, they represent a nice piece of quantum field theory.

In spite of this broad use, little is known about the Higgs models beyond perturbative corrections to the semiclassical approximation. This standard approach [3] is based on the assumption of the occurence of non-perturbative phenomenon of spontaneous symmetry breakdown (SSB) in the pure ϕ_4^4 theory, $\langle \phi \rangle \neq 0$. The coupling term between the scalar fields and the gauge fields in the unitary gauge gives rise to a mass term for the gauge field,

$$g^2 \; \phi^+\phi \; A_\mu A_\mu \rightarrow \tfrac{1}{2} \; m_A^2 \; A_\mu A_\mu \; . \tag{1.1}$$

The gauge fields are on the tree level only passive benefitors of this procedure and can be treated perturbatively.

However, SSB has never been demonstrated in the continuum ϕ^4

theory in 4 dimensions. Actually, in recent years several analytic [5] and numerical [6] results have given strong support to the old suspicion [7] that this theory might be trivial.

If this is the case, then only the interactions of the gauge fields themselves can save the Higgs phenomenon. If we cannot simply assume SSB in pure ϕ_4^4 theory, then we have to demonstrate the existence of the Higgs phenomenon from the very first principles. In analogy to SSB this requires a nonperturbative approach. Thus the Higgs phenomenon is an intrinsically nonperturbative effect in the gauge theory, closely similar to confinement. And it is a challenge for lattice gauge theories to show its existence and to understand its properties.

The one loop analysis in continuum by Coleman and E. Weinberg '73 [4] shows that the gauge fields can induce the Higgs mechanism on the level of effective potential and thus work in the right direction. Several later perturbative analyses using the renormalization group method indicate that, if the continuum ϕ_4^4 theory is trivial, the salvation of the Higgs phenomenon might be possible only in a certain range of parameters [8]. This could lead to a restriction on the value of the Higgs boson mass or perhaps even to its prediction. Such a result would be very valuable from the phenomenological point of view.

2. Formulation of Higgs models on the lattice

Instead of giving a sufficiently general definition of actions including all lattice Higgs models ever considered, I shall define a prototype U(1) lattice Higgs model with a particular choice of action and also of notation. All other models are more or less straightforward variations of this particular model.

The U(1) lattice Higgs model on a hypercubic lattice with lattice constant a is formulated by means of the following variables:

site variable $\phi_x = \rho_x \sigma_x$; $\rho_x \in R^+$, $\sigma_x \in U(1)$

link variable $U_\ell \in U(1)$; $\ell = x\mu$ (2.1)

plaquette product $U_p = \prod_{\ell \in \partial p} U_\ell$; $p = x\mu\nu$.

The action is

$$
\begin{aligned}
S = & -\frac{\beta}{2} \sum_p (U_p + U_p^*) \\
& + \lambda \sum_x (\phi_x^* \phi_x - 1)^2 + \sum_x \phi_x^* \phi_x \\
& - \kappa \sum_{x\mu}(\phi_x^* U_{x\mu} \phi_{x+\mu} + c.c.)
\end{aligned}
$$
(2.2)

and the measure is

$$
\rho_x \, d\rho_x \, d\mu(U_\ell) \, d\mu(\sigma_x),
$$
(2.3)

$d\mu$ being Haar measure.

The first term in the action (2.2) is the standard Wilson action [9] for the gauge field. It is a periodic function of the plaquette angle $F_{\mu\nu}(x)$ defined as

$$F_{\mu\nu}(x) = \nabla_\mu A_\nu(x) - \nabla_\nu A_\mu(x) \; ; \quad U_\ell = e^{iagA_\mu(x)} \; . \tag{2.4}$$

In some models a noncompact action of the type $F_{\mu\nu}^2$ is used instead of the Wilson action (see Secs. 3 and 8).

The λ term in (2.2) determines the quartic selfinteraction of the scalar field. For $\lambda = \infty$ the radial mode ρ_x of the scalar field ϕ_x is frozen at the value 1. The term proportional to κ is the hopping term arising from the covariant derivative. The action is symmetric under the reflection of κ, so that it is sufficient to consider $\kappa \geq 0$.

This parametrization of the action is connected to the parametrization usual in the continuum theory,

$$S_c = \int d^4x \; (\; |D_\mu \phi_c|^2 + m_c^2 \, |\phi_c|^2 + \lambda_c \, |\phi_c|^4 \;) \; , \tag{2.5}$$

by the relations

$$\phi_c = \phi \; \frac{\sqrt{\kappa}}{a} \; , \quad \lambda_c = \frac{\lambda}{\kappa^2} \; , \quad m_c^2 = \frac{1 - 2\lambda - 8\kappa}{\kappa a^2} \; . \tag{2.6}$$

I would like to comment on the parametrization of the lattice Higgs models. The possibility to rescale the ϕ field led to a variety of definitions of the coupling parameters associated with the scalar field. This makes a comparison of the results of different authors quite difficult. A standardization of our notation is very urgent. We should choose the notation which is consistent with the one used in lattice gauge theories with fermions, because we want to have one day both matter fields on the lattice simultaneously. Therefore I advocate the use of the parametrization (2.2) by means of the hopping parameter κ. This parametrization, introduced for $\lambda < \infty$ by Kuhnelt, Lang and Vones '84 [10] is suitable also in the very important limit case of the frozen radial mode ($\lambda = \infty$). The continuum-like parametrization (2.5) does not have a virtue of intuitive insight by means of the quasiclassical approximation, since for the reasons already disussed above this approximation is not trustworthy. The parameters m_c^2 and λ_c are anyhow unrenormalized, the consequence of which is that the Higgs phase transition occurs at unexpected negative values of m_c^2.

II. PROGRESS IN ANALYTIC UNDERSTANDING OF LATTICE HIGGS MODELS

3. How to define the Higgs mechanism

It is convenient to maintain the gauge invariance of the lattice gauge theories [9] and not to impose a gauge fixing condition. Then the well known theorem proved by Elitzur '75 [11] and others [12] forbids the existence of a nonzero expectation value $\langle\phi\rangle$ for the scalar field ϕ. It is less well known that $\langle\phi\rangle = 0$ has to hold also in the temporal gauge (Fröhlich, Morchio and Strocchi '80 [13]).

Recently, it has been demonstrated analytically for U(1) lattice Higgs models that there is no SSB also in the α-gauges for $\alpha \neq 0$ and $d \leq 4$. Kennedy and King '85 [14] show this for the noncompact version of the model in which the usual compact Wilson action for the gauge field (2.2) is replaced by the square of the plaquette angle,

$$1 - \cos F_{\mu\nu} \rightarrow \frac{1}{2} F^2_{\mu\nu} \ . \tag{3.1}$$

The α-gauges are defined by the gauge fixing function

$$\frac{1}{2\alpha} \sum_x [\nabla_\mu A_\mu(x)]^2 \ . \tag{3.2}$$

Borgs and Nill '85 [15] obtain for d = 4 the same result for the compact model (2.2) with the compact gauge fixing term

$$\frac{1}{\alpha} \sum_x \cos [\nabla_\mu A_\mu(x)] \ . \tag{3.3}$$

The presence of the Gribov copies allowed by (3.3) causes for large α $\langle \phi_x \phi_y \rangle = 0$ for $x \neq y$ (Nill '85 [16]).

Thus the simplest possible definition of the Higgs phenomenon by means of SSB does not seem to be available except in special cases. One might try to define this phenomenon directly by means of the vector boson mass m_A. In the semiclassical perturbative approach we would expect m_A behaving qualitatively as shown in Fig. 1. The vector boson mass m_A, being zero below the Higgs phase transition, should become nonzero above it. However, if below the Higgs PT nonperturbative effects are important, then m_A does not need to vanish there (Schwinger '62 [17]). The Monte Carlo results for m_A by Montvay '85 [18,19], Langguth and Montvay '85 [20] and by the Aachen–Graz group '85 [21,22] for SU(2) Higgs model confirm that the actual values of the vector boson mass behave quite differently to the expectation based on semiclassical perturbative analysis (Fig. 2).

At present a definition of the Higgs mechanism and a rigorous proof of its existence has been found, except for Z(2) [23], only for the noncompact U(1) lattice Higgs model. For this model it has been expected for d = 4 and known for d = 3 (Brydges, Fröhlich and Seiler '79 [24]) that the model consists of two phases: one with a massless photon (free charge or QED phase) and one with a massive photon (Higgs or screening phase). Recently, Balaban, Brydges, Imbrie and Jaffe '85 [25] found that it is possible to reformulate the noncompact U(1) model for d $\geq$ 3 as a compact one of Villain type:

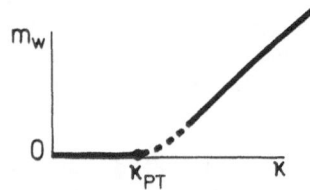

Fig. 1. Vector boson mass below and above the Higgs PT in semiclassical approximation.

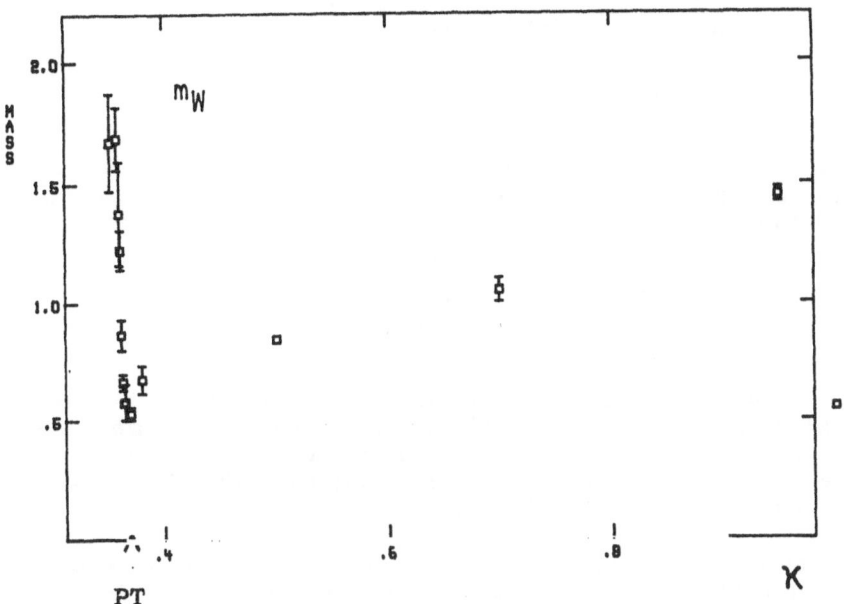

Fig. 2. Vector boson mass below and above the Higgs PT determined
by Monte Carlo simulation in SU(2) model for $\lambda = 3$ and
$\beta = 2.25$ [22].

$$e^{-\frac{1}{e^2} S^{comp}(\theta_p)} = \sum_{n_p} e^{-\frac{1}{2} \sum_p (\theta_p - \frac{2\pi}{e} n_p)^2} \qquad (3.4)$$

The summation over n_p on all plaquettes is constrained by the condition

$$\sum_{p \in \partial c} n_p = 0 , \qquad (3.5)$$

which means physically that there is no monopole in any 3-cube c. The
presence or absence of monopoles is the main physical difference between
the compact and noncompact abelian lattice gauge theories. Thus the model
is still of noncompact type. However, the reformulation in the form (3.4)
makes it possible to derive several nice analytic results. The same authors
show the existence of the phase with massive photon. Brydges and Seiler
'85 [26] and Borgs and Nill [27] demonstrate, by using different methods,
the existence of another phase with vanishing photon mass for $d \geqslant 3$. Thus
the noncompact U(1) lattice Higgs model consists of two phases and in this
model the Higgs mechanism operates.

Kennedy and King '85 [14] devised an order parameter for this model.
They define a gauge invariant two-point function of the scalar field which
contains a smeared string of the link products:

$$G_s(x,y) = \langle \phi_x^* e^{-ie \sum_{z\mu} A_\mu(z) h_\mu(z)} \phi_y \rangle ,$$

$$h_\mu(z) = \nabla_\mu V(z-x) - \nabla_\nu V(z-y) ,\qquad (3.6)$$

$$\triangle V(z) = -\delta(z) .$$

Using the action (3.4) and the cluster expansion they show for $d \gtrsim 3$ that there are two phases with different behaviour of $G_8(x,y)$ as $|x-y|$ approaches infinity,

$$G_8(x,y) \xrightarrow[\ |x-y| \to \infty\ ]{} \begin{cases} G_8^\infty \neq 0 \\ 0. \end{cases} \qquad (3.7)$$

Thus G_8^∞ is an order parameter. In the Landau gauge ($\alpha = 0$ in (3.2)) in which the smeared string in (3.6) vanishes, the phase with $G_8^\infty \neq 0$ even shows SSB. See also Ref. [28]. The identity of the two phases defined by the order parameter G_8^∞ with the Higgs and the QED phases has not yet been proved.

Unfortunately, these nice results have not been extended to the compact U(1) model, not to speak of the nonabelian ones. Actually, Seiler '85 [29] speculates that for large λ the compact U(1) model has only one confinement/Higgs/plasma phase with a massive vector boson inside the whole phase diagram.

The lack of a precise definition of the Higgs phenomenon in the lattice Higgs models is embarrassing, but not necessarily disastrous. As I shall describe later, it is easy to identify by means of MC simulation the phase transition manifold which contains the PT of the pure lattice ϕ_4^4 theory. The Higgs mechanism is expected to operate for κ above this manifold. This Higgs phase or Higgs region of a phase reduces in the limit $\beta \to \infty$ to the SSB phase of the pure lattice ϕ_4^4 theory .

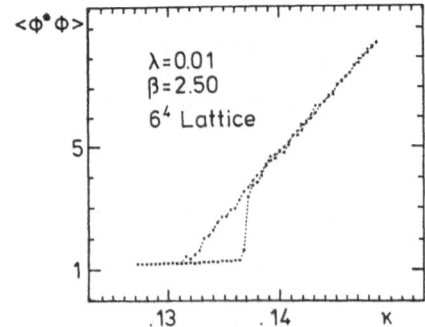

Fig.3. Gauge invariant polynomial of scalar field below and above the Higgs PT in U(1) model [30].

The expectation values of gauge invariant polynomials of scalar fields like $\langle \phi_x^\dagger \phi_x \rangle$ are larger in the Higgs region than below the Higgs PT manifold, where they are nonzero, too, but small. Furthermore, in the Higgs region these expectation values grow fast with increasing κ whereas they are nearly constant below the Higgs PT (Fig. 3). This quantitative difference substitutes for the qualitative difference of nonvanishing or vanishing $\langle \phi \rangle$ in the semiclassical approach using fixed gauge. Even without a genuine order parameter it is possible to distinguish the properties of vacuum in the Higgs region, where the scalar field condensates, from those below the Higgs PT, where the scalar field only fluctuates around zero.

I think that we should once and for all admit that there is no spontaneous symmetry breaking of gauge symmetries and avoid the use of this misleading term when gauge fields play an important dynamical role. Instead we can speak of condensate of scalar fields in vacuum in the Higgs region. This vacuum can have various patterns characterized by different expectation values of gauge invariant polynomials of condensed scalar fields. In Sec. 12 I shall discuss in more detail the characterization of the vacuum patterns in the Higgs region.

4. Higgs mechanism and confinement

An intriguing relation between the Higgs and confinement phenomena on the lattice follows from an observation by Osterwalder and Seiler '78 [31] and by Fradkin and Shenker '79 [32]. For the Z(2) and U(1) models with the Higgs fields in the fundamental representation, there exists, for large λ, an analytic connection C-H between the confinement and the Higgs regions of the same phase. This analytic result has been confirmed and extended to SU(2) (Fig.4) and SU(3) models by Monte Carlo studies [10,33,34]. The fact that for small and fixed λ the analytic connection is no more seen does not change the physical consequences of the result at large λ. Even for small λ both regions are connected via an analytic path going through the regions of large λ.

What is the meaning of the analytic connection between the confinement and Higgs regions? As the spectrum of the model varies continuously on the C-H path (Fig. 4), in the Higgs region we should expect a hadron-like spectrum analogous to that in the confinement region. This is a challenge for MC investigation of the masses in the Higgs models.

Furthermore, we have learned that the seemingly fundamental fields like vector or Higgs boson fields are actually in a gauge invariant formulation composite field aggregates, which are singlets with respect to gauge transformations. They are analogous to the composite hadrons in QCD. For

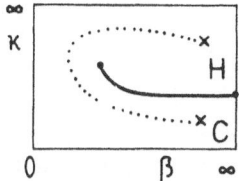

Fig. 4. Schematic phase diagram of the SU(2) model for λ large with the analytic connection between confinement (C) and Higgs (H) regions.

example, the following identities hold in the unitary gauge for the continuum SU(2) model with scalar field in the fundamental representation:

$$\phi(x) = \hat{\phi}^+(x)\phi(x) \ ,$$

$$\hat{\phi} = \phi/\rho \ , \qquad\qquad (4.1)$$

$$W_\mu^3(x) = \hat{\phi}^+(x) \ D_\mu \ \hat{\phi}(x) \ ,$$

etc.

This had been first noticed by Mack '77 [35] and later by 't Hooft '79 [36], Dimopoulos, Raby and Susskind '80 [37], Fröhlich, Morchio and Strocchi '81 [13] and by others [38,39]. It has been called "complementarity" in Ref. [37]. However, the relation between gauge invariant and gauge fixed expressions for the physical fields is analogous to a relation between the mass and energy of a particle in the special theory of relativity [35]. They coincide in a special Lorentz frame, but there is no doubt that one of the concepts is more fundamental.

With this aspect in mind one realizes that it might be useful to consider the lattice Higgs models also from the confinement point of view. A formulation of a confinement criterion in gauge theories with matter fields is a notoriously difficult problem. This is another similarity between the confinement and Higgs phenomena. I would like to mention some recent proposals.

Fredenhagen and Marcu '83 [40] suggested a criterion testing for the possibility of constructing a charged state. Such a state should be produced in a limit procedure, during which one of the charges of a gauge invariant dipole state is removed to infinity along the dipole axis. The two charges at a distance T are connected by a string of link variables on a path Γ shown in Fig. 5. Choosing $R = T/2$ the limit $T \to \infty$ is performed. The projection of the resulting state on the vacuum, regularized properly by dividing by the square root of the expectation value of a $T{\times}T$ Wilson loop, can be written symbolically as

$$\rho_{FM}^\infty = \lim_{T \to \infty} \frac{\langle \ \sqcap \ \rangle}{\langle \ \square \ \rangle^{1/2}} \ . \qquad\qquad (4.2)$$

Here the expectation value in the numerator is the gauge invariant two-point function of the ϕ field defined along the path Γ (Fig. 5),

$$G(R,T) = \langle \ \phi_x^* \prod_{\ell \subset \Gamma} U_\ell \ \phi_y \ \rangle. \qquad\qquad (4.3)$$

The parameter ρ_{FM}^∞ should vanish if a charged state is present. If instead quark fragmentation takes place, i.e. if confinement holds, then $\rho_{FM}^\infty \neq 0$.

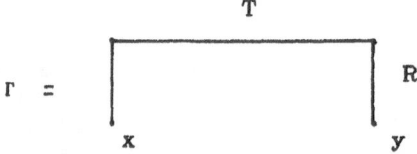

Fig. 5. Support Γ of the gauge invariant two-point function $G(R,T)$.

This criterion replaces an earlier proposal by Bricmont and Fröhlich '83 [41], who also used the gauge invariant two-point function, but with the string along the straight line connecting the charges. The function $G(0,T)$ is not directly testing for confinement [40,42].

Fredenhagen and Marcu tested these properties of ρ_{FM}^s in the Z(2) Higgs model by polymer expansion [40]. Subsequently Kondo '85 [43] and recently also Fredenhagen and Marcu [44] extended this test to the compact U(1) Higgs model. A different construction of the charged states has been also studied by Szlachanyi '84 [45].

Bricmont and Fröhlich '85 [42] suggested another criterion for confinement, testing whether color singlet particles can unbind and decompose into fundamental colored particles or not. Then the correlation function for a color singlet particle should decay as a two particle or as a one particle Green function, respectively. The difference between the two possibilities manifests itself in different Ornstein-Zernike power corrections of the form

$$|x - y|^{-(d-1)} \quad \text{or} \quad |x - y|^{-(d-1)/2} \, , \tag{4.4}$$

respectively, to the exponential decay of the Green function.

Bricmont and Fröhlich [46] also show that the gauge invariant two-point function $G(R,T)$ of the ϕ field calculated along the path Γ (Fig. 5) has the following asymptotic behavior:

$$G(R,T) \xrightarrow[\substack{T \to \infty \\ R = T/2}]{} < \begin{array}{ll} e^{-mT} & \text{in confinement phase} \\ \\ e^{-mT} \dfrac{1}{T^{(d-1)/2}} & \text{in free charge phase.} \end{array} \tag{4.5}$$

III. PROPERTIES OF VARIOUS LATTICE HIGGS MODELS

In the following sections I shall briefly describe the established properties of various lattice Higgs models in four dimensions. They have been obtained mostly by the Monte Carlo simulation combined with plausible conjectures. The mean field method is used only rarely and its results have to be verified by Monte Carlo simulations, too.

I have tried to set up a list of references as complete as possible. I apologize for eventual unintended omissions. When quoting the Aachen group I mean various subsets of the following set of physicists: H.G. Evertz, V. Grösch, K. Jansen, H.A. Kastrup, T. Neuhaus and myself.

5. Z(2) Higgs model

The Z(2) model introduced by Wegner '71 [47,23] is a prototype of gauge theories with matter which has been studied in many analytic investigations and for various purposes since the advent of lattice gauge theories. I shall concentrate on the studies of the phase diagram done since the fundamental work by Fradkin and Shenker '79 [32]

Their prediction of the analytic connection between the confinement and Higgs regions for $\lambda = \infty$ has been confirmed in the Monte Carlo study by Creutz '80 [48]. Bhanot and Creutz '80 [49] also found an interesting phase structure (two frustration phases) for negative ß. Horn with various

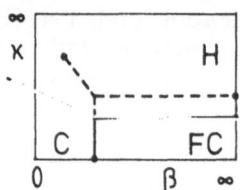

Fig. 6. Schematic phase diagram
of the Z(2) model for
$\lambda = \infty$ (dashed line) and
λ small. FC is the free
charge phase.

collaborators [50] investigated both phases of the Z(2) model for $\lambda = \infty$ by means of Hamiltonian formalism.

A qualitative step forward has been made by Munehisa and Munehisa '82 [51]. They allowed the radial mode of the scalar field to fluctuate by choosing λ finite. It turned out that the phase diagram changes substantially for small λ. The analytic connection vanishes as long as λ is kept fixed. Similar results were obtained later in Z(N) for several values of N by Gerdt, Ilchev and Mitrjushkin '84 [52]. The confinement-Higgs phase transition extends into the region of negative ß where it eventually joins a frustration phase transition [53].

The phase diagrams for large and small λ thus differ qualitatively as indicated schematically in Fig. 6. This difference had not been to my knowledge anticipated by analytic investigations. The change of the phase diagram for small fixed λ does not invalidate the conclusion that the confined and Higgs regions are analytically connected, however, as this connection persists in a three dimensional phase diagram via the regions of large λ.

6. U(1) Higgs model, Higgs field charge q = 1

This model has been studied intensively for $\lambda = \infty$ by analytic methods. By means of duality transformation in the spirit of Ref. [54] the model can be rewritten in terms of topological excitations - monopoles [55,56,39]. Fradkin and Shenker '79 [32] and Banks and Rabinovici '79 [39] pointed out the existence of an analytic connection between the confinement and Higgs regions in one of the phases of the model, in the confinement-Higgs phase. For large ß and small κ a Coulomb or free charge phase has been expected [55,32,39]. Its properties correspond to the continuum scalar

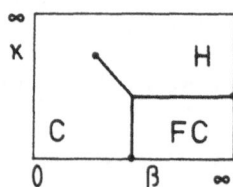

Fig. 7. Schematic phase diagram of
the U(1) model for $\lambda = \infty$.
FC is the free charge or
Coulomb phase.

QED for weak coupling. Peskin '78 [55] pointed out the relation of the Higgs PT between both phases at large ß to the Coleman–Weinberg phenomenon [4]. Katznelson '84 [57] considered also the phase structure of the U(1) Higgs model with a mixed action for gauge field.

The first Monte Carlo study of U(1) Higgs model had been performed by Bowler et al. '81 [58] and by Callaway and Carson '82 [59]. They found both phases and the analytic connection in the confinement–Higgs phase (Fig. 7). The Higgs PT for $\lambda = \infty$ seems to be smooth on small lattices and thus consistent with being of second order. But the question of the order is still open. Later, Kripfganz, Ranft and Ranft '83 [60] and Labastida et al. '85 [61] detected also magnetic monopoles in both phases. The monopole density drops during both the confinement–Coulomb and confinement–Higgs PT's. But monopoles do not seem to play any role in the Higgs PT.

Similarly to Z(2) the phase diagram changes when the radial mode fluctuates, i.e. for λ finite and small. Munehisa '84 [62] was the first to find that for small λ the Higgs PT is clearly of first order, and that the analytic connection disappears. This has been confirmed by several further studies [30,63–65]. The Aachen group, Lang and Vones '85 [30,65] found also frustration phases for negative ß and determined the whole three-dimensional phase diagram (Fig. 8).

The first order of the Higgs PT for small λ can be understood as non-perturbative indication of validity of the Coleman–Weinberg mechanism [4].

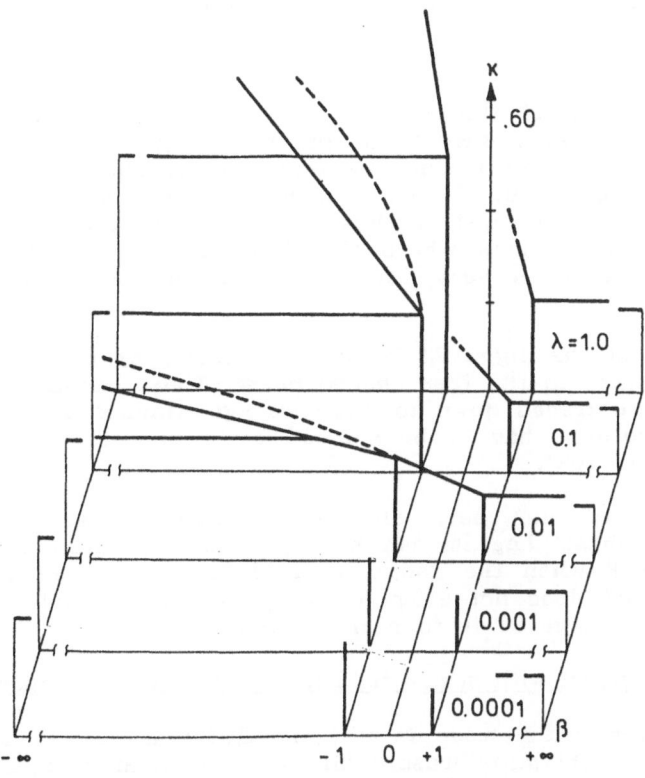

Fig. 8. Three dimensional phase diagram of the U(1) model [30].

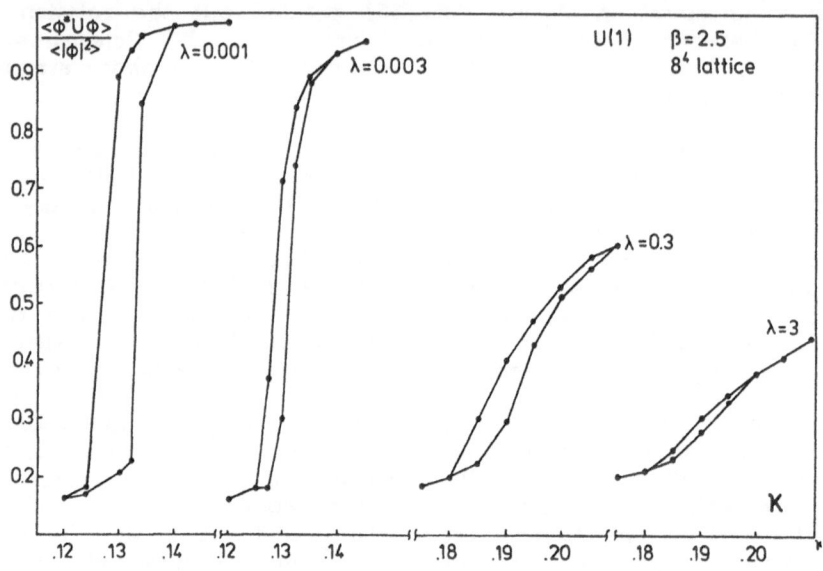

Fig. 9. Weakening of the Higgs PT with increasing λ in the U(1) model [65].

But Kleinert '83 [66] pointed out that with growing λ the Higgs PT might change its order. Monte Carlo studies revealed a tendency of the Higgs PT to weaken with growing λ or ß (Fig. 9). It is an open question whether and where the order of the Higgs PT changes.

The points at which the order of a PT changes are usually called tri-critical points. The experience from spin models [67] and from pure U(1) lattice gauge theory [68] teaches us that the systems can behave in the vicinity of tricritical points in a confusing way, and that the order of the PT can be easily misinterpreted. Thus it will be very difficult to determine with certainty the regions where the Higgs PT is of higher order. Of course, this question is very important from the point of view of the continuum limit.

The study of the Higgs PT is not made easier by spurious excitations occurring sometimes in the free charge phase. These excitations have been in the limit κ = 0 traced down to Dirac strings winding around the lattice [69]. Their occurence has to be avoided by a proper choice of thermalization procedure [65].

The Higgs PT has been also investigated by mean field method [64,70,71]. The phase diagram can be reproduced qualitatively for low or large values of ß when the gauge is fixed or not, respectively. But the mean field method does not seem to be sufficiently reliable for study of properties of the Higgs PT in four dimensions.

7. SU(2) Higgs model with scalar field in the fundamental representation

For a frozen radial mode (λ = ∞) the SU(2) model was studied quite early by Lang, Rebbi and Virasoro '81 [72]. Again an analytic connection between the confinement and Higgs regions was found. This means that the

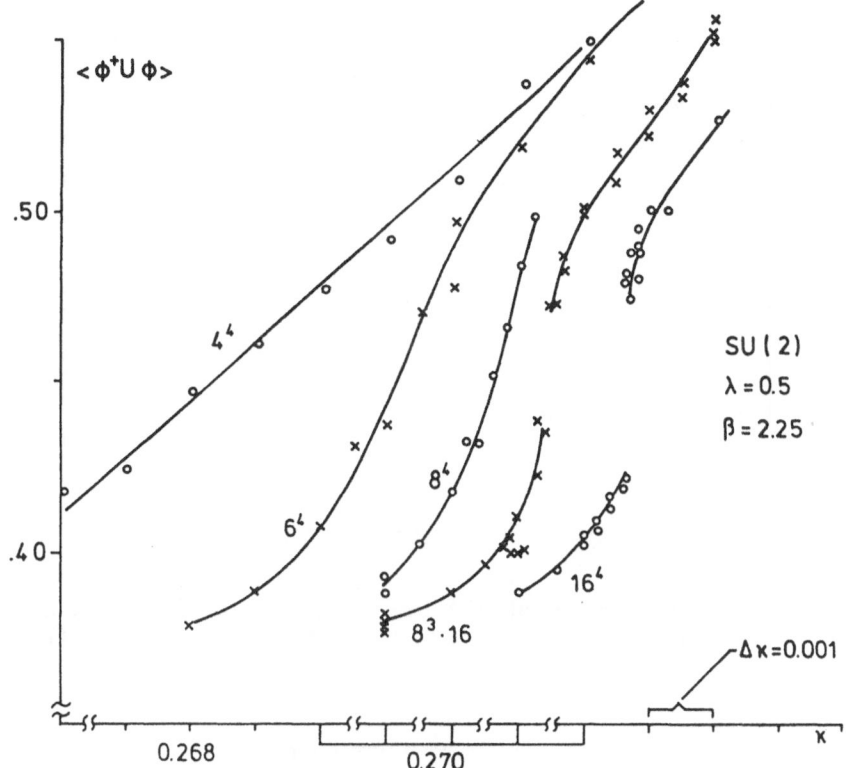

Fig. 10. Change of the Higgs PT in the SU(2) model with growing
lattice size [33,75]. The κ-axis is shifted by Δκ = 0.001
for consecutive lattice sizes.

model exists for ß ⪰ 0 only in one phase. Nevertheless, there is a Higgs PT
line emerging from the PT point of the pure φ⁴ theory on the lattice at
ß = ∞. It has an end point at a certain low value of ß. The order of this
PT line remains an open question [20,72,73].

Kuhnelt, Lang and Vones '84 [10] observed a shift of the end point of
the Higgs PT line towards lower values of κ when λ decreases. It also
shifts to lower ß and soon achieves the ß = 0 axis. Further investigations
dislosed that for low λ the Higgs PT is of first order and that it weakens
with growing λ or ß [33,74].

Again we face the question whether the Higgs PT changes its order
when λ and/or ß are sufficiently large. The Aachen group, Lang and Vones
'85 [33] studied in detail the PT for λ = 0.5 and ß = 2.25. On small lattices
of sizes 4⁴ to 8⁴ the system behaves smoothly (Fig. 10) so that one is
tempted to assume that the Higgs PT at this point is already of second
order. Also the correlation length, corresponding to the inverse Higgs
boson mass, calculated on an 8³*16 lattice (I shall discuss the masses in
more detail in Section 14) attains at this point the value of five lattice
spacings [22,33], which is approximately a maximum value one can obtain on
the lattice of this size. However, the latest calculation by the Aachen
group, Landau and Xu [75] reveals a clear two-state signal on a 16⁴ lattice
for the same values of λ and ß (Fig. 11) and a discontinuity in ⟨φ⁺Uφ⟩
across the PT (Fig. 10). Langguth and Montvay '85 [20] found a histogram

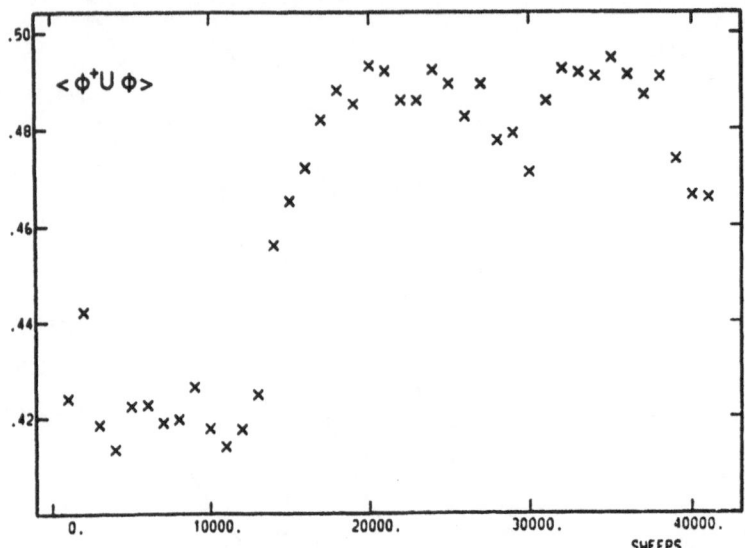

Fig. 11. A two-state signal with one phase flip in the SU(2) model on a 16^4 lattice for $\lambda = 0.5$, $\beta = 2.25$ and $\kappa = 0.27062$ [75].

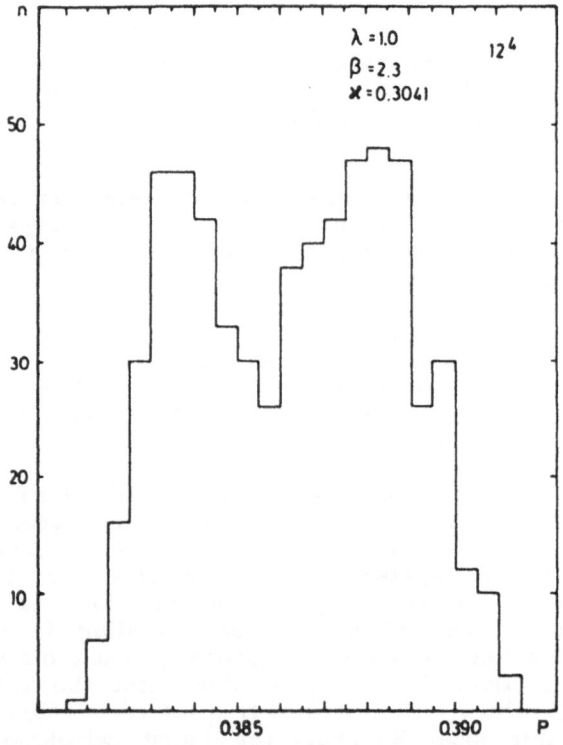

Fig. 12. A double peaked distribution of the mean plaquette P in the SU(2) model on a 12^4 lattice for $\lambda = 1$, $\beta = 2.3$ and $\kappa = 0.3041$ [20].

with two peaks in the Higgs PT at $\lambda = 1$ and $\beta = 2.3$ (Fig. 12).

These results give us a warning that a second order-like behaviour (inclusive usual finite size effects [33]) observed on smaller lattices can change into a first order-like behaviour on larger lattices. I am afraid that for some time to come we still will have to live with an uncertain order of the Higgs PT. Various possible positions of the second order transitions on the manifold of Higgs PT are indicated in Fig. 13.

Mean field studies of the Higgs PT [70,76] deserve similar remarks as in the case of U(1) model. Decker, Montvay and Weisz '85 [77] suggested another analytic approach to the SU(2) Higgs model. They expanded it for large λ around the same model with frozen radial mode ($\lambda = \infty$). Such an expansion might be valuable if it turns out that λ is in some sense an irrelevant parameter, as suggested by Montvay '85 [19] and Langguth and Montvay '85 [20]. I shall return to this question and to the results for spectrum, static potential and for other quantities in the SU(2) model in later sections.

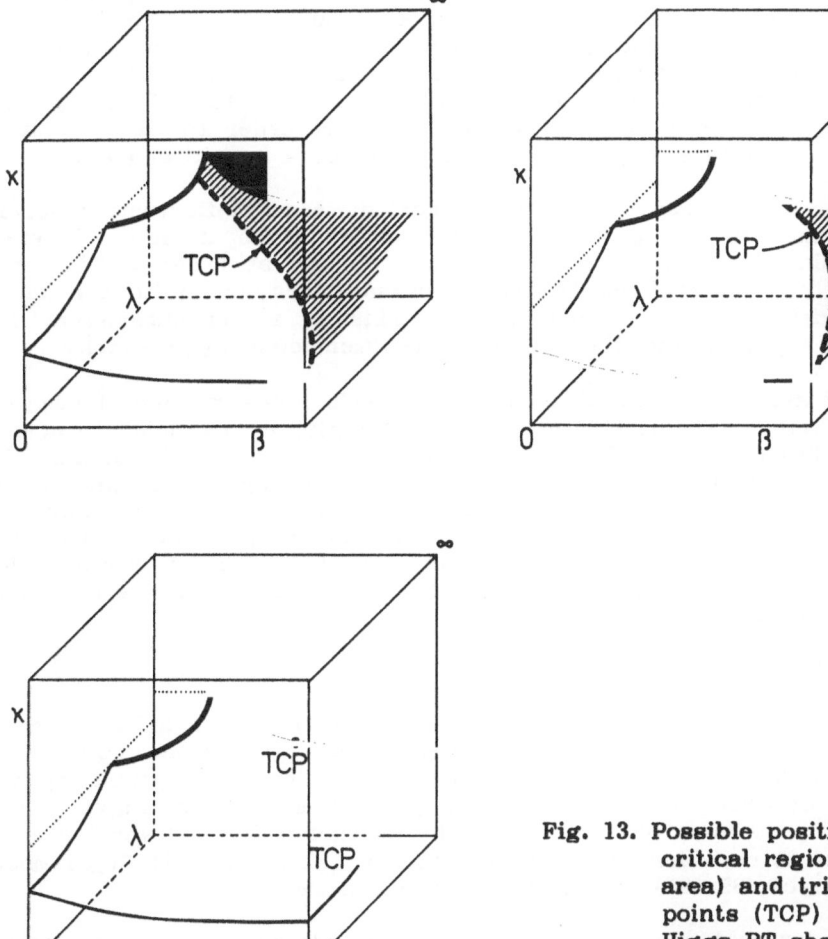

Fig. 13. Possible positions of critical regions (shaded area) and tricritical points (TCP) on the Higgs PT sheet of the SU(2) model.

8. Phase transition in lattice superconductor

We should keep in mind that the Higgs models in four dimensions are an analogue to the Ginzburg-Landau theory of superconductivity. The relevance of this theory to the problem of masses of gauge bosons had been pointed out by Anderson already in 1963 [2]. In particular he already then realized that the Goldstone boson transforms in gauge theory into a component of the massive vector boson. (I thank H.A. Kastrup for bringing the content of the paper by Anderson to my attention.) The three-dimensional lattice superconductors might teach us something even today.

In analogy to Coleman and Weinberg [4] it was realized by Halperin, Lubensky and Ma '74 [78] that, due to the radiative corrections, the phase transition in three-dimensional superconductor should be weakly of first order. The order of this PT was studied on a lattice. The lattice super-conductors are usually chosen with noncompact action for the U(1) gauge field in order to approximate the continuum theory as closely as possible. It turned out in Monte Carlo simulation (Dasgupta and Halperin '81 [79]) that for such a model with frozen radial mode (i.e. $\lambda = \infty$, extreme type II superconductor case) the PT is smooth, presumably of second order. Lawrie '82 [80] suggested on the basis of the ε expansion that the arguments for the first order of the PT fail for ß below a certain value. Bartholomew '83 [81] investigated the model by Monte Carlo method for various values of λ and found that the transition is of first order for small λ. He even attempted to estimate for what value of λ the order of the PT changes by extrapolating a discontinuity in $\langle |\phi| \rangle$ on the first order PT as a function of λ to the point where it gets zero. This would be a tricritical point.

Kleinert '82 [82] rewrote the same model in terms of fluctuating vortex lines and found that the order of PT changes when the attraction between the vortex lines at small λ changes into repulsion at large λ. Thus the results for three-dimensional lattice superconductors are quite consistent with the existence of a tricritical line. This is an encouragement for investigating the same question also in four-dimensional Higgs models.

I would also like to mention two Monte Carlo investigations of compact lattice superconductors performed, not surprisingly, by particle physicists. Bhanot and Freedman '81 [83] found that the Higgs PT might actually be only a cross-over at finite ß for a singly charged Higgs field and $\lambda = \infty$. This would be an interesting difference to the above mentioned results by Dasgupta and Halperin [79]. Could it be that the compact and noncompact formulations of a three-dimensional abelian Higgs model differ as much? For finite λ Munehisa '85 [84] found a first order PT in agreement with Bartholomew [81].

9. SU(2)*U(1) Higgs model

The above studies of the U(1) and SU(2) Higgs models in four dimensions do not take into account the mixing of both groups as required by the Standard Model of electroweak interactions. This is justified by the necessity to understand first the general properties of lattice Higgs models and to develop methods for performing the continuum limit. Nevertheless, it is instructive to consider the phase diagram of the SU(2)*U(1) Higgs model found by Shrock '85 [85] for the frozen radial mode.

The scalar field has charge one and transforms as a fundamental representation of SU(2). There are two gauge field couplings $ß_1$ for U(1) and $ß_2$ for SU(2) and the hopping parameter $\kappa = ß_h/2$. Thus the simplest phase diagram is already three-dimensional.

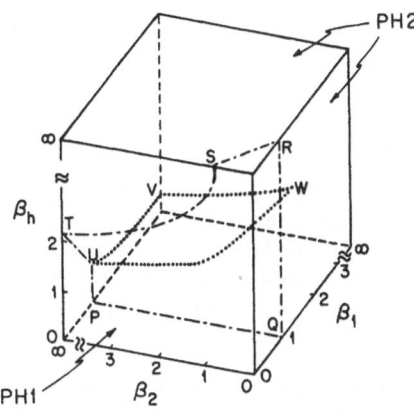

Fig. 14. Phase diagram of the
SU(2)*U(1) model for
λ = ∞ [85].

The Shrock cube (Fig. 14) consists of two phases. Phase PH1 for low β_1 is the phase with confinement of both SU(2) and U(1) charges. It is separated by a sheet PQRSTU of phase transitions from phase PH2 in which the electromagnetic U(1) is in the Coulomb phase. The face $\beta_1 = \infty$ of the cube corresponds to the SU(2) Higgs model. The face $\beta_2 = \infty$ is the U(1) Higgs model with additional global SU(2) symmetry. This gives rise to the phase transition line TU, and the total confinement and Higgs regions are not analytically connected in spite of frozen radial mode.

10. U(1) model with multiple charged scalar field

As was pointed out by Fradkin and Shenker '79 [32], if the charge of scalar field q is greater than 1, the model reduces for λ = ∞ and κ = ∞ to the pure Z(q) gauge theory. This theory exhibits one PT for q from 2 to 4 and two PT's for q ⩾ 5. Thus there is no analytic connection between the confinement phase at low ß and the Higgs (or frozen) phase at high ß and κ.

The phase diagrams have been investigated numerically by Creutz '80 [48] (with U(1) substituted by Z(6)), by Kripfganz, Ranft and Ranft '83 [60] and recently by Labastida et al. '85 [86]. They are shown schemati-

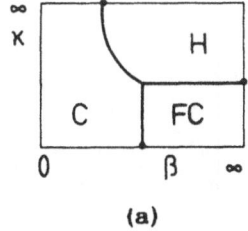

(a)

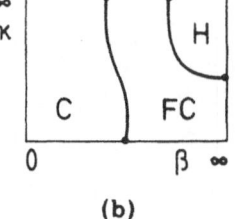

(b)

Fig. 15. Schematic phase diagram of the U(1) model for λ = ∞ with
multiple charged scalar field: (a) for 2 ⩽ q ⩽ 4,
(b) for q ⩾ 5.

cally in Fig. 15. For q growing to infinity the Coulomb-Higgs PT line shifts to ß infinite. But one should perhaps use the rescaled coupling ßq² instead of ß [24].

11. SU(2) with Higgs field in adjoint representation

The Georgi-Glashow model enjoys much attention in the lattice gauge theory as it is a prototype of the Higgs mechanism in grand unified theories requiring scalar field in adjoint representation of a nonabelian group. For κ = ∞ the lattice model reduces to the pure U(1) lattice gauge theory, and the Higgs phase is expected to contain a massless U(1) gauge field.

The phase diagram for λ = ∞ had been determined by Lang, Rebbi and Virasoro '81 [72] and by Brower et al. '82 [87], and is shown schematically in Fig. 16a. The Higgs PT line, separating the confinement and Higgs phases connects the second order PT of the O(3) spin model at ß = ∞ with a probably weakly first order [68] PT of the pure U(1) gauge theory at κ = ∞. Thus a tritritical point should be expected somewhere on this line.

Karsch, Seiler and Stamatescu '83 [88] investigated the phase diagram of the model with frozen radial mode at finite temperature. They held temperature constant in lattice units using a lattice of one size only (6^3*3). A line of deconfining PT's connecting the deconfining transitions in pure SU(2) and U(1) gauge theories has been found (Fig. 16b). It is interesting that no unambiguous signal of the Higgs PT has been seen for ß < ∞. Could it be that the PT of the O(3) spin model is not stable against coupling to gauge field at finite temperature? In any case this paper reminds us that the results for Higgs models on small lattices for large ß might be distorted by finite temperature effects.

Lee and Shigemitsu '85 [89] studied the Higgs PT for finite λ by looking for the massless U(1) field in the Higgs phase. They calculated two quantities. One of them is the energy $E(\vec{p})$ of a state with the lowest nonzero momentum $|\vec{p}|$ = $2\pi/N_S$ contributing to the correlation function of the U(1) field B_μ. Here N_S is the size of the lattice in space directions, and correlations are calculated in one (larger) time direction. For a massless

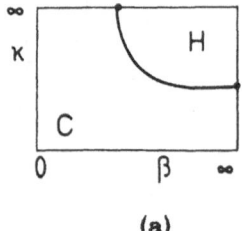

(a)

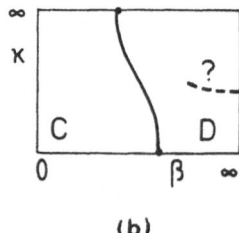

(b)

Fig. 16. Schematic phase diagram of the SU(2) model with octet scalar field (a) at zero temperature and (b) at finite temperature.

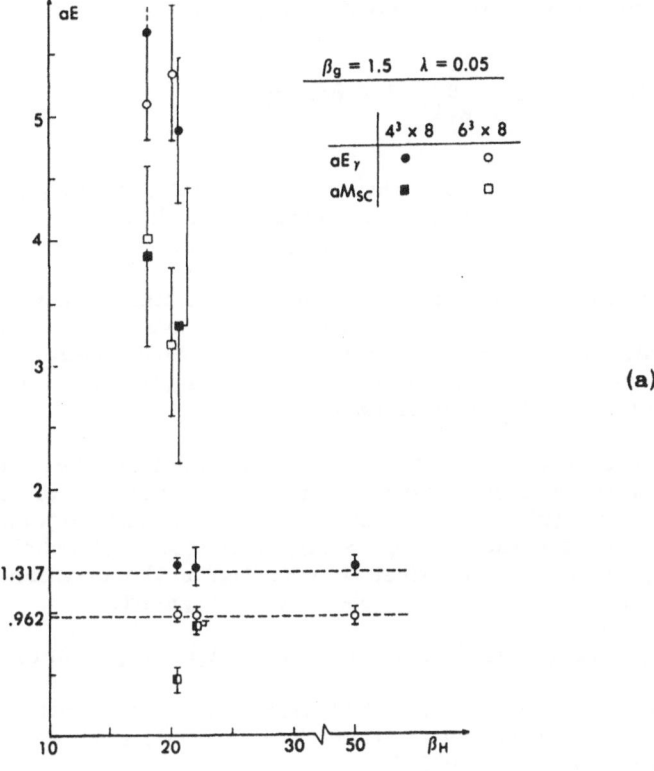

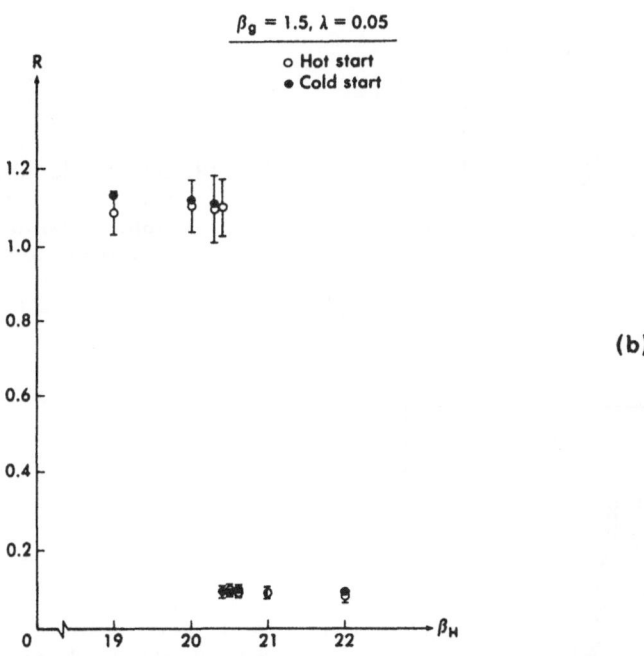

Fig. 17. Vanishing (a) of the photon mass and (b) of the longitudinal field in the Higgs phase of the adjoint SU(2) model [89].

state the energy should satisfy the dispersion relation

$$\cosh E(\vec{p}) = 1 + \sum_{k=1}^{3} (1 - \cos p_k) .\tag{11.1}$$

The data are in agreement with this relation in the Higgs phase (Fig. 17a). This method for detecting massless states on small lattices had been used earlier successfully in pure U(1) lattice gauge theory [90].

Another method is to calculate the longitudinal component of the B_μ field for $\vec{p} \neq 0$ and to compare it with the transverse one. The ratio R of both components squared is plotted in Fig. 17b. Both results are consistent with the presence of a massless field in the Higgs phase, whereas in the confinement phase the B_μ field is massive.

In the framework of the Georgi–Glashow model on the lattice it is also possible to study nonperturbative quantum properties of the 't Hooft-Polyakov monopoles [91]. This classical solution of field equations plays an important role in the theories of early Universe. Schierholz, Seixas and Teper '85 [92] have been the first to investigate the creation and stability of the monopoles as well as their role in the Higgs PT.

12. SU(3) Higgs model with scalar field in various representations

The phase diagram of the SU(3) model for $\lambda = \infty$ with Higgs field in fundamental or adjoint representations has been determined by a group in Hiroshima (Kikugawa et al. [34,93]).

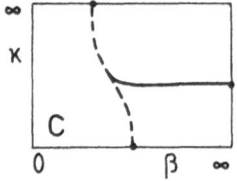

Fig. 18. Schematic phase diagram of the SU(3) model with scalar field in fundamental representation for $\lambda = \infty$.

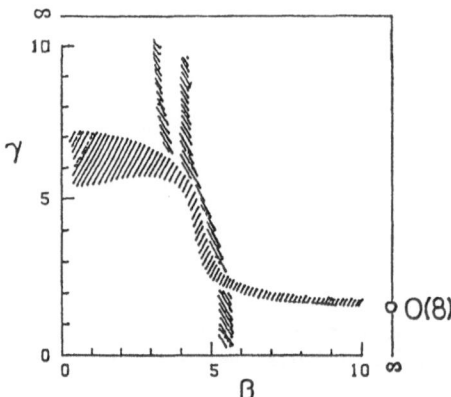

Fig. 19. Phase diagram of the SU(3) model with scalar field in adjoint representation for $\lambda = \infty$ at finite temperature [93].

152

For a triplet Higgs field the limit cases $\beta = \infty$, $\kappa = 0$ and $\kappa = \infty$ correspond to O(6) spin model, pure SU(3) and pure SU(2) lattice gauge theories, respectively. Since both limit gauge theories have no PT for Wilson actions, the Higgs and confinement regions are again analytically connected. The cross-overs at $\kappa = 0, \infty$ are connected by a line of cross-overs for all values of κ (Fig. 18). The Higgs PT seems to have its end point on this cross-over line [34]. It would be interesting to know how this picture changes for $\lambda < \infty$. The endpoint might move to lower β as in SU(2) or the cross-over might change into a genuine PT as in pure SU(2) or SU(3) models with mixed actions.

The SU(3) model with an octet Higgs field is of relevance for GUT. This makes its phase diagram at finite temperature of much interest. For $\lambda = \infty$ such a diagram is shown in Fig. 19 [93].

The adjoint model with fluctuating radial mode ($\lambda < \infty$) was studied by Gupta and Heller '84 [94] and by Olynyk and Shigemitsu '85 [95] with the aim to determine the vacuum pattern in the Higgs phase. A global SU(3) symmetry of adjoint scalar field can be broken into SU(2)*U(1) or U(1)*U(1) symmetries. Olynyk and Shigemitsu [95] suggest to distinguish between the vacuum patterns characterized on the basis of SSB in pure ϕ^4 theory by looking at the following ratio of the gauge invariant polynomials:

$$6 \left\langle \frac{(\operatorname{tr} \phi^3)^2}{(\operatorname{tr} \phi^2)^3} \right\rangle \propto \left\langle \begin{array}{l} 1 \quad [\text{like SU(2)*U(1) in } \phi^4 \text{ theory}] \\[6pt] 0 \quad [\text{like U(1)*U(1) in } \phi^4 \text{ theory}] . \end{array} \right. \qquad (12.1)$$

Because of the gauge field fluctuations these relations are expected to hold only approximately. The first vacuum pattern in (12.1) is preferred by the

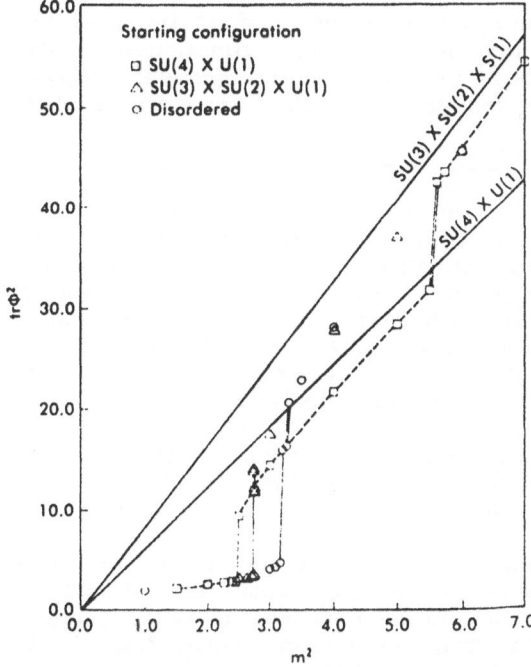

Fig. 20. Vacuum patterns in the SU(5) model [99].

Monte Carlo data at several points of the phase diagram [93-95]. Suranyi '85 [96] finds the same vacuum pattern also by means of strong coupling expansion. This, of course, does not exclude the possibility that the vacuum pattern is a function of coupling constants and that it may be different in regions which have not yet been investigated. As we do not know which regions are more relevant for a continuum limit, we should be very cautious in physical interpretation of the results for vacuum patterns on the lattice.

There are two investigations of the SU(3) model with scalar field described by a 3*3 matrix. Karsch, Seiler and Stamatescu '85 [97] consider also finite temperature effects in such a model. They interpret the scalar field as a matter field with several flavours. Drouffe, Jurkiewicz and Krzywicki '84 [98] studied both SU(2) and SU(3) models of this type.

13. SU(5) model with scalar field in adjoint representation

This model is of direct relevance for GUT. Therefore it deserves extensive attention. The vacuum pattern suggested by SSB of global symmetry of pure ϕ^4 theory may be changed by radiative corrections due to gauge fields, as indicated by perturbative estimates in continuum theory. Olynyk and Shigemitsu '85 [99] found that similar change occurs also on the lattice. They have found a (possibly metastable) SU(4)*U(1)-like vacuum pattern instead of SU(3)*SU(2)*U(1) obtained on the tree level (Fig.20). The same comment on the possible dependence of the vacuum patterns on the coupling parameters I have written for SU(3) model is applicable here, too.

IV. HIGH STATISTICS CALCULATIONS OF OBSERVABLES IN SU(2) MODEL

At present only the SU(2) lattice Higgs model with scalar field in the fundamental representation is beyond the exploratory investigation of phases and phase transitions. Results of high statistics calculations of the Higgs boson and vector boson masses, of static potential, of screening effects and of gauge invariant two-point function of scalar field are now available. They have been obtained by two groups: Langguth and Montvay [18-20,100] and the Aachen group, Landau, Lang and Xu [21,22,33,101].

14. Higgs boson and vector boson masses

Both groups use for the calculation of the Higgs boson and the vector boson masses, m_H and m_W, respectively, correlation functions between gauge invariant field aggregates with suitable quantum numbers [18,19,22]. Montvay et al. [18,20] work in unitary gauge on 8^4 and 12^4 lattices. The Aachen group et al. [21,22,33,75] perform Monte Carlo calculation also over gauge degrees of freedom using 8^3*16 and 16^4 lattices. In the vicinity of the Higgs PT the correlation functions are measurable up to distance 8. But long time fluctuations require high statistics and long thermalization [100]. The aim of these first calculations of m_H and m_W is to estimate the dependence of the masses on the coupling parameters. Both groups fix λ and β and determine the masses in the vicinity of the Higgs PT as a function of κ. This has been done for several fixed values of λ and β.

The κ dependences of both masses show the same characteristic pattern at all λ,β points investigated until now (Figs.21-22). Higgs boson mass m_H has a dip in the interval where the Higgs phase transition takes place on finite lattices. The correlation length $1/m_H$ in the Higgs boson channel achieves values about 5 lattice spacings which is close to the spatial size of the lattices used. The dip, first observed by the Aachen group, Lang and Vones [21,22,33] is very narrow and difficult to find. The

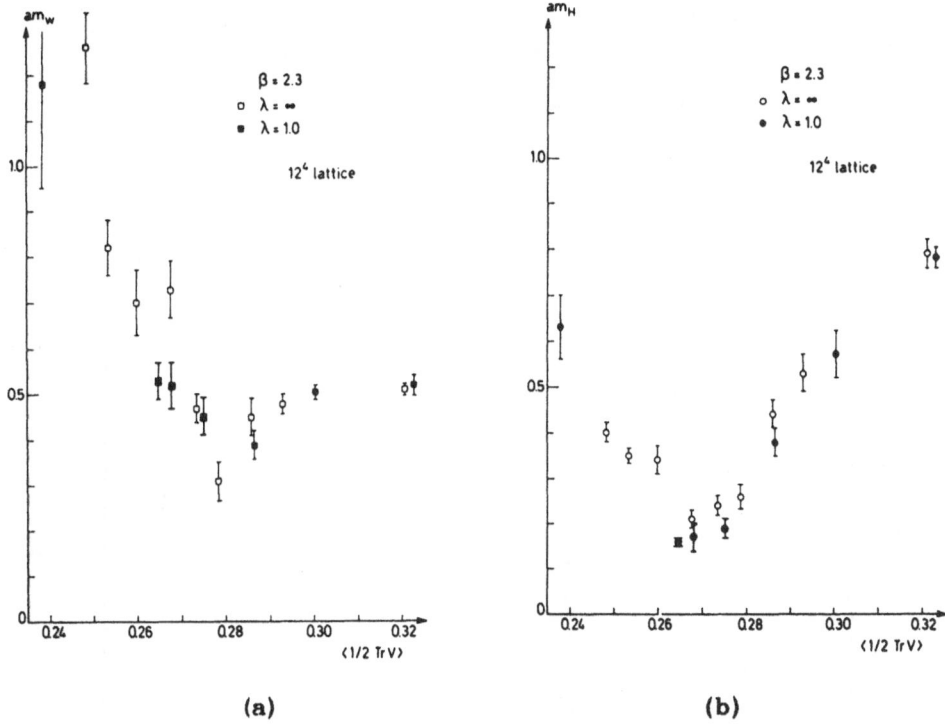

Fig. 21. The vector boson (a) and Higgs boson (b) masses in the SU(2) model for $\lambda = \infty$ and $\lambda = 1$ at $\beta = 2.3$ [20].

vector boson mass m_W does not show any dip. It decreases during the Higgs PT from very high (unmeasurable) values in the confinement region to lower values in the Higgs region monotonously. Thus we observe a behaviour quite opposite to the naive expectations based on the semiclassical tree level treatment of Higgs mechanism (Fig. 1). For κ above the Higgs PT the vector boson mass starts to rise again, indicating that the quasiclassical picture might be correct qualitatively at large κ. Below and on the Higgs PT m_W is greater than m_H. Thus the model is in the regime of superconductors of type I. The critical-like behaviour is due to the increase of the Higgs correlation length $1/m_H$ whereas $1/m_W$ remains much smaller than the lattice size. For κ above the Higgs PT the relation is reverse. The Higgs region corresponds to a type II superconductor for κ immediately above the Higgs PT.

Langguth and Montvay [18-20] concentrated on the λ dependence of the spectrum. For β fixed at 2.3 they compared the masses at $\lambda = \infty$ and $\lambda = 1$ (Fig. 21). When the masses are plotted not as a function of κ but as a function of the variable

$$ L = \langle \frac{\phi_x^+}{\rho_x} U_{x\mu} \frac{\phi_{x+\mu}}{\rho_{x+\mu}} \rangle , \qquad (14.1) $$

the mass curves are equal within the error bars. If this kind of universality holds for all β and λ for which the Higgs PT is of higher order, it might mean that λ is an irrelevant parameter. Then the continuum limit might be independent of λ and the Higgs boson mass should be predictable (provided the limit is nontrivial) [19,20]. The Aachen group and Lang [22] did not calculate the quantity L. Thus their data for $\beta = 2.25$ and $\lambda = 3$

155

and λ = 0.5 cannot be used for testing the universality suggested by Montvay [19,100]. But the strong dependence of the properties of the Higgs PT on λ for small λ [33] indicates that the λ-universality in the above simple sense might hold only in a restricted region.

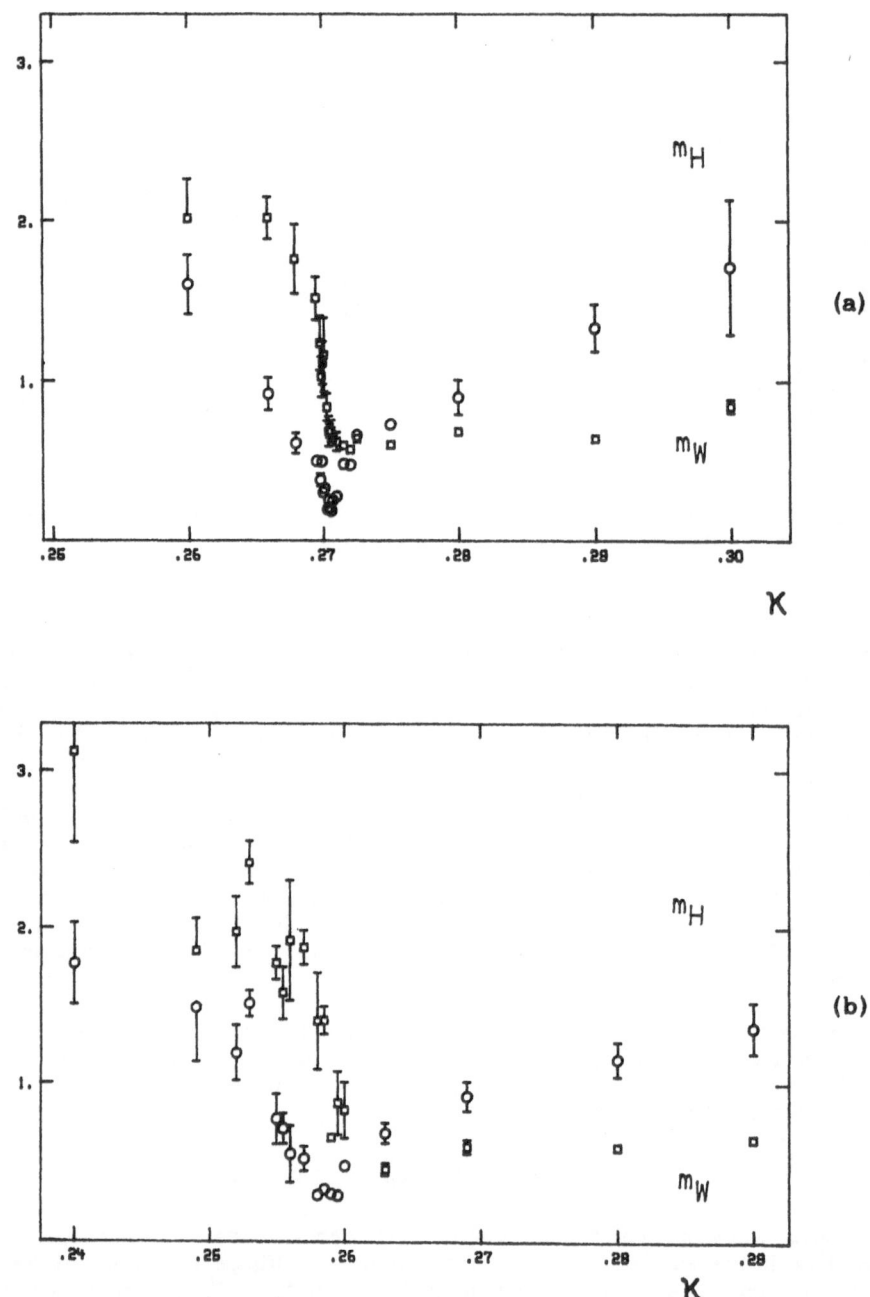

Fig. 22. The Higgs boson and vector boson masses in the SU(2) model for λ = 0.5 and (a) ß = 2.25 and (b) ß = 2.4 [22].

The Aachen group and Lang '85 [22] concentrated on the investigation of the ß dependence of the mass curves. For $\lambda = 0.5$ the data at ß = 2.25 and ß = 2.4 are quite comparable (Fig. 22). When the PT points are superposed, the Higgs mass curves $m_H(\kappa)$ are equal within error bars. The same seems to be true for $m_W(\kappa)$ at $\kappa < \kappa_{PT}$. In the Higgs region ($\kappa > \kappa_{PT}$) the values of $m_W(\kappa)$ might be slightly lower for ß = 2.4 than for ß = 2.25 at the same distance from the Higgs PT. The difference, which would mean that the lines of constant m_H/m_W approach the Higgs PT with growing ß, is only marginal, however. The convergence of these lines to the PT line will have to be studied in a much broader ß interval.

15. Static potential

Montvay [19] collected a lot of data on the static potential calculated by means of Wilson loops on a 12^4 lattice. For $\kappa > \kappa_{PT}$ he observed a flattening of the potential due to the screening of static sources by matter field. In the confinement region ($\kappa < \kappa_{PT}$) the screening length seems to be too large to be observed on lattices of accessible sizes.

Montvay fitted the potential by Yukawa or Hulthén analytic expressions

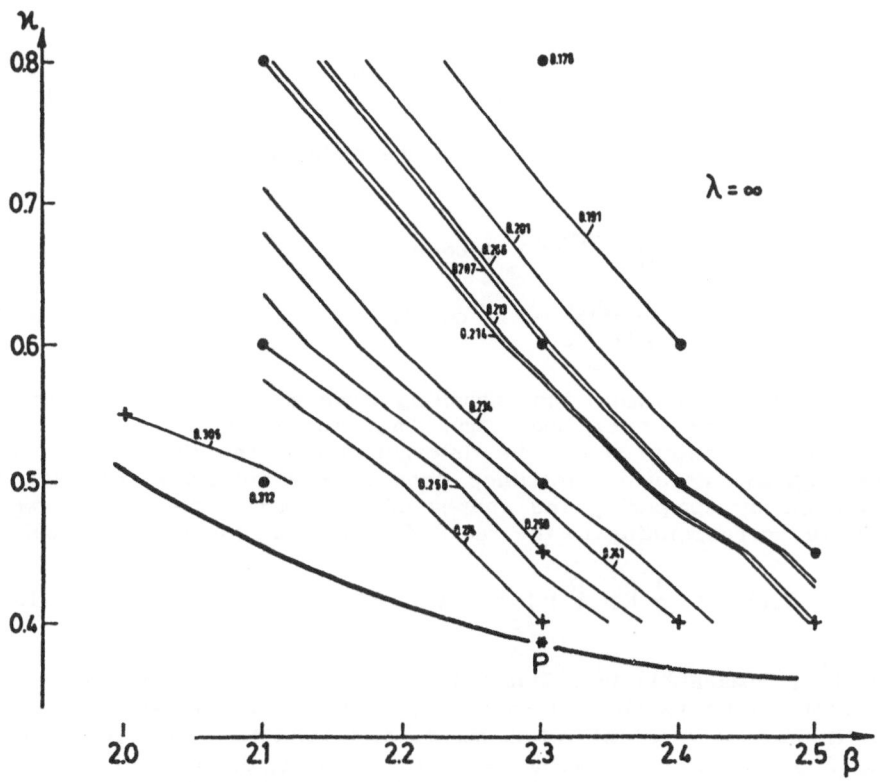

Fig. 23. Lines of constant α in the SU(2) model for $\lambda = \infty$ [19].

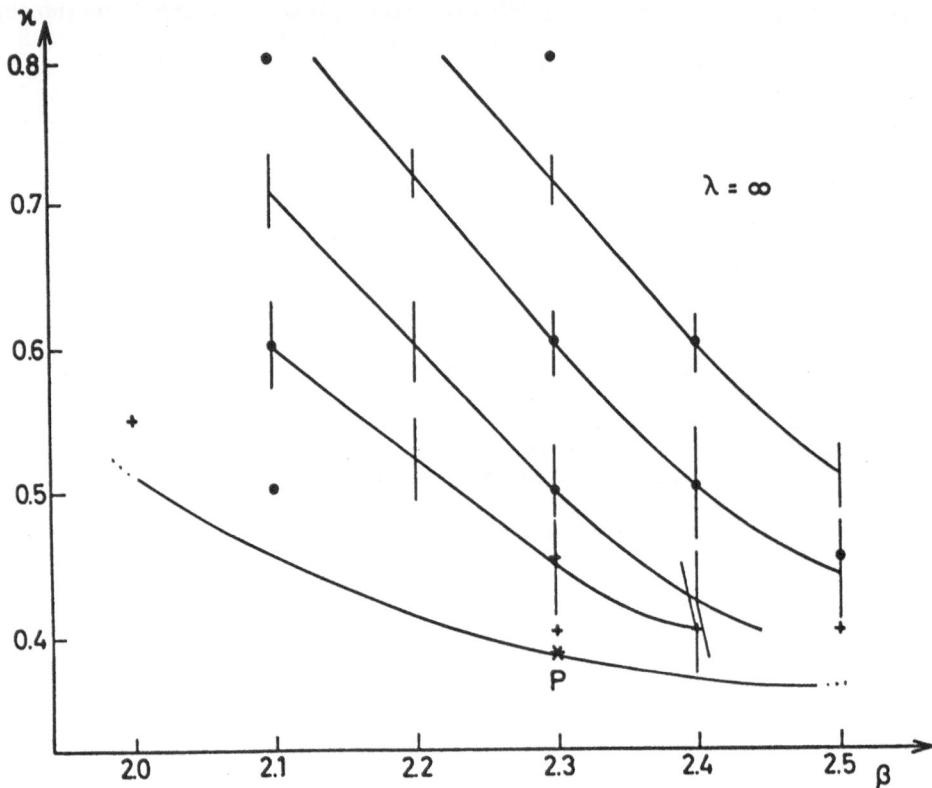

Fig. 24. Lines of constant force in the SU(2) model for $\lambda = \infty$ [19].

$$V(R) \;=\; \left\langle \begin{array}{l} - \dfrac{\alpha}{R}\, e^{-m_{sc}R} \quad + \text{ const} \\[3mm] - \dfrac{\alpha m_{sc}}{e^{m_{sc}R} - 1} \quad + \text{ const.} \end{array} \right. \tag{15.1}$$

This allows him to estimate the screening mass m_{sc} which turns out to be consistent with the vector boson mass m_W. Furthermore, the renormalized fine structure constant α can be determined and the lines of constant α plotted (Fig. 23). He also determined lines of constant force (Fig. 24). They are obtained by comparing the shapes of the forces after their rescaling by a factor corresponding to the ratio of lattice constants in physical units

$$F_1(R) \;=\; \frac{1}{\xi}\, F_2\!\left(\frac{R}{\xi}\right); \qquad \xi = \frac{a_2}{a_1}. \tag{15.2}$$

It is a remarkable fact that both lines of constant α and of constant forces seem to be parallel. It would be very interesting to compare them also with lines of constant ratios m_H/m_W and thus to verify the scaling of various physical observables. However, this ratio is difficult to calculate for κ far above κ_{PT} where the data in Figs.23-24 have been obtained.

The static potential has been also used as a test of the eventual λ universality mentioned in the preceeding section. Montvay [19] finds that the potential $V(R,\lambda,\beta,\kappa)$ is independent of λ if plotted as a function of L (14.1) instead of κ.

16. Gauge invariant two-point function of the scalar field

The Aachen group, Landau and Xu '85 [101] calculated the gauge invariant two-point function G(R,T) of the ϕ field, Eq. (4.3), in the SU(2) model with doublet scalar field. The data have been obtained on 16^4 and $8^3 \ast 16$ lattices for R = 0,...,8 and R = 0,...,4, respectively, and for T = 1,...,8. The two-point function decays for all R and for T $\geq$ 4 exponentially,

$$G(R,T) = C_G(R) \, e^{-m_\phi(R)T} \, . \tag{16.1}$$

It turns out that at all points of the phase diagram studied $m_\phi(R)$ is independent of R and thus defines a new observable with dimension of mass,

$$m_\phi(R) = m_\phi \, . \tag{16.2}$$

The mass m_ϕ calculated on a 16^4 lattice is shown for $\lambda = 0.5$ and $\beta = 2.4$ in Fig. 25a as a function of κ. It decreases monotonously with increasing κ and its slope increases in the Higgs PT region. The data for a $8^3 \ast 16$ lattice obtained in a much broader range of κ indicate similar behaviour (Fig. 25b). The mass m_ϕ behaves as a function of κ differently from the masses m_H and m_W (Figs.21-22).

It is instructive to compare m_ϕ with the values of the static potential V(R) at large distances R. For $\kappa \gtrsim \kappa_{PT}$ the potential becomes constant already at distances smaller than the sizes of lattices used. For these values of R one finds

$$m_\phi = \frac{1}{2} \, V(R) \, , \qquad R \text{ large.} \tag{16.3}$$

The data for V(R)/2, R = 7 and 4 are indicated in Fig. 25 by circles.

The physical meaning of the quantity V(R = ∞)/2 is quite obvious: It is the energy in fields screening the external static charge. In a sense, it is thus the energy of the lowest bound state of a quantum of matter field ϕ with a static charge. (This quantum is not the Higgs boson since it carries SU(2) color and is thus confined, see Sec.4.) Similar interpretation for m_ϕ determined by the exponential decay (16.1) of the two-point function G(R,T) has been suggested in Refs. [40,44]. Thus the approximate equality (16.2), noticed first in Monte Carlo data by T. Neuhaus, has got a good theoretical explanation.

For $\kappa \lesssim \kappa_{PT}$ the values of V(R)/2 are lower than m_ϕ. This is expected since V(R) is not yet the asymptotic value of the screened potential V(R = ∞) in the confinement region.

17. Test for confinement in gauge theory with matter field

As I have explained in Sec.4, the Fredenhagen-Marcu criterion (4.2) [40] tests for the possibility to construct a charged state. The SU(2) Higgs model with Higgs field in the fundamental representation has for $\beta \gtrsim 0$ only one confinement-Higgs phase. Thus both above and below the Higgs PT, we should expect ρ_{FM}^∞ to be nonzero.

The Aachen group, Landau and Xu '85 [101] calculated the quantities

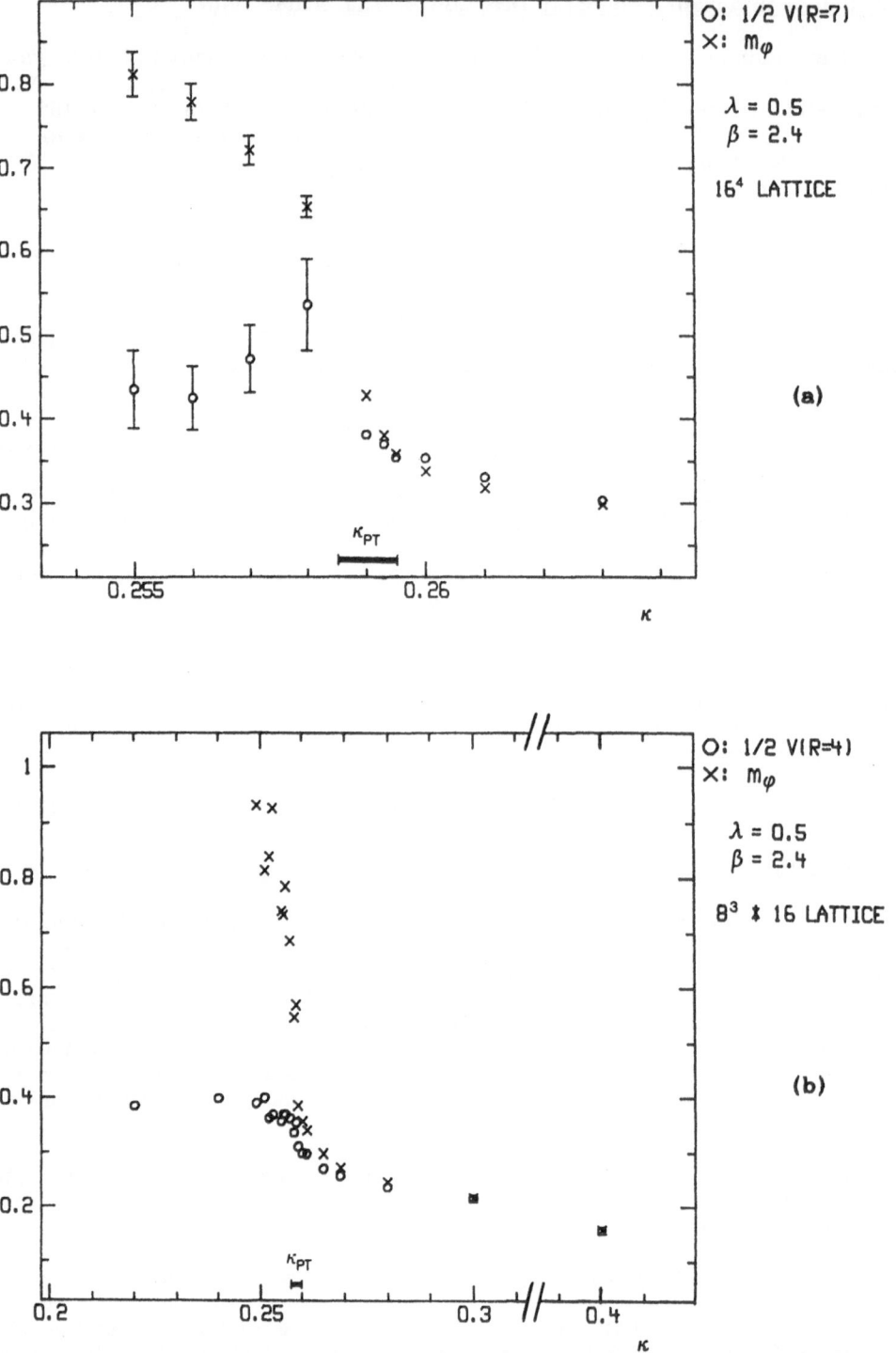

Fig. 25. The mass m_ϕ denoted by crosses (a) in the vicinity of the
Higgs PT on a 16^4 lattice and (b) in a broader κ-interval on
a 8^3*16 lattice in the SU(2) model for $\lambda = 0.5$ and $\beta = 2.4$.
Circles indicate values of 1/2V(7) and 1/2V(4) [101].

$$\rho_{FM}(R_1,R_2,T) \;=\; \sqrt{\frac{G(R_1,T)\;G(R_2,T)}{W(R_1+R_2,T)}}\;, \qquad\qquad (17.1)$$

where $G(R,T)$ is given by Eq.(4.3) and $W(R_1+R_2,T)$ is the expectation value of the rectangular Wilson loop $(R_1+R_2)*T$. Since there is no necessity to consider only $R = T/2$ in (4.2), an increase of R proportional to T being sufficient, the ratios (17.1) are an inessential generalization of the ratios in (4.2) allowing us to consider more data. In the vicinity of the Higgs PT and above it the ratios (17.1) are easily calculated on a 16^4 lattice even for $T = 7$ and $R_1+R_2 = 7$. For $T = 6$ the error bars are still quite small. Fig. 26 shows the results for $\rho_{FM}(3,3,6)$ as a function of κ. They are apparently different above and below the Higgs PT.

The crucial question is to what degree the ratios $\rho_{FM}(3,3,6)$ represent the asymptotic values ρ_{FM}^{∞} for $T = \infty$ as required by (4.3). In Fig. 27 the values of $\rho_{FM}(R_1,R_2,T)$ are shown as a function of one half of the perimeter of the Wilson loop, $\tau = R_1+R_2+T$. Here R_1 and R_2 are chosen within the constraints $|R_1+R_2-T| \leqslant 2$, $|R_1-R_2| \leqslant 1$ and $R_1+R_2 \leqslant 7$. For $\kappa < \kappa_{PT}$ (Fig. 27a) the ratios rapidly approach zero, whereas for $\kappa > \kappa_{PT}$ they stay finite and after a decrease for small τ they indicate a slight increase for $\tau \geqslant 12$.

Thus for $\kappa > \kappa_{PT}$ the data (Fig. 27b) suggest that ρ_{FM}^{∞} is nonzero, as expected, though it might be too early to determine the asymptotic value ρ_{FM}^{∞} from the result at $\tau = 12$.

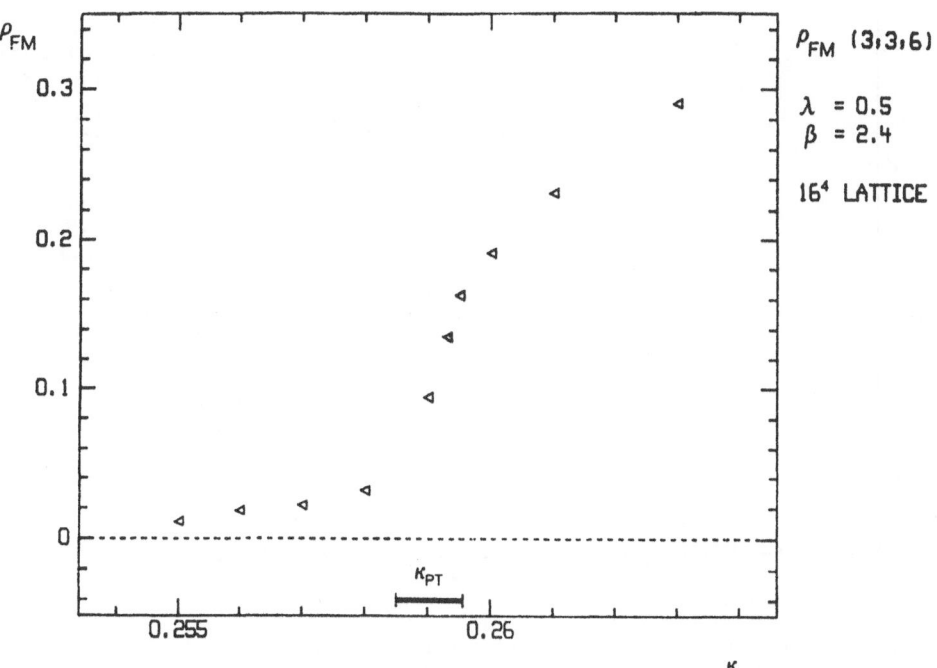

Fig. 26. The ratio $\rho_{FM}(3,3,6)$ in the vicinity of the Higgs PT [101].

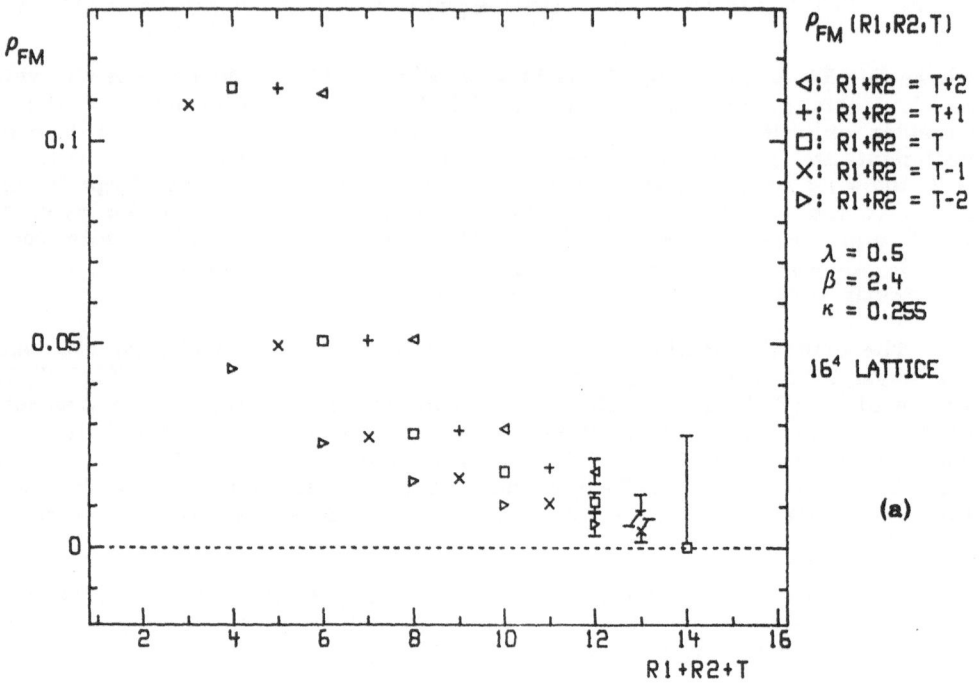

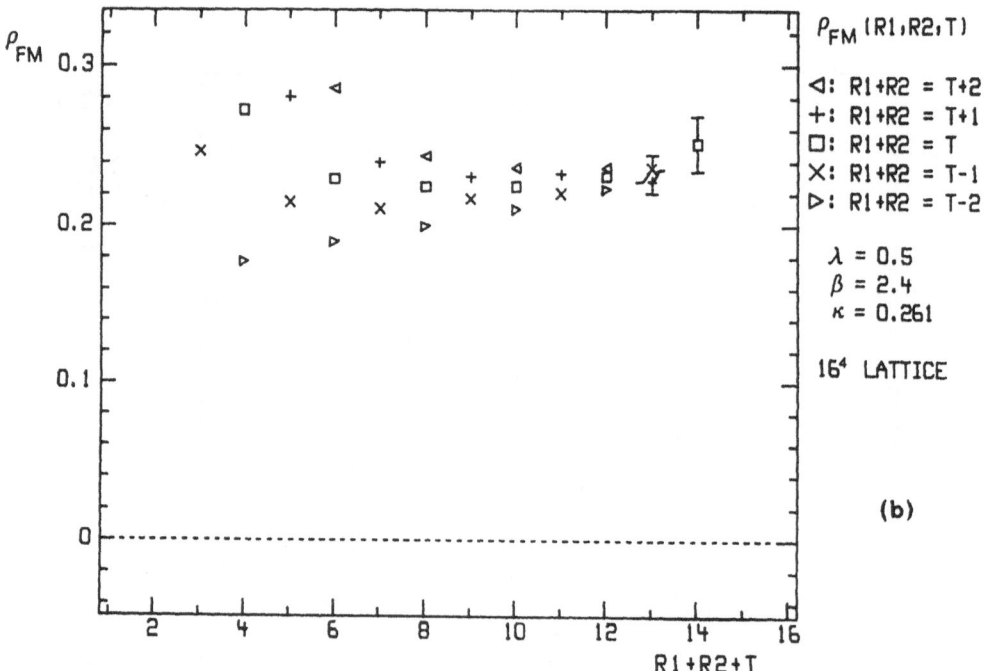

Fig. 27. The ratios $\rho_{FM}(R_1,R_2,T)$ in the SU(2) model for
$\lambda = 0.5$ and $\beta = 2.4$ closely (a) below and (b) above the
Higgs PT [101].

For $\kappa < \kappa_{PT}$ the explanation of the data might be as follows: The Fredenhagen–Marcu parameter ρ°_{FM} (4.2) tests for the absence of charged states by testing for hadronization during the process of removing one of the charges of a dipole to infinity [40]. But at distances T smaller than the screening length of the confinement potential, no hadronization is expected and the overlap between the dipole state and hadron states should be very small. One can expect, however, that ρ_{FM} is not a monotonous function of τ (as indicated also for $\kappa > \kappa_{PT}$ in Fig. 27b) and that it will get nonzero for T above the screening length. Since for $\kappa < \kappa_{PT}$ the screening length is larger than the lattice size, ρ_{FM} is still zero. If this interpretation is correct, then the parameter ρ°_{FM} is a good indicator of confinement in principle, but of limited use for small κ in practice. This trouble is obviously caused by the fact that ρ°_{FM} does not smoothly generalize the Wilson criterion valid for $\kappa = 0$, since it is based on a physical picture different from that of the rising static potential.

18. Outlook

All recent analytic and Monte Carlo results leave the question of continuum limit open. We do not know where and exactly how to construct the continuum theory and whether the result will be nontrivial. The lack of unique continuum strategy is caused by the absence of reliable perturbative renormalization group calculations in continuum theory, i.e. by absence of asymptotic freedom. This is a big difference from lattice QCD.

Most probably the continuum strategy will have to be developed on the basis of Monte Carlo experience with the lattice Higgs models. Numerical study should influence theoretical thinking and vice versa.

As first steps we should localize the regions where the Higgs PT is critical and determine the renormalization group flows. The RG trajectories determined by means of constant mass ratios and constant forces should be compared with the results of Monte Carlo renormalization group method. First studies of lattice Higgs models based on RG methods have been discussed at this conference by P. Hasenfratz [102] and R. Petronzio [103].

Before we really know how to perform the continuum limit it might be useful to develop some optimistic scenaria, within which it might be possible to obtain phenomenologically relevant results. This has been attempted for SU(2) by Montvay '85 [100] and by Langguth, Montvay and Weisz [104]. The idea is to find points where the renormalized fine structure constant α has the value required for the electroweak interactions and to calculate the ratio m_H/m_W there. The results for the masses in lattice units and for the coupling constants are at $\beta = 8$ [104]

$$m_W = 0.19(1) \qquad g_{HHH} = 6.9 \pm 2.7$$

$$m_H = 1.18(14) \qquad g_{HWW} = 0.28 \pm 0.08.$$

The result $m_H \approx 6m_W$ is interesting. However, we do not know whether the assumed scenario is correct, what the finite temperature effects are and whether the Monte Carlo method does work for such large β properly. Therefore it would be premature to suggest this value for the Higgs boson mass to our colleagues experimentalists as a prediction of lattice gauge theory.

ACKNOWLEDGEMENTS

I am grateful to G. Bhanot, J. Bricmont, D. Brydges, P. Damgaard,

H.G. Evertz, J. Fröhlich, V. Grösch, D. Horn, K. Jansen, K. Kanaya, H.A. Kastrup, H. Kleinert, K.-I. Kondo, D.P. Landau, W. Langguth, I.D. Lawrie, G. Mack, M. Marcu, I. Montvay, Y. Munehisa, T. Neuhaus, G. Roepstorff, G. Schierholz, L. Sehgal, J. Shigemitsu, R.E. Shrock and I.O. Stamatescu for discussions and informations. In particular I am indebted to C.B. Lang and E. Seiler for sharing their knowledge with me and for various suggestions. I would also like to thank V. Grösch and J. Smižanská-Jersák for help with the manuscript.

REFERENCES

[1]   L.B. Okun, "Particle Physics Prospects: August 1981",
        Proceedings of the 10. International Symposium on Lepton
        and Photon Interactions at High Energies, Bonn 1981.

[2]   P.W. Anderson, Phys. Rev. 130 (1963) 439.

[3]   P.W. Higgs, Phys. Lett. 12 (1964) 132;
        Phys. Rev. Lett. 13 (1964) 508;
        Phys. Rev. 145 (1966) 1156.
        F. Englert and R. Brout, Phys. Rev. Lett. 13 (1964) 321.
        G.S. Guralnik, C.R. Hagen and T.W.B. Kibble,
        Phys. Rev. Lett. 13 (1964) 585.
        T.W. Kibble, Phys. Rev. 155 (1967) 1554.

[4]   S. Coleman and E. Weinberg, Phys. Rev. D7 (1973) 1888.

[5]   M. Aizenman, Phys. Rev. Lett. 47 (1981) 1;
        Commun. Math. Phys. 86 (1982) 1.
        G.A. Baker and J.M. Kincaid, J. Stat. Phys. 24 (1981) 469.
        D.C. Brydges, J. Fröhlich and T. Spencer,
        Commun. Math. Phys. 83 (1982) 123.
        J. Fröhlich, Nucl. Phys. B200 (1982) 281.
        A.D. Sokal, Ann. Inst. H. Poincare A37 (1982) 317.
        M. Aizenman and R. Graham, Nucl. Phys. B225 [FS9] (1983) 261.
        C. Aragao de Carvalho, S. Caracciolo and J. Fröhlich,
        Nucl. Phys. B215 (1983) 209.
        K. Gawedzki and A. Kupiainen, Phys. Rev. Lett. 54B (1985) 92.

[6]   B. Freedmann, P. Smolensky and D. Weingarten,
        Phys. Lett. 113B (1982) 481.
        D.J.E. Callaway and R. Petronzio,
        Nucl. Phys. B240 [FS12] (1984) 577.
        C.B. Lang, Phys. Lett. 155B (1985) 399;
        Tallahassee preprint FSU-SCRI-85-6
        (to be published in Nucl. Phys. B).

[7]   K.G. Wilson, Phys. Rev. B4 (1971) 3184.
        K.G. Wilson and J. Kogut, Phys. Rep. 12C (1974) 76.

[8]   R. Dashen and H. Neuberger, Phys. Rev. Lett. 50 (1983) 1897.
        M.A. Beg, C. Panagiotakopoulos and A. Sirlin,
        Phys. Rev. Lett. 52 (1984) 883.
        D.J.E. Callaway, Nucl. Phys. B223 (1984) 189.
        E. Ma, Phys. Rev. D31 (1985) 1143.
        M. Lindner, Munchen preprint MPI-PAE/PTh 52/85.

[9]   K.G. Wilson, Phys. Rev. D10 (1974) 2445.

[10] H. Kühnelt, C.B. Lang and G. Vones,
Nucl. Phys. B230 [FS10] (1984) 16.

[11] S. Elitzur, Phys. Rev. D12 (1975) 3978.

[12] G.F. De Angelis, D. de Falco and F. Guerra,
Phys. Rev. D17 (1978) 1624.
M. Lüscher, Preprint DESY 77-16 (1977) (unpublished).

[13] J. Fröhlich, G. Morchio and F. Strocchi,
Phys. Lett. 97B (1980) 249;
Nucl. Phys. B190 [FS3] (1981) 553.

[14] T. Kennedy and C. King, Phys. Rev. Lett. 55 (1985) 776;
Princeton preprint 1985.

[15] C. Borgs and F. Nill, München preprint MPI-PAE/PTh 59/85.

[16] F. Nill, München preprint MPI-PAE/PTh 39/85.

[17] J. Schwinger, Phys. Rev. 125 (1962) 397.

[18] I. Montvay, Phys. Lett. 150B (1985) 441.

[19] I. Montvay, Preprint DESY 85-005.

[20] W. Langguth and I. Montvay, Phys. Lett. 165B (1985) 135.

[21] J. Jersák, Talk given at the Conference "Advances in the
Lattice Gauge Theory", April 1985, Tallahassee,
Aachen preprint PITHA 85/12.

[22] H.G. Evertz, J. Jersák, C.B. Lang and T. Neuhaus,
Aachen preprint PITHA 85/23.

[23] R. Marra and S. Miracle-Sole, Commun. Math. Phys. 67 (1979) 233.

[24] D. Brydges, J. Fröhlich and E. Seiler,
Nucl. Phys. B152 (1979) 521.

[25] T. Balaban, D. Brydges, J. Imbrie and A. Jaffe,
Ann. Phys. 158 (1984) 281.

[26] D.C. Brydges and E. Seiler, Princeton Univ. preprint (1985).

[27] C. Borgs and F. Nill (in preparation).

[28] C. Borgs and F. Nill, München preprint MPI-PAE/PTh 73/85.

[29] E. Seiler, München preprint MPI-PAE/PTh 47/85.

[30] K. Jansen, J. Jersák, C.B. Lang, T. Neuhaus and G. Vones,
Phys. Lett. 155B (1985) 268.

[31] K. Osterwalder and E. Seiler, Ann. Phys. 110 (1978) 440.
E. Seiler, "Gauge Theories as a Problem of Constructive
Quantum Field Theory and Statistical Mechanics",
Lecture Notes in Physics 159 (Springer 1982).

[32] E. Fradkin and S. Shenker, Phys. Rev. D19 (1979) 3682.

[33] J. Jersák, C.B. Lang, T. Neuhaus and G. Vones,
      Phys. Rev. D32 (1985) 2761.

[34] M. Kikugawa, T. Maehara, T. Minazuki, J. Saito and
      H. Tanaka, Prog. Theor. Phys. 69 (1983) 1207.

[35] G. Mack, DESY report 77/58 (1977) (unpublished).

[36] G. 't Hooft, Cargése Lecture, in "Recent Developments
      in Gauge Theories", Ed. G. 't Hooft et al., Plenum 1980.

[37] S. Dimopoulos, S. Raby and L. Susskind,
      Nucl. Phys. B173 (1980) 208.

[38] L.F. Abbot and E. Farhi, Phys. Lett. 101B (1981) 69.

[39] T. Banks and E. Rabinovici, Nucl. Phys. B160 (1979) 349.

[40] K. Fredenhagen and M. Marcu,
      Commun. Math. Phys. 92 (1983) 81;
      DESY preprint 85-008.

[41] J. Bricmont and J. Fröhlich, Phys. Lett. 122B (1983) 73.

[42] J. Bricmont and J. Fröhlich,
      Nucl. Phys. B251 [FS13] (1985) 517.

[43] K.-I. Kondo, Prog. Theor. Phys. 74 (1985) 152.

[44] K. Fredenhagen and M. Marcu (in preparation).

[45] K. Szlachanyi, Phys. Lett. 147B (1984) 335.

[46] J. Bricmont and J. Fröhlich (in preparation).

[47] F.J. Wegner, J. Math. Phys. 12 (1971) 2259.

[48] M. Creutz, Phys. Rev. D21 (1980) 1006.

[49] G. Bhanot and M. Creutz, Phys. Rev. B22 (1980) 3370.

[50] D. Horn and S. Yankielowicz, Phys. Lett. 85B (1979) 347.
      D. Horn and E. Katznelson, PL 91B (1980) 397.
      D. Horn, Phys. Rep. 67 (1980) 103.
      D. Horn and M. Karliner, Phys. Lett. 109B (1982) 288.

[51] T. Munehisa and Y. Munehisa, Phys. Lett. 116B (1982) 353.

[52] V.P. Gerdt, A.S. Ilchev and V.K. Mitrjushkin,
      Yad. Fiz. 40 (1984) 1097.

[53] T. Munehisa and Y. Munehisa,
      Nucl. Phys. B215 [FS7] (1983) 508.

[54] T. Banks, R. Myerson and J. Kogut, Nucl. Phys. B129 (1967) 493.

[55] M. Peskin, Ann. Phys. 113 (1978) 122.

[56] M.B. Einhorn and R. Savit, Phys. Rev. D17 (1978) 2583;
      Phys. Rev. D19 (1979) 1198.

[57] E. Katznelson, Nucl. Phys. B240 [FS12] (1984) 19.

[58] K.C. Bowler, G.S. Pawley, B.J. Pendleton, D.J. Wallace and
      G.W. Thomas, Phys. Lett. 104B (1981) 481.

[59] D.J.E. Callaway and L.J. Carson, Phys. Rev. D25 (1982) 531.

[60] J. Ranft, J. Kripfganz and G. Ranft,
      Phys. Rev. D28 (1983) 360.

[61] J.M.F. Labastida, E. Sanchez-Velasco, R.E. Shrock and P. Wills,
      Stony Brook preprint ITP-SB-85-40.

[62] Y. Munehisa, Phys. Rev. D30 (1984) 1310;
      Phys. Rev. D31 (1985) 1522.

[63] V.P. Gerdt, A.S. Ilchev and V.K. Mitrjushkin,
        Dubna preprint E2-85-59.
      G. Koutsoumbas, Phys. Lett. 140B (1984) 379.
      R.L. Sugar and W.D. Toussaint, Phys. Rev. D32 (1985) 2061.

[64] D. Espriu and J.F. Wheater,
      Nucl. Phys. B258 (1985) 101.

[65] K. Jansen, J. Jersák, C.B. Lang, T. Neuhaus and G. Vones,
      Nucl. Phys. B265 [FS15] (1986) 129.

[66] H. Kleinert, Phys. Lett. 128B (1983) 69;
      San Diego preprint (1985).

[67] D.P. Landau and R.H. Swendsen,
      Phys. Rev. Lett. 46 (1981) 1437.

[68] J. Jersák, T. Neuhaus and P.M. Zerwas,
      Phys. Lett. 133B (1983) 103.
      H.G. Evertz, J. Jersák, T. Neuhaus and P.M. Zerwas,
      Nucl. Phys. B251 [FS13] (1985) 279.

[69] V. Grösch, K. Jansen, J. Jersák, C.B. Lang, T. Neuhaus and
      C. Rebbi, Phys. Lett. 162B (1985) 171.

[70] Y. Sugiyama and K. Kanaya, Prog. Theor. Phys. 73 (1985) 176.

[71] B.J. Pendleton, Phys. Lett. 118B (1982) 121.
      Y. Sugiyama and T. Yokota, Nagoya preprint DPNU-85-38.

[72] C.B. Lang, C. Rebbi and M. Virasoro,
      Phys. Lett. 104B (1981) 294.

[73] M. Tomiya and T. Hattori, Phys. Lett. 140B (1984) 370.

[74] V.P. Gerdt, A.S. Ilchev, V.K. Mitrjushkin, I.K. Sobolev
        and A.M. Zadorozhny, Nucl. Phys. B265 [FS15] (1986) 145.
      V.P. Gerdt, A.S. Ilchev, V.K. Mitrjushkin,
      and A.M. Zadorozhny, Z. Phys. C29 (1985) 263;
        Dubna preprint E2-85-104.

[75] H.G. Evertz, J. Jersák, D.P. Landau, T. Neuhaus and J.-L. Xu
      (in preparation).

[76]   P.H. Damgaard and U.M. Heller, Phys. Lett. 164B (1985) 121.

[77]   K. Decker, I. Montvay and P. Weisz, Preprint DESY 85-123 (1985).

[78]   B.I. Halperin, T.C. Lubensky and S.K. Ma,
       Phys. Rev. Lett. 32 (1974) 292.

[79]   C. Dasgupta and B.I. Halperin, Phys. Rev. Lett. 47 (1981) 1556.

[80]   I.D. Lawrie, J. Phys. C15 (1982) L879.

[81]   J. Bartholomew, Phys. Rev. B28 (1983) 5378.

[82]   H. Kleinert, Nuovo Cimento Lett. 35 (1982) 405.

[83]   G. Bhanot and B.A. Freedman,
       Nucl. Phys. B190 [FS3] (1981) 357.

[84]   Y. Munehisa, Yamanashi preprint 85-01.

[85]   R.E. Shrock, Phys. Lett. 162B (1985) 165;
       Stony Brook preprint ITP-SB-85-46.

[86]   J.M.F. Labastida, E. Sanchez-Velasco, R.E. Shrock and P. Wills,
       Stony Brook preprint ITP-SB-85-42.

[87]   R.C. Brower, D.A. Kessler, T. Schalk, H. Levine and
       M. Nauenberg, Phys. Rev. D25 (1982) 3319.

[88]   F. Karsch, E. Seiler and I.O. Stamatescu,
       Phys. Lett. 131B (1983) 138.

[89]   I.H. Lee and J. Shigemitsu, Ohio preprint DOE/ER/01545-363.

[90]   B. Berg and C. Panagiotakopoulos,
       Phys. Rev. Lett. 52 (1984) 94.

[91]   G. 't Hooft, Nucl. Phys. B79 (1974) 276.
       A.M. Polyakov, Sov. Phys. JETP Lett. 20 (1974) 194.

[92]   G. Schierholz, J. Seixas and M. Teper,
       Phys. Lett. 151B (1985) 69;
       Phys. Lett. 157B (1985) 209.

[93]   M. Kikugawa, T. Maehara, J. Saito, R. Sasaki, H. Tanaka,
       and Y. Yamaoka, Prog. Theor. Phys. 74 (1985) 553.

[94]   S. Gupta and U.M. Heller, Phys. Lett. 138B (1984) 171.

[95]   K. Olynyk and J. Shigemitsu,
       Nucl. Phys. B251 [FS13] (1985) 472.

[96]   P. Suranyi, Phys. Lett. 164B (1985) 342.

[97]   F. Karsch, E. Seiler and I.O. Stamatescu,
       Phys. Lett. 157B (1985) 60.
       I.O. Stamatescu, Talk at the Conference "Advances in the
       Lattice Gauge Theory", April 1985, Tallahassee,
       Free University Berlin preprint.

[98] J.M. Drouffe, J. Jurkiewicz and A. Krzywicki,
          Phys. Rev. D29 (1984) 2982.

[99] K. Olynyk and J. Shigemitsu,
          Phys. Rev. Lett. 54 (1985) 2403;
          Phys. Lett. 154B (1985) 278.
      K. Olynyk, Talk given at the Conference "Advances in the
          Lattice Gauge Theory", April 1985, Tallahassee,
          Ohio Univ. preprint DOE/ER/01545-365.

[100] I. Montvay, Talk at the Conference "Advances in the
          Lattice Gauge Theory", April 1985, Tallahassee,
          Preprint DESY 85-050.

[101] H.G. Evertz, V. Grösch, J. Jersák, H.A. Kastrup, D.P. Landau,
          T. Neuhaus and J.-L. Xu, Aachen preprint PITHA 85/24.

[102] A. Hasenfratz and P. Hasenfratz (in preparation).

[103] D.J.E. Callaway and R. Petronzio, Preprint CERN-TH 4270/85.

[104] W. Langguth, I. Montvay and P. Weisz, Preprint DESY 85-138 (1985).

RELAXATION AND CORRELATIONS IN TIME IN A FINITE VOLUME

J. Zinn-Justin
Service de Physique Théorique
CEN-Saclay, 91191 Gif-sur-Yvette Cedex, France

INTRODUCTION

To describe relaxation towards equilibrium and correlating in time in stochastic numerical simulations, it is necessary to examine dynamical stochastic evolution equations from the renormalization group (R.G.) point of view, and then finite volume effects since all numerical simulations take place of course in a finite volume. The stochastic evolution equation which describes in the continuum space and time limit the time behavior of numerical simulations is the Langevin equation and in a first part we shall discuss briefly its algebraic properties. We shall show how it is possible to associate with the Langevin equation a conventional effective action in such a way that its renormalization and R.G. properties can be discussed in the language of standard field theory. However in this discussion it is essential to recognize that this effective action has a B.R.S. [1] type symmetry of the form encountered in the quantization of gauge theories.

It is then also possible to study finite size effects by adapting techniques developped for the static case in the cylindrical geometry [2].

Some small scale numerical simulations have been done [3] in the case of the 2D non linear σ model to verify R.G. predictions with finite size effects. More analytical calculations are in preparation [4].

I. LANGEVIN EQUATION AND EFFECTIVE ACTION

Given an euclidean action $\beta \mathcal{A}(\varphi_\alpha)$ for example :

171

$$\mathcal{H}(\varphi) = \int d^d x \left[\frac{1}{2} \left[\partial_\mu \varphi_\alpha(x) \right]^2 + V(\varphi(x)) \right] \qquad (1)$$

it is possible to construct stochastic dynamics which at equilibrium lead to the field distribution

$$\exp - \beta \mathcal{H}(\varphi)$$

They can be described by a Langevin equation and therefore lead also to a Fokker-Planck for the time dependent field distribution. The simplest form of the Langevin equation is :

$$\varphi_\alpha(x, t) = -\beta \, \frac{\Omega}{2} \, \frac{\delta \mathcal{H}}{\delta \varphi_\alpha(x, t)} + \nu_\alpha(x, t) \qquad (2)$$

in which ν_α is a gaussian white noise centered around zero, and defined by its two point function :

$$\langle \nu_\alpha(x, t) \, \nu_\beta(x', t') \rangle = \Omega \delta(x-x') \, \delta(t-t') \, \delta_{\alpha\beta} \qquad (3)$$

However associated with the same equilibrium distribution different Langevin equations can be written which fall in distinct universality classes[5]. For example with the same noise the equation :

$$\varphi_\alpha(x, t) = \frac{\Omega}{2} F_\alpha(\varphi) + \nu_\alpha(x, t) \qquad (4)$$

with the condition :

$$\int dx \, dt \sum_\alpha \frac{\delta}{\delta \varphi_\alpha(x, t)} \left[e^{-\beta \mathcal{H}} \left(\beta \, \frac{\delta \mathcal{H}}{\delta \varphi_\alpha(x, t)} + F_\alpha(\varphi) \right) \right] = 0 \qquad (5)$$

To study the renormalization and R.G. properties of equation (4) it is convenient to introduce a functional representation for the generating functional $Z(J)$ of time dependent correlation functions :

$$Z(J) = \langle \exp \int dx \, dt \, J_\alpha(x, t) \, \varphi_\alpha(x, t) \rangle_\nu \qquad (6)$$

To impose equation (4) a standard trick is to write a functional δ-function with the correct weight factor :

$$Z(J) = \int [d\rho(\nu)] \int [d\varphi] \, \det M_{\alpha\beta} \, \delta(\varphi_\alpha(x, t) - \frac{\Omega}{2} F_\alpha(\varphi)$$

$$- \nu_\alpha(x, t)) \exp \int dx \, dt \, J_\alpha(x, t) \, \varphi_\alpha(x, t) \qquad (7)$$

in which $[d\rho(\nu)]$ is the probability distribution for the noise and $M_{\alpha\beta}$ the functional derivative of Langevin equation (4) :

$$M_{\alpha\beta}(x,t;x',t';\varphi) = \delta_{\alpha\beta}\frac{d}{dt}\delta(t-t')\delta(x-x') - \frac{\Omega}{2}\frac{\delta F_\alpha(x,t)}{\delta\varphi_\beta(x',t')} \quad (8)$$

Both the determinant and the δ function can be represented by functional integrals :

$$Z(J) = \int [d\rho(\nu)][d\varphi][d\lambda][d\overline{C}][dC] \exp\int dx\,dt\left[-\lambda_\alpha(x,t)\varphi_\alpha(x,t)\right.$$

$$\left. + \frac{\Omega}{2}\lambda_\alpha(x,t)F_\alpha(\varphi;x,t)+\lambda_\alpha\nu_\alpha+\overline{C}_\alpha M_{\alpha\beta}C_\beta\right] \quad (9)$$

The quantities C and $\overline{C}$ represent fermion fields.

Introducing the generating functional of connected noise correlation functions $w(\lambda)$:

$$\exp w(\lambda) = \int [d\rho(\nu)] \exp\int dx\,dt\; \nu_\alpha(x,t)\;\lambda_\alpha(x,t) \quad (10)$$

we can write expression (9) as :

$$Z(J) = \int [d\varphi][d\lambda][d\overline{C}][dC]\exp-S(\varphi,\lambda,C,\overline{C})+\int J_\alpha(x,t)\varphi_\alpha(x,t)dxdt$$
$$(11)$$

in which S is the effective action :

$$S = -w(\lambda) + \int dxdt\left[\lambda_\alpha\varphi_\alpha - \frac{\Omega}{2}\lambda_\alpha F_\alpha - \overline{C}_\alpha M_{\alpha\beta}C_\beta\right] \quad (12)$$

If $w(\lambda)$ and $F(\varphi)$ are local functionals, we can use standard methods of field theory to discuss the renormalization of action (12). In particular in the case of Langevin equation (2) with the gaussian noise (3), it is possible to show that the effective action (12) is renormalizable in the same space dimension as the static action $\mathcal{H}(\varphi)$. However even there power counting alone does not by itself imply conservation of the special form (12) necessary to insure the connection with a Langevin equation. This is a consequence of a BRS symmetry.

II. B.R.S. SYMMETRY

Langevin equations are a special example of a class of problems[1] which by construction have a similar algebraic structure which implies an anticommuting global symmetry closely related to the B.R.S. symmetry[6] which appears in the quantization of gauge theories. The transformation law has here the form :

$$\delta \varphi_\alpha (x, t) = \bar{\epsilon} \ C_\alpha (x, t)$$

$$\delta \ C_\alpha (x, t) = 0 \qquad\qquad (13)$$

$$\delta \ \bar{C}_\alpha (x, t) = \bar{\epsilon} \ \lambda_\alpha (x, t)$$

$$\delta \ \lambda_\alpha (x, t) = 0$$

in which $\bar{\epsilon}$ is an anticommuting C-number. The verification is elementary and relies only on the observation that $M_{\alpha\beta}(\varphi)$ is a functional gradient term and thus :

$$\frac{\delta M_{\alpha\beta}}{\delta \varphi\gamma} = \frac{\delta M_{\alpha\gamma}}{\delta \varphi_\beta} \qquad\qquad (14)$$

Actually in the special case of Langevin equation (2) with noise (3) the effective action is even supersymmetric. Introducing a superfield notation :

$$\Phi_\alpha = \varphi_\alpha (x, t) + \bar{\theta} \ C_\alpha (x, t) + \bar{C}_\alpha (x, t) \theta + \bar{\theta}\theta \ \lambda_\alpha (x, t) \qquad\qquad (15)$$

in which θ and $\bar{\theta}$ are anticommuting coordinates, the action becomes[7] :

$$S(\Phi) = \int dxdt \ d\bar{\theta}d\theta \left[\frac{\partial \Phi_\alpha}{\partial t} \ \theta \ \frac{\partial \Phi_\alpha}{\partial \theta} + \frac{\Omega}{2} \left(\frac{\partial \Phi_\alpha}{\partial \theta} \ \frac{\partial \Phi_\alpha}{\partial \bar{\theta}} - \beta \ \mathcal{A}(\Phi) \right) \right] \qquad (16)$$

The B.R.S. symmetry corresponds to :

$$\Phi(\bar{\theta}, \theta) \mapsto \Phi(\bar{\theta}+\bar{\epsilon}, \theta) \qquad\qquad (17)$$

while the second symmetry corresponds to

$$\Phi(t, \bar{\theta}, \theta) \mapsto \Phi(t - \frac{2}{\xi} \ \bar{\theta}\epsilon, \bar{\theta}, \theta+\epsilon) \qquad\qquad (18)$$

III. RENORMALIZATION AND RENORMALIZATION GROUP

These symmetries are preserved under renormalization. As a consequence in addition to the usual static renormalizations, only one additional time scale renormalization is needed :

174

$$\Omega_0 = Z_\Omega \, \Omega \qquad\qquad (19)$$

with which is associated a new R.G. function and one time scale in the long distance, long time limit, the relaxation or correlation time τ. In A.F. field theories τ behaves like :

$$\tau \sim \xi^2 (\ell n \; \xi)^\alpha \qquad\qquad (20)$$

in which ξ is the correlation length. In gauge theories for example[3] :

$$\alpha_{gauge} = -\frac{35}{88} \qquad\qquad (21)$$

For a non trivial fixed point instead :

$$\tau \sim \xi^{2-z} \qquad\qquad (22)$$

in which z is in general a new exponent (related to Z_Ω).

IV. FINITE SIZE EFFECTS[2]

I shall just mention the main ingredients needed in the analysis of finite size effects in a finite volume system with periodic boundary conditions. More details can be found in reference [2].

1) Divergences of perturbation series are only due to short distance effects and thus independent of size of the system. The linear size L of the system does not enter in the R.G. equations, but only in the solution because a new dimensionless quantity L/ξ_∞, in which ξ_∞ is the infinite volume correlation length, can be formed. For example now the relaxation time τ can be written :

$$\tau(L) = \tau_\infty \; f(L/\xi_\infty) \qquad\qquad (23)$$

For A.F. field theories, if we neglect the logarithmic dependence we have approximatively

$$\tau(L) \not\sim L^2 \; g(L/\xi_\infty) \qquad\qquad (24)$$

which is similar of course to the free field behavior.

2) For a temperature T very close to T_c or for $T < T_c$ in the presence of massless modes, and with periodic boundary conditions, it is necessary to isolate the zero momentum mode of the field in the functional integral :

$$\varphi(t, x) = \varphi_0(t) + \sum_{\substack{\vec{q} \\ \vec{q} = \frac{2\pi m}{L}}} \varphi_{\vec{q}}(t) \, e^{i\vec{q}\vec{x}} \qquad (25)$$

because this mode would lead to I.R. divergences in Feynman graphs.

One integrates at $\varphi_0(t)$ fixed on all other modes using perturbation expansion and renormalization group. This leads to an effective path integral for the last mode $\varphi_0(t)$:

$$Z = \int [d\varphi_0(t)] \, \exp - L^d \, S_{ef}(\varphi_0, L) \qquad (26)$$

which has been to be treated exactly for example by going over to a Schrödinger equation.

REFERENCES

1) Zinn-Justin J., Renormalization and Stochastic Quantization, Saclay preprint in preparation
2) Brézin E. and Zinn-Justin J., Nucl. Phys. B257 (FS14) (1985) 867
3) Lainee F., Thèse de 3ème cycle, Paris 1985
4) Niel J.C., Finite Size Effect Calculations for Dynamical Systems (Saclay)
5) Hohenberg P.C. and Halperin B.I., Rev. Mod. 49 (1977)
6) Becchi C., Rouet A. and Stora R., Ann. Phys. (NY) 98 (1976) 287
7) Nakazato H., Namiki M., Ohba I. and Oknano K., Progr. Theor. Phys. 70 (1983) 298
   Egorian E.S. and Kalitsin S., Phys. Lett. 129B (1983) 320

TOPOLOGY IN LATTICE GAUGE THEORY

G. Schierholz

Institut für Theoretische Physik der Universität, Keil
and
Deutsche Elektronen-Synchroton DESY, Hamburg

ABSTRACT

We have evidence that the vacuum of (pure) SU(2) and SU(3) gauge theories
possesses an underlying instanton structure. Starting from Monte Carlo
generated equilibrium gauge field configurations representing the physical
vacuum, we obtain by systematically freezing the quantum fluctuations by
successive relaxation (approximate) solutions of the classical equations
of motion, which turn out to have discrete values of the action

$$S \approx \beta 2\pi^2 N, \quad N = 0, 1, 2, \ldots$$

in close agreement with the continuum (multi-) instanton solutions. We show
that these lattice (multi-) instantons are localized in space time, that
they carry a topological charge $|Q| = N$ and that they give rise to a number
of fermion zero modes in accordance with the Atiyah-Singer index theorem.

In order to study the role played by topology in the physics of the vacuum
of QCD and SU(N) gauge theories quantitatively, we need a fast algorithm
for reliably computing the topological charge Q on large lattices. Two
such algorithms exist now for gauge group SU(2). Using recently derived
explicit formulae for the 2- and 3-cochains, we are able to integrate the
Chern-Simons density analytically and arrive at a local algebraic expression

for the topological charge which is relatively easy to implement and, since it does not resort to numerical integration, fast to compute on the lattice. The other algorithm is due to Philips and Stone which is based on simplicial lattices and a geometrical interpolation of the transition functions.

This talk will not be reprinted here, since by the time these proceedings appear the numerical results - on which this workshop was mainly focussed - that were presented will be superseded by new results on large lattices and with high statistics being in progress now. The material presented can be found in

I.A. Fox, J.P. Gilchrist, M.L. Laursen and G. Schierholz: Phys. Rev. Lett. $\underline{54}$, 749 (1985)

E.-M. Ilgenfritz, M.L. Laursen, M. Müller-Preußker, G. Schierholz and H. Schiller: Nucl. Phys. $\underline{B268}$, 693 (1986)

M.L. Laursen, G. Schierholz and U.-J. Wiese: DESY 85-062 (1985), to be published in Comm. Math. Phys.

M. Göckeler, M.L. Laursen, G. Schierholz and U.-J. Wiese: DESY 85-142 (1985)

LARGE N QCD: THE EGUCHI KAWAI APPROACH

O. Haan

Physics Department
University of Wuppertal
Gauss-Str. 20, D-5600 Wuppertal 1

ABSTRACT: Large N QCD is the zeroth order of the 1/N expansion, which is a candidate for a sensible approximation scheme for strong interaction phenomena. Monte Carlo simulations of reduced models reveal properties of this "bare" QCD. Results for static potential, string tension and deconfinement are reviewed, showing that large n QCD confines and that the variation from N=3 to N=∞ is small for $T_c/\sqrt{\sigma}$. The prospect of measuring properties of glueballs and mesons at large N in reduced models is discussed.

INTRODUCTION

At the beginning of this survey on nonperturbative results for large N QCD, I would like to recall the motivation for studying a theory seeming so remote from the reality of three colors.

The complex phenomena of strong interactions cannot be understood from a perturbative treatment of the gauge coupling parameter g. In 1974 't Hooft introduced a more promising expansion parameter[1]: 1/N, where N is the number of colors. Large N QCD is the zeroth order in this 1/N expansion.

Although the large N theory has not yet been solved, some of its properties follow from the fact that only planar Feynman diagrams without closed quark loops survive the large N limit (with gauge coupling $g/\sqrt{N}$, g fixed). If large N QCD confines color at all then it describes an infinite number of stable, noninteracting mesons and glueballs. Many phenomenological aspects of strong interactions can be understood as 1/N perturbations to this zeroth order description of hadrons[2].

Thus the parameter 1/N in QCD is analoguous to the finestructure constant α in QED, and a theoretical understanding of QCD requires an understanding of large N QCD, just as the analysis of QED presupposes the solution of the Dirac equation. In particular, confinement must be due to effects which survive the large N limit[3]. Also the meson and glueball spectrum of QCD should be classified at large N where no ambiguities due to finit width and/or mixing occurs. Of course, the properties of physical N=3 particles and processes can be obtained only after systematically adding 1/N corrections.

Baryons play a special role, their properties at large N emerge from a Hartree-like treatment of the N quark system[2]. They behave like solitons in the theory of weakly (1/N) coupled mesons and gluons. The effective Lagrangian approximation to QCD has this kind of baryonic solutions[4], so the old idea of Skyrme[5] finds a justification from the large N picture.

These results from an 1/N analysis of QCD motivated great efforts to find ways for a quantitive study of the large N limit. With the Eguchi-Kawai[6] reduction Monte Carlo simulations of large N lattice gauge theories became possible. The analysis of QCD as a theory of weakly interacting mesons and glueballs now can proceed in three steps: i) Study of confinement at large N, ii) calculation of glueballs and meson masses at large N, iii) inclusion of 1/N corrections.

Simulations of reduced models indeed show confinement[7-9], as well as a deconfining phase transition at finite temperature[10-12], and for the first time a physical quantity, $T_c/\sqrt{\sigma}$, has been determined for large N[9]. Comparison to N=2 and 3 shows little N dependence. These results of course strengthen the confidence in the large N approach and stimulate efforts to treat the second point of the list. I will describe the inclusion of fermions and a way to handle correlations in reduced model. Preliminary results from the lightest glueball have been obtained with this method. I can say nothing concerning 1/N corrections, they have to be studied in the future.

Before giving some details of these results, I will explain shortly the mechanism of reduction and some features of Monte Carlo simulation in reduced models.

2. REDUCED MODELS

In 1982 Eguchi and Kawai[6] showed, that for large N the gauge theory on a single point is equivalent to the infinite volume theory. This reduction was established by the fact that the Schwinger Dyson equations for Wilson loops coincide in both theories, if the additional $U(1)^d$ symmetry of the reduced model is unbroken.

For weak coupling this symmetry must be protected from spontaneous breakdown[13], either by quenching the eigenvalues of the link matrices[14] or by imposing twisted boundary conditions on the reduced model[15]. I will discuss this twisted Eguchi Kawai (TEK) model, because it is best suited for numerical simulation and most nonperturbative results for large N QCD have been obtained within this model.

Twisted boundary conditions in SU(N) gauge theories are given by:

$$U_\mu(x+Le_\nu) = \Gamma_\nu U_\mu(x) \Gamma_\nu^+ \quad , \quad \Gamma_\nu \in SU(N) \tag{1}$$

The matrices Γ_μ have to obey 't Hoofts consistency condition[16]

$$\Gamma_\mu \Gamma_\nu = z_{\nu\mu} \Gamma_\nu \Gamma_\mu \ , \ z_{\mu\nu} = exp-\frac{2\pi i}{N} n_{\mu\nu} \in Z_N$$

In the one point model, eq. (1) with L=1 can be used to construct translation matrices

$$T(x) = \prod_{\mu=1}^{4} \Gamma_\mu^{x_\mu} \tag{3}$$

180

such that the link variables at point x of an extended lattice are represented by $T(x)\, U_\mu\, T^+(x)$ in the one point model.

Wilson loops in the reduced model are given by

$$W(c) = \langle T_r \prod_i T_{(x_i)}\, U_{\mu_c}\, T^+_{(x_i)} \rangle = Z(c) \langle T_r \prod_i U_{\mu_i}\, \Gamma_{\mu_i} \rangle \qquad (4)$$

the twist factor $Z(c) \in Z_N$ comes from commuting the Γ_μ's according eq. (2). Changing variables $U_\mu \Gamma_\mu \rightarrow U_\mu$ we find as a special case of eq. (4) the action of the TEK model:

$$S = \beta \cdot N \cdot \sum_{\mu,\nu} Z_{\nu\mu}\, T_r\, U_\mu\, U_\nu\, U_\mu^+\, U_\nu^+ \qquad . \qquad (5)$$

The groundstate of (5) in the weak coupling limit is a solution of eq. (2). Of particular importance are twist factor $Z_{\mu\nu}$, which have solutions with the properties[7],[17],[18]

$$T_r\, \Gamma_\mu^m = 0 \;,\; m = 1,\ldots, L_\mu - 1$$
$$T_r\, \Gamma_\mu^{L_\mu} = N \;,\; L_1 \cdot L_2 \cdot L_3 \cdot L_4 = N^2 \;, \qquad (6)$$

because the Wilson loops, cf eq. (4), in this groundstate reproduce exactly the weak coupling values for Wilson loops on an extended $L_1 \times L_2 \times L_3 \times L_4$ lattice. The Feynman diagrams for Wilson loops in this kind of one point model are closely related to the diagrams in the SU(N) Wilson theory on a lattice of equivalent size $L_1 \times L_2 \times L_3 \times L_4$[7],[18],[19]: Planar diagrams in both models coincide, nonplanar diagrams differ, but are suppressed by powers of $1/N^2$ in both models. This shows that at large N TEK model and lattice gauge theory are equivalent, at least perturbatively.

With symmetric[7] and hot[17],[18] twists symmetric and asymmetric equivalent lattices with N^2 points can be constructed. The (N^2-1) color degrees of freedom of the SU(N) matrices are converted into (N^2-1) momentum modes of nonzero momentum, therefore, the problem of constant modes in lattice gauge theories[20] is avoided in twisted reduced models.

For many applications, as we will see, it is convenient to construct generalized reduced models, in which either not all color degrees of freedom are converted into momentum modes, or the lattice is reduced not to one point, but to a larger set of points, or both. The first kind of model can be used to study the nonplanar corrections in the TEK model[21] or to include fermions into the reduction scheme[22]. A partially reduced model[12] which lives on two points in every time slice of the unreduced time direction proves to be very convenient from several aspects which will be mentioned below.

It has the action

$$S = \beta \cdot N \sum_{t=1}^{T} \Big\{ \sum_{i,j} \big(z_{i_1} z_{j_i} z_{i_j}\, T_r\, V_i(t) W_j(t) W_i^+(t) V_j^+(t)$$
$$+ \tfrac{2}{3i} T_r\, W_i(t) V_j(t) V_i^+(t) W_j^+(t) \big)$$
$$+ \sum_i \big(T_r V_i(t) W_0(t) V_i^+(t+1) V_0^+(t) + c.c.$$
$$+ T_r\, W_i(t) V_0(t) W_i^+(t+1) W_0^+(t) + c.c. \big) \Big\} \qquad (7)$$

The twist is a hot twist[18] with $L_0 = 1$ ($Z_{10} = 1$) and the equivalent lattice for T timeslices of the SU(N) partial-reduced model has the volume $V = 2TN^2$. Partical reduction to more general sets of points is possible[22], [23].

Twisted reduction is a very efficent way to generate planar diagrams: only one degree of freedom per link on the equivalent lattice is needed as compared to (N^2-1) degrees of freedom in the usual Wilson theory.

3. MONTE CARLO SIMULATION OF REDUCED MODELS

Monte Carlo methods are applicable to reduced models in the usual way apart from some peculiarities which I will discuss now.

The effects from treating finite N instead of an infinite number of colors are twofold: i) The equivalent lattice has a finite volume, N^2, with the usual finite size effects; ii) nonplanar contributions are not completely suppressed, their influence can be controlled with the methods of ref. 21). Compared to the SU(3) gauge theory on an equivalent lattice. we find that the advantage of a smaller number of degrees of freedom (by a factor of 8) is overbalanced by the disadvantage of the long range coupling in color space.

Updating of SU(N) matrices can be done by successive multiplication with elements of different SU(2) subgroups[24]. The SU(2) elements are generated with the Metropolis algorithm in the one point model which is bilinear in the link matrices. Our partial reduced model allows the use of the more efficient heatbath method because every plaquette in the action is the product of four different link variables. The a_0-component of the SU(2) matrix

$$\Delta = a_0 \cdot 1 + i \, \vec{a} \cdot \vec{\sigma} \tag{8}$$

has to be generated with the probability

$$da_0 \; \theta(1-a_0^2) \; \sqrt{1-a_0^2} \; e^{\lambda a_0} \; , \tag{9}$$

where λ is determined by the neighbouring links. Instead of correcting the easily generated $da_0 \, \theta(1-a_0^2) \exp \lambda a_0$ distribution for the factor $\sqrt{1-a_0^2}$ [25] (which has high rejection rate for large λ), we generate, with three gaussian random numbers, the probability[26]

$$da_0 \; \theta(1-a_0) \sqrt{1-a_0} \; e^{\lambda a_0} \tag{10}$$

and correct for $\theta(1+a_0) \sqrt{1+a_0}$ with an extremely small rejection rate in our large N calculations. Recently it has been shown that this modified heatbath has some advantages even for N=3[27]

4. CONFINEMENT AT LARGE N

The simulations of all reduced models reveal a strong first order phase transition at $\beta_c^* = 0.35$, which can be viewed as the 4-dimensional manifestation of the large N transition in the Gross-Witten solution[28] of 2-dimensional lattice gauge theory and/or the large N version of the singularity in the plane of fundamental and adjoint couplings[29]. To lowest order in the 1/N expansion, $\beta_c^* = 0.35$ corresponds to a value $6/g^2 = 6.3$ in the

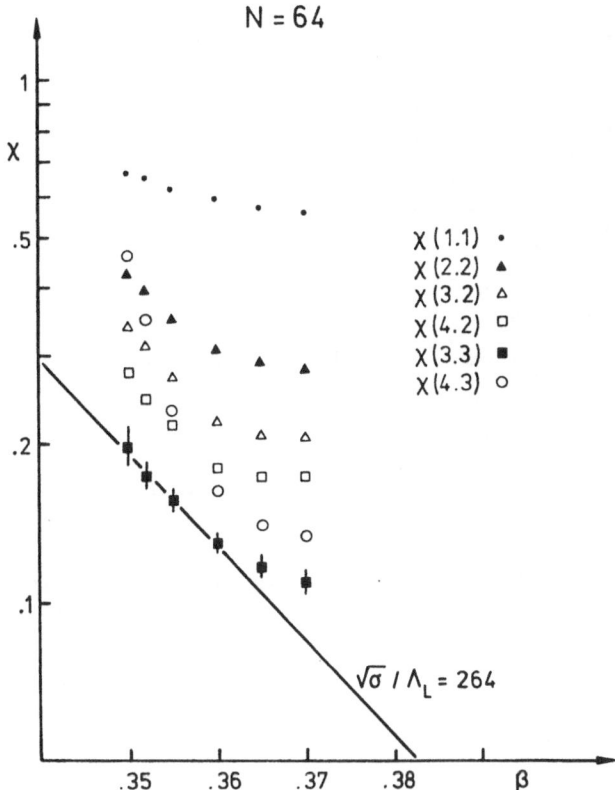

Fig. 1: Creutz ratios for N = 64.

SU(3) theory, where scaling behaviour has been observed.

Measurements of Creutz ratios for N = 36[7] and N = 64[8] indeed indicate scaling behaviour for $\beta_c^* < \beta < 0.36$ (fig.1). The ratios decrease with increasing size because of the small lattice spacing for these β-values. An upper bound on the string tension can be deduced: $\sqrt{\sigma}/\Lambda_L < 264$
This number should be compared to early results[30] for the SU(3) string tension from 6^4 and 8^4 lattices $\sqrt{\sigma}/\Lambda_L < 200$.

For a more precise evaluation of σ one has to use reduced models of larger equivalent sizes. Therefore, we studied Wilson loops in a partial reduced model with N = 30 and 24 timeslices, corresponding to a 24 x 10 x 12 x 15 lattice[9]. We fixed the gauge by choosing all timelike matrices equal to one. By this we speed up the simulation by a factor of 2 and the Wilson loop measurements by an even greater factor. This model allows the propagation of states with nonzero external charges. These unwanted states are suppressed by a large timelike extension of the lattice. From Wilson loop measurements at $\beta = 0.37$ in 1500 consecutive configurations, V(R), R = 1, ..., 5 was extracted (fig. 2) and fitted to the form (fig. 3).

$$V(R) = -A/R + V_0 + \sigma \cdot R \ . \tag{11}$$

The result for the 1/R-term, $A = 0,298 \pm 0,020$, is close to the value $\pi/12 = 0,262$ from the string picture[31]. The string tension from this fit is $= 0.0336 \pm 0.0078$. Assuming scaling at $\beta = 0.37$ we find

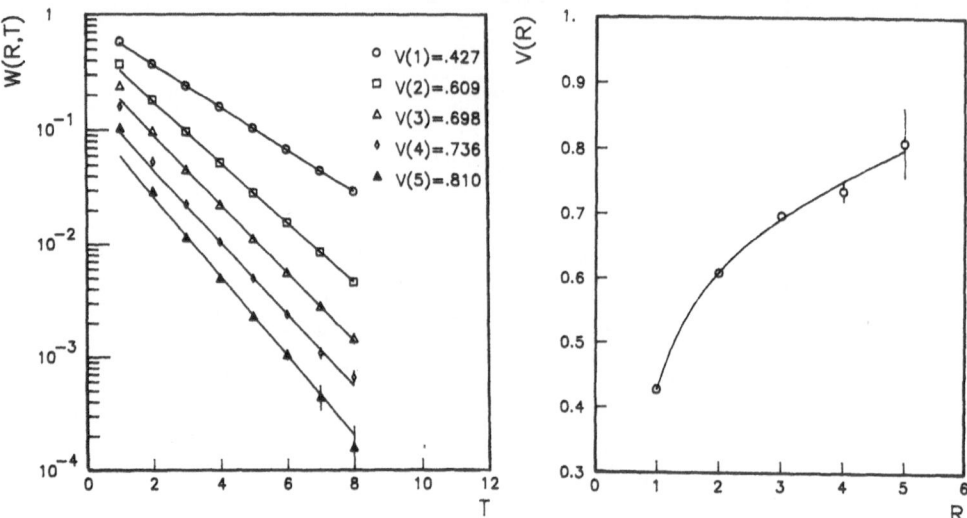

Fig.2: Wilson loops with exponential
fits.

Fig.3: The static potential, fit
according to eq. (11).

$$\sqrt{\sigma}/\Lambda_L = 162 \pm 25 \qquad (12)$$

With a string tension = $(450 \text{ MeV})^2$, we find a lattice spacing = 0.08 fm.
The same spacing in the SU(3) case is obtained for $6/g^2 = 6.1$. As can be
seen from fig. 3 with R = 5 at this lattice spacing, we have not yet reached
the asymptotic linear rising part of the potential. Measurements for larger
R are desirable. Also scaling has to be checked by simulating at different
ß-values.

Our results support very strongly the confinement property of large N
QCD, and thereby all features of the 1/N-expansion which have been deduced
from the confinement assumption[2]. The question of the N-dependence cannot
be answered by comparing to the SU(3) value for the string tension, $\sqrt{\sigma}/\Lambda$ =
90 $\pm$ 5[32] because of the undetermined relation between the lattice scales
at different N. Only by measuring a second observable, the deconfinement
temperature, we will be able to judge the variation of physical QCD results
with N.

The early attempts of locating the deconfinement temperature in large
N QCD used the 1-point TEK-model with hot twist[17,18] and very asymmetric
equivalent spatial volume V = $2L_0$ x $3L_0$ x $6L_0$ where L_0 is the extension of
the lattice in time direction. This model describes large N lattice QCD with
temperature T = $1/a.L_\nu$. For values of L_0 from 2 to 5[10, 11, 33] the critical
β_c coincides with the $\beta^* = 0.35$ of the first order bulk transition which is
independent of temperature (cf. fig. 5). The separation of the two transitions occurs at larger values of L_0. $L_0 > 5$ could not be reached within the
hot twist 1-point model, which wastes many degrees of freedom to produce the
very asymmetric spatial volume.

More economic ways to produce the finite temperature use the symmetric
twist with different couplings in spatial and temporal directions[34], or the
partial reduced model with L_0 time slices[12]. With the first method an isolated deconfining transition is found for N = 81 at a temperature

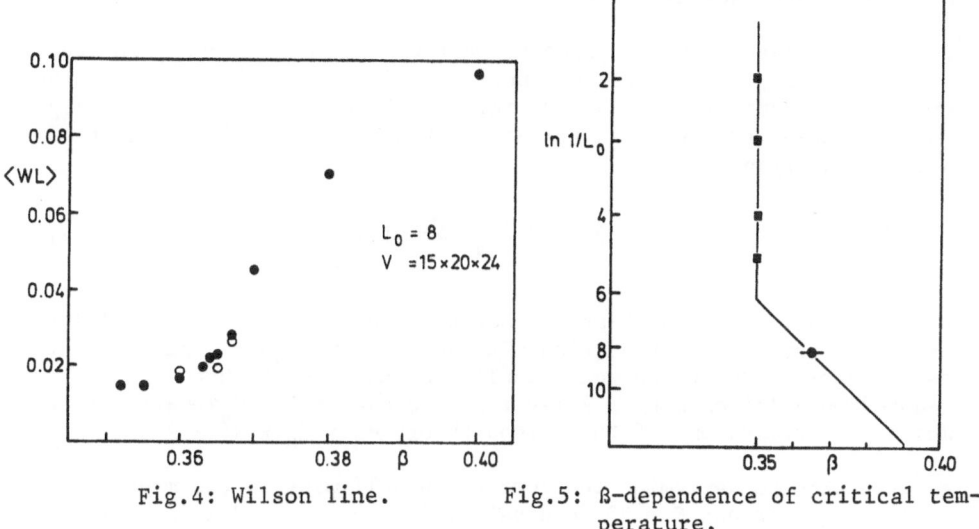

Fig.4: Wilson line.   Fig.5: ß–dependence of critical tem-
                                        perature.

$$T_c/\Lambda_L = 105 \pm 5 \;\;, \; a/a_t = 1.50$$

$$T_c/\Lambda_L = 110 \pm 6 \;\;, \; a/a_t = 1.75 \tag{13}$$

The results for different ratios of spatial and temporal lattice spacings
are consistent with asymptotic scaling behaviour[34].

   In the partial reduced model with N = 60, corresponding to a spatial
volume V = 15 x 20 x 24, the deconfinement transition for L_0 = 8 is clearly
separated from the bulk transition[12]. From the behaviour of the averaged
Wilson-line (fig.4) WL,

$$WL = \frac{1}{N} \cdot Tr \prod_{t=1}^{L_0} U_0(t) \tag{14}$$

and from an analysis of the distribution of its value in several hundred
sweeps[12] at each ß-value we find a critical $ß_c = 0,365 \pm 0,005$ (fig.5).
If asymtotic scaling holds for this ß (as the results for Creutz-ratios
indicate) the corresponding temperature is

$$T_c/\Lambda_L = 101 \pm 4 \tag{15}$$

   From (15) and (12) we can eliminate the unknown lattice scale   and
obtain

$$T_c/\sqrt{\sigma} = 0.62 \pm 0.13 \tag{16}$$

which coincides within the large errors with the value of $T_c/\sqrt{\sigma}$ in the
SU(3) theory[35]. This is the first quantitative confirmation of the common
believe that large N QCD is a good zeroth order approximation to the N = 3
theory.

5. THE PARTICLE SPECTRUM AT LARGE N

In the Euclidean lattice approach, particle masses are obtained from the asymptotic decay of correlation functions. The one point TEK model does not reproduce the correlations of the infinite lattice Wilson theory. E.g. the two-plaquette-correlation in the Wilson theory becomes independent of distance by reduction

$$\langle \frac{1}{N} Tr\, U(P_x)\, \frac{1}{N} Tr\, U(P_o)\rangle \;\rightarrow\; \langle (\frac{1}{N} Tr\, U(P))\rangle \;. \tag{17}$$

Particle propagation in Eguchi-Kawai models is only possible, if at least one dimension survives the reduction. It can be shown, comparing the Schwinger Dyson equations[36] or the weak coupling expansions[22] that e.g. our partial reduced model[12] on a L_o x 2 lattice, reproduces for $N \to \infty$ exactly the zero momentum correlations of the Wilson theory on a L_o x N^2 x 2 lattice.

We attempted to determine the lowest glueball mass in the model of ref 12) with L_o = 12 and N = 30 (equivalent lattice 10 x 12 x 15) for ß = 0,355, 0,360, 0,370. A source at timeslice 0 (realized by fixed spacelike plaquettes) creates glueballs which are detected by 1 x 1, 1 x 2 or 2 x 2 Wilson loops at time t . The signals at times t = 2, ..., 10 are fitted by

$$\langle \frac{1}{N} Tr\, U(C_t)\rangle = A + B\, cosh\, M\cdot(t-6) \tag{18}$$

and give the following results:

$$\begin{aligned}
M &= 0.87 \pm 0.20 \quad (ß=0.355), \\
M &= 1.7 \pm 0.4 \quad (ß=0.360), \\
M &= 1.4 \pm 0.4 \quad (ß=0.370).
\end{aligned} \tag{19}$$

It is very probable that exited glueball states contribute to the correlation function, especially for the larger ß values. The low mass value at ß = 0.355 could be an indication that here the lowest glueball dominates the t-dependence. Conversion into physical units, using the two loop ß-function and our string tension result, yields for the lowest glueball mass

$$M = 1450\ MeV \tag{20}$$

This has the expected order of magnitude but of course a more reliable mass estimate deserves further studies.

The meson spectrum at large N can be obtained within the Eguchi-Kawai approach if the fermion fields can be reduced. Reduction with translation matrices (cf. eq (3)) is possible only for fields in the adjoint representation. This requires to send to infinity the number of flavours, N_f , simultaneous with $N \to \infty$, keeping fixed the ratio N_f/N [22],[38].

A different reduction scheme is possible, if only properties of mesons, i.e. fermion bilinears, are required[39]. The following partial reduced model, with fermions in the fundamental representation, reproduces all zero momentum mesonic correlation functions of the large N Wilson theory on an infinite lattice:

$$S = S_g + S_q \tag{21}$$

where S_g is the action of a partial reduced TEK model and

$$S_q = \frac{1}{2} \sum_{t,\mu} \left\{ \overline{\Psi}(t)(1-\gamma_\mu) U_\mu(t) \Psi(t+e_\mu)_o) + \overline{\Psi}(t+e_\mu)_o)(1+\gamma_\mu) U_\mu^+(t) \Psi(t) \right\} \quad (22)$$

The fermion matrix of this reduced model is smaller by a factor $(3N)^2$ than the fermion matrix of the SU(3) theory on a lattice of equivalent spatial size $V_s = N^2$. Thus the inversion of the reduced matrix can be done faster and with less computer power than in the SU(3) case.

6. CONCLUSIONS

Properties of large N QCD, the bare theory of strong interactions, can be calculated nonperturbatively in the Eguchi-Kawai reduction approach. Up to now, the confining properties of this theory have been established. For $T_c / \sqrt{6}$ it has been shown quantitatively that $1/N = 1/3$ is only a small perturbation of the $1/N = 0$ bare theory. The possibility to determine the particle spectrum for $N \to \infty$ has been discussed, including preliminary results for the lowest glueball mass and, for the first time, reduction for mesons which are composed of quarks in the fundamental representation. A more detailed study of particle properties at large N may be helpful to understand better the complicated phenomena of the hadronic world.

REFERENCES

1. G. 't.Hooft, Nucl. Phys. B72, 461 (1974)
2. E. Witten, Nucl. Phys. B160, 57 (1979)
3. J. Greensite, Nucl. Phys. B249, 263 (1985)
4. E. Witten, Nucl. Phys. B223, 422, 433 (1983)
5. T.H.R. Skyrme, Proc. Soc. London, Ser. A260, 127 (1961)
6. T. Eguchi and H. Kawai, Phys. Rev. Lett. 48, 1063 (1982)
7. A. Gonzales-Arroyo and M. Okawa, Phys. Lett. 133B, 415 (1983)
8. K. Fabricius and O. Haan, Phys. Lett. 139B, 293 (1984)
9. O. Haan and K. Meier, the Static Potential for large N, Preprint, University of Wuppertal
10. S.R. Das and J.B. Kogut, Phys. Lett. 141B, 105(1984), Phys. Lett. 145B, 375 (1984), Nucl. Phys. B257 [FS14], 141 (1985)
11. K. Fabricius, O. Haan and F. Klinkhamer, Phys. Rev. D30, 2227 (1984)
12. K. Fabricius and O. Haan, Nucl. Phys. B260, 285 (1985)
13. G. Bhanot, M. Heller and H. Neuberger, Phys. Lett. 113B, 47 (1982)
14. G. Bhanot, M. Heller and H. Neuberger, Phys. Lett. 115B, 237 (1982)
15. A. Gonzales-Arroyo and M. Okawa, Phys. Rev. D27, 2397 (1983)
16. G. 't.Hooft, Comm. Math. Phys. 81, 267 (1981)
17. F. Klinkhamer and P. van Baal, Nucl. Phys. B237, 274 (1984)
18. K. Fabricius and C.P. Korthals-Altes, Nucl. Phys. B240 [FS12],237 (1984)
19. T. Eguchi and R. Nakayama, Phys. Lett. 122B, 59 (1983)
20. A. Coste et al., Preprint, Marseille CPT 85/P1777
21. K. Fabricius, T. Filk and O. Haan, Phys. Lett. 144B, 240 (1984)
22. K. Fabricius and C.P. Korthals-Altes, Preprint, CERN-TH 4260/85, Nucl. Phys. B, in Press
23. H. Hasegawa and A. Nakamura, Decreasing the finite size effects through the twisted boundary conditions, Preprint
24. N. Cabbibo and E. Marinari, Phys. Lett. 119B, 387 (1982)
25. M. Creutz, Phys. Rev. D21, 2308 (1980)
26. K. Fabricius and O. Haan, Phys. Lett. 143B, 459 (1984)
27. A.D. Kennedey and B.J. Pendleton, Phys. Lett. 156B, 393 (1985)
28. D. Gross and E. Witten, Phys. Rev. D21, 446 (1980)
29. Yu. M. Makeenko and M.I. Polikarpov, Nucl. Phys. B205 [FS5], 386 (1982)
30. M. Creutz and K.J.M. Moriarty, Phys. Rev. D26, 2166 (1982)

31. M. Lüscher, Nucl. Phys. B180, [FS2], 317 (1981)
32. C. Micheal, Potentials, contribution in these proceedings
33. K. Fabricius and O. Haan, unpublished
34. S.R. Das and J.B. Kogut, Phys. Rev. D31, 2704 (1985)
35. S.A. Gottlieb et. al., Phys. Rev. Lett. 55, 1958 (1985)
36. H. Levine and H. Neuberger, Phys. Rev. Lett. 49, 1603 (1982)
37. K. Fabricius and O. Haan, unpublished
38. S.R. Das, Phys. Lett. 132B, 155 (1983)
39. O. Haan, Mesons in the Eguchi-Kawai Model, Preprint NORDITA-85/30

HADRON MASS CALCULATION ON A 24^3 x 48 LATTICE[*]

Ph. de Forcrand[1], A. Konig[2], K.H. Mutter[3],
K. Schilling[3], and R. Sommer[3]

ABSTRACT

Preliminary results on a hadron mass calculation with Wilson fermions at ß=6.3 on a 24^3 x 48 lattice are presented. The quenched approximation was used in the updating of the gluonic fields and the approximate block diagonalization scheme for the computation of quark propagators.

1. INTRODUCTION

In the frame work of lattice gauge theory hadron masses are extracted from the exponential decay of hadron-hadron correlation functions. In previous computations[1] on rather small lattices (with less than 32 sites in time direction) it was found that hadron hadron correlation functions do not yet show a clean signal for exponential decay - at least for reasonably small quark masses. In the quenched approximation the main limitation on the lattice size comes from the computation of quark propagators at large time separations. To overcome this difficulty, Klaus Schilling[2] and myself have proposed an approximation scheme for the computation of quark propagators, which we call "approximate block diagonalization".

In the first part of my talk, I will review this method. Then I will show you preliminary results on hadron hadron correlation functions obtained from a Monte Carlo study on a 24^3 x 48 lattice at ß=6.3. Finally, I shall present an argument that the block diagonalization scheme might be helpful in the updating with dynamical quarks. For this purpose, I will discuss the QCD action in terms of the spectral density of the fermion matrix. From this representation one expects that the fluctuations of the small eigenvalues have the most important influence on the updating of the fermion determinant.

2. THE WILSON ACTION ON A BLOCKED LATTICE WITH SPACING 2a.

The basic idea of the approximate block diagonalization scheme is

---

[*] Talk presented by K.H. Mütter

1) Cray Research
2) PRACLA-SEISMOS, Hannover
3) Physics Department, University of Wuppertal
   Supported by Deutsche Forschungsgemeinschaft

easily explained:
The large distance behaviour of the quark propagators is governed by the small eigenvalue modes of the fermion matrix. Therefore, we look for an effective fermion action with effective fermionic and gluonic fields, defined on a coarser "blocked" lattice, which keeps the small eigenvalue modes almost unchanged. The blocking proceeds as follows: We denote the sites on the original lattice with spacing a by x

$$x = a\,(n_0, n_1, n_2, n_3)$$

the sites on the blocked lattice with spacing 2a by $\tilde{x}$

$$\tilde{x} = 2a\,(\tilde{n}_0, \tilde{n}_1, \tilde{n}_2, \tilde{n}_3)$$

The 16 sites x within one block B ($\tilde{x}$) with lower left corner $\tilde{x}$ are given by 16 internal vectors

$$\tau = (\tau_0, \tau_1, \tau_2, \tau_3) \quad , \quad \tau_\mu = (-1)^{n_\mu} = \pm 1 \qquad (2.1)$$

such that

$$x = \tilde{x} + \frac{a}{2}\,(1 - \tau_0, 1 - \tau_1, 1 - \tau_2, 1 - \tau_3)$$

The new fermionic fields on the blocked lattice are introduced by means of a parallel transporter, which brings the fields $\Psi(x)$ on the original lattice to the lower left corner x:

$$\chi_\tau(\tilde{x}) = \mathcal{U}\,(\tilde{x}, x)\,\Psi(x) \qquad (2.2)$$

The parallel transporter is the product of link variables along a standard path from $\tilde{x}$ to x. Note, that on the blocked lattice, we have 16 fermionic modes specified by τ. The nearest neighbour couplings on the original lattice are expressed in terms of couplings of the blocked lattice making use of a geometrical relation between Kronecker δ's on the original and blocked lattices:

$$\delta_{x', x+\mu} = \frac{1}{2}\left(T_\mu^{(+)} + T_\mu^{(-)}\right)_{\tau' \tau}\,\delta_{\tilde{x}', \tilde{x}+2\mu} + \frac{1}{2}\left(T_\mu^{(+)} - T_\mu^{(-)}\right)_{\tau' \tau}\,\delta_{\tilde{x}', \tilde{x}} \qquad (2.3)$$

The $T_\mu^{(\pm)}$ are matrices in a 16 dimensional space, the basis of which is labelled by the internal vectors τ. They obey the following commutation and anti-commutation relations

$$\left[T_\mu^{(+)}, T_\nu^{(-)}\right]_- = \left[T_\mu^{(-)}, T_\nu^{(-)}\right]_- = 0, \quad \left[T_\mu^{(+)}, T_\nu^{(+)}\right]_- = 0 \;\; \mu \neq \nu, \quad \left[T_\mu^{(+)}, T_\mu^{(+)}\right]_+ = 0 \qquad (2.4)$$

If we express the Wilson action on the original lattice in terms of the 16 fermionic fields on the blocked lattice:

$$\sum_{x' x} \bar{\Psi}(x')\,\Delta^{-1}(x', x)\,\Psi(x) = \sum_{\tilde{x}' \tau'} \sum_{\tilde{x} \tau} \bar{\chi}_{\tau'}(\tilde{x}')\,\Delta^{-1}(\tilde{x}'\tau', \tilde{x}\tau)\,\chi_\tau(\tilde{x}) \qquad (2.5)$$

we are led in a quite natural way to new gauge fields as well. One finds two types: The <u>first</u> one mediates the interaction between two neighbouring points on the blocked lattice

$$\mathcal{U}_\mu^{(\tau)}(\tilde{x}, \tilde{x} \pm 2\mu) = \mathcal{U}(\tilde{x}, x)\,\mathcal{U}(x, x \pm \mu)\,\mathcal{U}(x \pm \mu, \tilde{x} \pm 2\mu) \quad , \quad \tau_\mu = \mp 1 \qquad (2.6)$$

It can be represented as a parallel transporter from $\tilde{x}$ to $\tilde{x} \pm 2\mu$ along a path which is specified by the internal vector τ. The <u>second</u> type of gauge fields describes a self-interaction at $\tilde{x}$

$$\mathcal{U}_\mu^{(\tau)}(\tilde{x}, \tilde{x}) = \mathcal{U}(\tilde{x}, x)\,\mathcal{U}(x, x \mp \mu)\,\mathcal{U}(x \mp \mu, \tilde{x}) \quad , \quad \tau_\mu = \mp 1 \qquad (2.7)$$

and represents a closed Wilson loop starting and ending at $\tilde{x}$. In terms of the new fields on the blocked lattice, the Wilson action acquires a rather lengthy form, for which I refer you to ref (2).

The crucial point of the block diagonalization scheme can be seen already in the free case. Here, you see the free fermion matrix, Fourier-transformed on the blocked lattice:

$$\Delta^{-1}(\tilde{p}\tau', \tilde{p}\tau) = -M + \frac{1}{2a}\sum_\mu T_\mu^{(+)}\left(i\gamma_\mu \sin\tilde{p}_\mu 2a + \cos\tilde{p}_\mu 2a - 1\right) \tag{2.9}$$

$$+ \frac{1}{2a}\sum_\mu T_\mu^{(-)}\left((1-\cos\tilde{p}_\mu 2a)\gamma_\mu - i\sin\tilde{p}_\mu 2a\right)$$

where

$$M = m_q \mathbf{1} + \frac{1}{a}\sum_\mu\left(1 - T_\mu^{(+)}\right) \tag{2.10}$$

stands for the mass matrix of the 16 fermionic modes. If we transform to a basis with the T_μ and M diagonal, we find
one light mode $\quad M_0 = m_q$

and fifteen heavy modes

$$M_\mu = m_q + g_\mu/a \;,\; \mu = 1,\cdots,15, \quad g_\mu = 2, 4, 6, 8$$

In this basis the fermion matrix eq. (2.9) is diagonal, except for the terms in the last line, which involve the remaining interactions between the fermionic modes. If we switch off these interactions it is clear, that the infrared behaviour of the quark propagator is dominated by the light mode with mass M_0. The interaction between the light mode and the heavy modes can be taken into account perturbatively with the lattice spacing a as an expansion parameter. In our computations, we took into account the first order correction in a, which amounts to reproduce the leading order of the Wilson term.

All the considerations, demonstrated here for the   free case only, can be performed as well for the interacting case. The resulting effective fermion matrix for the light fermionic mode on the blocked lattice with spacing 2 a looks as follows:

$$\Delta^{-1}(\tilde{x}', \tilde{x}) = -m_q \delta_{\tilde{x}',\tilde{x}} + \frac{1}{4a}\sum_\mu \gamma_\mu S_{2\mu}(\tilde{x}',\tilde{x})$$

$$+ \frac{1}{4a}\sum_\mu C_{2\mu}(\tilde{x}',\tilde{x})$$

$$\tag{2.11}$$

$$+ \frac{1}{4a}\sum_\mu \gamma_\mu W_\mu(\tilde{x})\delta_{\tilde{x}',\tilde{x}} - \frac{1}{32a}\sum_\mu C_{4\mu}(\tilde{x}',\tilde{x})$$

where I have used these abbreviations:

$$S_{2\mu}(\tilde{x}',\tilde{x}) = W(\tilde{x},\tilde{x}+2\mu)\delta_{\tilde{x}',\tilde{x}+2\mu} - W(\tilde{x},\tilde{x}-2\mu)\delta_{\tilde{x}',\tilde{x}-2\mu}$$

$$C_{\mu\mu}(\tilde{x}',\tilde{x}) = W(\tilde{x},\tilde{x}+n\mu)\delta_{\tilde{x}',\tilde{x}+n\mu} + W(\tilde{x},\tilde{x}-n\mu)\delta_{\tilde{x}',\tilde{x}-n\mu} - 2\delta_{\tilde{x}',\tilde{x}} \;, n=2,4$$

The effective gluon fields $W(\tilde{x}, \tilde{x}+2\mu)$, $W_\mu(\tilde{x})$ are averages over the fields defined above (ref. eqs. (2.6), (2.7)):

$$W(\tilde{x},\tilde{x}+2\mu) = \frac{1}{16}\sum_\tau U^{(\tau)}(\tilde{x},\tilde{x}+2\mu)$$

$$W_\mu(\tilde{x}) \quad = \frac{1}{16}\sum_\tau \gamma_\mu U_\mu^{(\tau)}(\tilde{x},\tilde{x})$$

Let me mention that this scheme can also be applied on the $\sqrt{3}$ blocked lattice introduced by R. Gupta and his coworkers.[3] In that case you find on the blocked lattice nine fermionic modes, one light and 8 heavy. The resulting effective fermion matrix for the light mode looks interesting in so far, as you find in addition to nearest neighbour couplings interactions along diagonals in three-dimensional cubes. This might help to accelerate the restoration of rotational invariance.

3. NUMERICAL RESULTS

We used 25 gluonic background fields, computed by one of us (Ph.d.F.) in the quenched approximation at ß=6.3 on a 24^3 x 48 lattice. The thermalization on this large lattice was achieved as follows: First fields were produced on a 24^3 x 6 lattice. The final configuration was copied 8 times in the time direction. 510 sweeps were performed over the resulting 24^3 x 48 lattice. Finally, 30 configurations were stored, two successive ones being separated by 100 sweeps.

For the computation of quark propagators we applied the block diagonalization scheme twice, which reduces the rank of the matrix to be inverted by a factor 256. Up to now, we have computed 136 quark propagators on 25 background fields and for these values of the hopping parameter:

$$\varkappa = .13, .1308, .1315, .1325$$

We find for the critical value of the hopping parameter:

$$\varkappa_c = .13285 \ (10)$$

In the following I will show you first preliminary results for the hadron propagators obtained on the Cyber 205 in Karlsruhe.

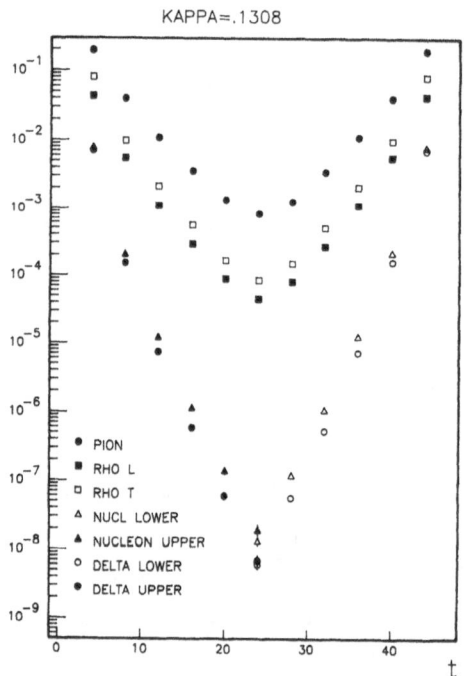

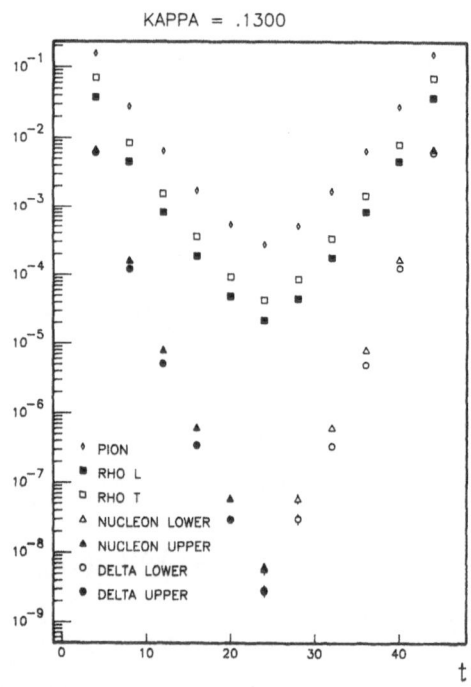

Fig. 1: The pion, ρ-, nucleon and propagators for $\varkappa$ =.130 ($m_q a$ =.081).

Fig. 2: Same as fig. 1 for $\varkappa$ =.1308 ($m_q a$ =.059).

Figures 1 and 2 show the pion, ρ, nucleon and Δ propagators for the lowest κ -largest quark mass values:

$$\varkappa = .130 \qquad m_q a = \frac{1}{2}\left(\varkappa^{-1} - \varkappa_c^{-1}\right) = .081$$

$$\varkappa = .1308 \qquad m_q a = .059$$

ρ_L and ρ_T refer to two different polarizations for the ρ. The open (closed) triangels and circles are data points for the upper (lower) components in the Dirac spinor for the nucleon and Δ respectively.

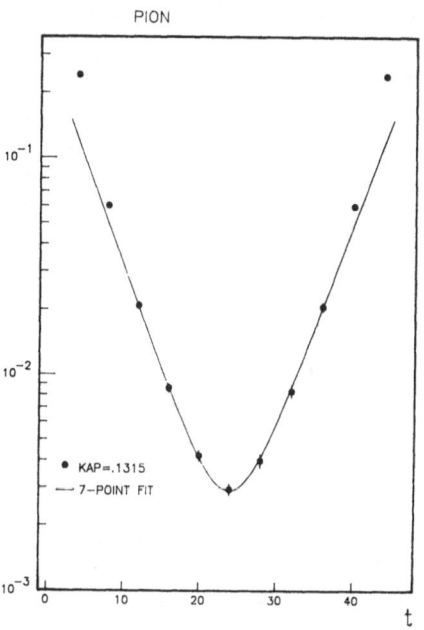

Fig. 3: The pion propagator.

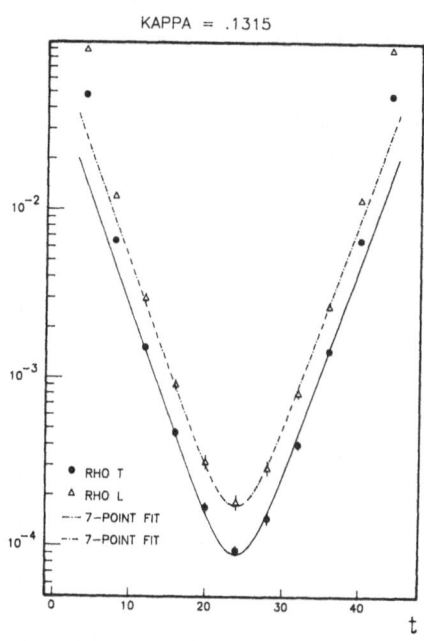

Fig. 4: The ρ-propagator.

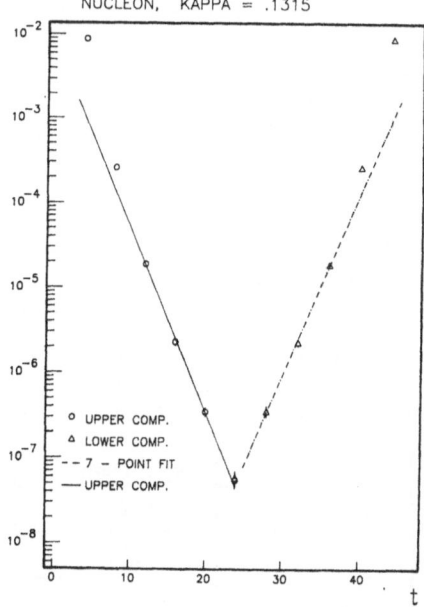

Fig. 5: The nucleon propagator.

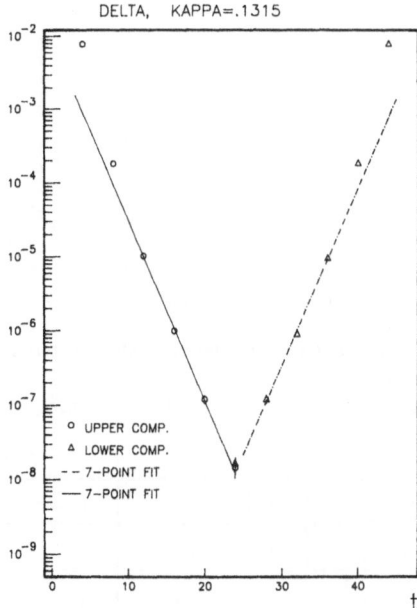

Fig. 6: The Δ - propagator.

In figure 3 you can see the pion propagator for $\varkappa = .1315$, $u_q a = .038$. The solid line is a cosh fit c cosh $u_q a(t-T/2)$ to the large distance points $12 \leq t \leq 36$. The small distance points at t = 4,8 deviate from the cosh fit. In a previous calculation on a 16^3 x 56 lattice[4] at ß = 6 we observed deviations from a cosh fit only for t = 4. Fig. 4 shows the ρ propagator for longitudinal and transverse polarizations and cosh fits to the large distance points for $\varkappa = .1315$. Figs. 5 and 6 show the nucleon and Δ propagators with exponential fits to the large distance points. Summarizing one can say: The propagators for $12 \leq t \leq 36$ nicely follow the single cosh and exponential fits. Therefore, it looks as if the large distance limit is reached in that interval. Fig. 7 shows our preliminary hadron masses as a function of the quark mass

$$u_q a = \tfrac{1}{2}\left(\varkappa^{-1} - \varkappa_c^{-1} \right)$$

Though the statistical errors are still very large for the smallest value of the quark mass ($u_q a = .01$) you readily realize by simple linear extrapolation to small quark masses, that the nucleon to ρ mass ratio is still very large, about 1.9 ± .2. In table 1 you can see a comparison of our new results (in column A) with two previous calculations on large lattices. In

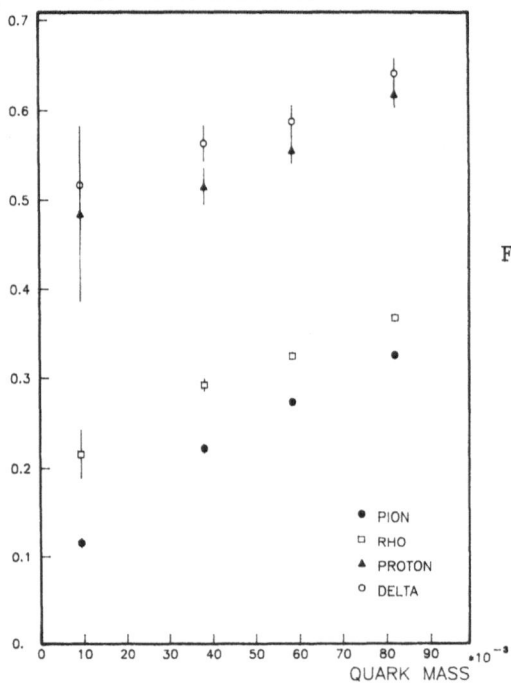

QUARKMASSDEPENDENCE

Fig. 7: Hadron masses versus quark mass.

Table 1: Comparison of hadron mass ratios:
A: This computation (Wilson fermions twice blocked, at ß=6.3 on a 24^3 x 48 lattice).
B: Ref. 4 (Wilson fermions twice blocked, at ß=6.0 on a 16^3 x (2 x 28) lattice).
C: Ref. 5 (Kogut Susskind fermions, at ß=6.0 on a 16^3 x 32 lattice).

	A	B	C	Experiment
m_N/m_ρ	1.9 ± .2	1.7 ± .1	1.4 ± .14	1.25
m_Δ / m_N	1.2 ± .15	1.2 ± .1	-	1.31
$a_\rho^{-1}(\beta)\,[GeV]$	3.9 ± .15	2.5 ± .06	1.66 ± .13	-
$a_N^{-1}(\beta)\,[GeV]$	2.2 ± .22	1.8 ± .2	-	-

column B, I have listed the results of the aforementioned computation on a 16^3 x 56 lattice at ß = 6 (also with Wilson fermions on a twice blocked lattice)

The third column C) contains the results of Barkai, Moriarty and Rebbi[5], you heard about in the preceding talk. Concerning the nucleon to ρ mass ratio, the computation with Kogut-Susskind fermions is closer to the experimental value. Of course, there is no a priori reason, that the experimental value of 1.25 should be reproduced correctly in the quenched approximation. However, in the continuum limit both computations with Wilson and Kogut-Susskind fermions should yield the same result. In other words: At least in one of these two computations at ß = 6 the continuum limit is not yet reached. In the last two lines of table 1 I have listed the values of the inverse lattice spacing a^{-1} (β) which you get if you insert for the ρ or the nucleon its experimental value. In both cases the ratio of the lattice spacings at ß = 6.3 and ß = 6

$$\frac{a(\beta=6.3)}{a(\beta=6.0)} = \begin{cases} .64 \pm .04 & \text{from } \rho \\ .82 \pm .13 & \text{from nucleon} \\ .713 & \text{from asymptotic scaling} \end{cases}$$

is roughly compatible with asymptotic scaling within errors. Finally, let me mention that Oswald Haan and Klaus Schilling have started a large lattice computation of the hadron mass spectrum with Wilson fermions, but without blocking, in order to study the systematic errors possibly introduced by the approximate block diagonalization scheme. They use our configurations at ß = 6 on a 16^3 x 28 lattice. The computation will be performed with the assistance of the Siemens company on a Fujibu VP200 in Munich.

4. THE SPECTRAL REPRESENTATION OF THE LATTICE QCD ACTION

Let me consider the Wilson action with two flavoured quarks with degenerate bare mass. The fermionic part is given by log det M, where M is the positive hermitean matrix:

$$M = \left(1 - 2\varkappa \sum_\mu Q_\mu\right)\left(1 - 2\varkappa \sum_\mu Q_\mu^+\right) \tag{4.1}$$

$\varkappa$ is the hopping parameter and Q_μ the hopping matrix in μ direction with matrix elements:

$$Q_\mu(x',x) = \frac{1}{2}(1-\gamma_\mu)\,\mathcal{U}(x,x+\mu)\,\delta_{x',x+\mu} + \frac{1}{2}(1+\gamma_\mu)\,\mathcal{U}(x,x-\mu)\,\delta_{x',x-\mu} \tag{4.2}$$

The gluonic part of the action can be written in terms of M as well. For this purpose we compute:

$$TR\,M^2 = const. - 16\varkappa^4 \sum_P tr\left(\mathcal{U}(P) + \mathcal{U}^+(P)\right) \tag{4.3}$$

Obviously, the whole action

$$S = \log \det M + S_g \tag{4.4}$$

can also be expressed in terms of the spectral density $P(\lambda, \mathcal{U}, \varkappa)$ of the positive hermitean matrix M:

$$S = -TR1 \int_{\lambda_0}^{\lambda_1} d\lambda\, P(\lambda, \mathcal{U}, \varkappa)\, g(\lambda) + const. \tag{4.5}$$

where I have chosen the normalization

$$\int_{\lambda_0}^{\lambda_1} d\lambda\, P(\lambda, \mathcal{U}, \varkappa) = 1 \tag{4.6}$$

On a lattice with N_t sites in time direction N_s sites in space direction, the rank of the matrix M is:

$$TR1 = 12 \cdot N_t \cdot N_s \tag{4.7}$$

Note, that the dependence of the action on the background field $\mathcal{U}$ enters only in the spectral density. The weight factor $g(\lambda)$ turns out to be

$$g(\lambda) = -\log \lambda + \hat{\beta} \lambda^2 \tag{4.8}$$

with

$$\hat{\beta} = \beta / 6.16 \cdot x^4 \tag{4.9}$$

The eigenvalues λ of M are constrained to the interval

$$\lambda_o \leqslant \lambda \leqslant \lambda_1 \tag{4.10}$$

where

$$\lambda_o \geqslant \begin{cases} (1 - 8x)^2 & \text{for } x \leqslant 1/8 \\ 0 & \text{for } x > 1/8 \end{cases} \tag{4.11}$$

$$\lambda_1 \leqslant (1 + 8x)^2 \qquad \text{for all } x \tag{4.12}$$

The bounds follow from the observation that the hopping matrices Q_μ (eq. (4.2)) are unitary

$$Q_\mu \, Q_\mu^+ = 1 \qquad \mu = 0, 1, 2, 3 \tag{4.13}$$

It is quite instructive to compare the spectral representations (4.5) in the quenched and unquenched case. They differ only in the field independent weight function $g(\lambda)$, which is given by eq. (4.8) in the unquenched case and which is obviously

$$g_{qu}(\lambda) = \hat{\beta} \lambda^2 \tag{4.14}$$

in the quenched case.
For a comparison of the quenched and unquenched case let me discuss the relative difference of the two weight functions

$$\gamma(\lambda) = \frac{g(\lambda)}{g_{qu}(\lambda)} - 1 = -\frac{\log \lambda}{\hat{\beta} \lambda^2} \tag{4.15}$$

It has a minimum at $\log \lambda = \frac{1}{2}$, from which one gets the bound

$$|\gamma(\lambda)| \leqslant \frac{1}{2 e \hat{\beta}} \qquad \lambda \geqslant 1 \tag{4.16}$$

For a typical choice of parameters $\beta=6$, $x=.15$ the upperbound (4.16) turns out to be very small (.0015). Therefore, the most important effect of the fermion determinant in the updating of the gluonic fields arises from the fluctuations of the spectral density for small eigenvalues $\lambda_o \leqslant \lambda \leqslant 1$. Since the approximate block diagonalization scheme treats those modes correctly, it should be sufficient to compute the determinant of the effective fermion matrix on the coarser blocked lattice. This suggests a novel approximation scheme to compute the effect of dynamical fermions on a blocked lattice in the updating procedure. Work along this line is in progress.

REFERENCES

1. H. Lipps, G. Martinelli, R. Petronzio and F. Rapuano; Phys. Lett. 126B (1983) 250
   A. Billoire, E. Marinari and R. Petronzio; Nucl. Phys. B251 [FS13](1985) 141
   A. König, K.H. Mütter and K.Schilling; Phys. Lett. 147B (1984) 145 and Nucl. Phys. B259 (1985) 33
2. K.H. Mütter and K. Schilling; Nucl. Phys. B230 [FS10] (1984) 275
3. R. Cordery, R. Gupta and M. Novotny; Phys. Lett. B128 (1983) 425
   A. Patel and R. Gupta; Nucl. Phys. B251 [FS13] (1985) 789
4. A. König, K.H. Mütter, K. Schilling and J. Smit; Phys. Lett. 157B (1985) 421
5. D. Barkai, K.J.M. Moriarty and C. Rebbi, Phys. Lett. 156B (1985) 385.

CALCULATION OF WEAK MATRIX ELEMENTS: SOME TECHNICAL ASPECTS[*]

Claude Bernard, Terrence Draper, George Hockney, and A. Soni

University of California, Los Angeles, CA 90024
and University of California, Irvine, CA 92717

INTRODUCTION

Since a report of the first results of this project has recently been published,[1] we will confine this description of the conference talk to those points which are not covered or not covered fully in the published work. Therefore, the following should be read in conjunction with Ref. 1.

CHIRAL BEHAVIOR

The behavior of the $\Delta I = 1/2$ weak amplitudes which is seen on our lattice is not what would be predicted by lowest order Chiral Perturbation Theory (CPTh). Rather than falling as $m_M{}^2$ (where m_M is the meson mass), the amplitudes are actually found to be rising with decreasing m_M in this mass range, that is, for $m_M \sim 1$ GeV to 500 MeV. Now, it is perfectly possible that this is the actual physical behavior of the amplitudes for these heavy masses. There is no guarantee that lowest order CPTh is good for m_M this large: for the K_0-$\bar{K}_0$ mixing amplitude (but not, it is true, for the $\Delta I = 1/2$ amplitudes) independent evidence exists[2] that CPTh may break down when $m_M \sim m_K$. However, the possibility that the mass dependence seen is a lattice artifact must be considered very seriously, especially because we are using Wilson fermions, with chiral invariance explicitly broken on the lattice.

There have been several ways proposed to investigate this question. The most straightforward, in principle, would simply be to extend the numerical studies down to lower values of meson mass. If the lowest order CPTh behavior did not set in by the time $m_M \sim m_\pi$, then one would be sure that lattice artifacts were responsible; conversely, if the expected chiral behavior was observed for low masses, one would have confidence that the higher mass results were indeed physical. However, practical considerations make it extremely difficult to work at very low masses, though work at masses somewhat lower than those used in Ref. 1 is in progress.[3] Some more indirect tests that we have used will be described below. These include both perturbative and non-perturbative estimates of the magnitude of chirally non-invariant terms generated by the lattice

---

[*]Presented by C. Bernard.

regularization, and an analysis based solely on the "figure-eight" graphs which are not expected to have chiral problems. For left-left operators (including $O_\pm$ and Q_1, Q_2, and Q_3) each of these tests has the same result: We find no evidence that the mass dependence seen is an artifact. This gives us some confidence that the results we have computed in the "charm in" case (which involves only left-left operators) are reliable. In particular, we have no reason at this point to doubt our conclusion that the dominance of the $\Delta I = 1/2$ amplitude is due to the dominance of the eye graphs.

However, for left-right operators the situation is less clear. As explained below, there are some indications that the corrections to left-right operators due to mixing with two-quark operators may be quite important (at least 40% for our values of the parameters), which makes suspect our results for quantities involving left-right operators such as ξ (related to ε'/ε). Weighed against this is the qualitative agreement between results calculated in the "charm in" and "charm out" approaches (in particular the mass dependences are very similar), which seems to indicate that the special problems with left-right operators do not change the qualitative picture. This of course could be numerical coincidence, so at the very best the conclusion is that more work needs to be done.

An alternative approach proposed by Bochicchio et al.[4] involves a systematic development of current algebra on the lattice. This is an elegant method which does not, in principle, rely on perturbation theory. It remains to be seen, however, whether it can be implemented in a practical manner for this problem. Further, even if the chiral properties are completely controlled, the "K $\to$ 0" techniques (described below and previously[5]) must still be used to remove s-d mixing terms which are chirally correct but unphysical.

Finally, one could switch to Kogut-Susskind fermions, which have better chiral properties (but much more complicated flavor structure) than Wilson fermions. The proper use of Kogut-Susskind fermions in this problem is rather subtle, but at least one group[6] is reportedly close to finishing numerical work with them. We are also (in collaboration with K. Barad) studying these fermions in this problem but are probably much further from any conclusions. It will be very interesting to compare the chiral behavior of the $\Delta I = 1/2$ amplitudes with both types of fermions.

Mixing with Two-Quark Operators: Perturbation Theory

Through diagrams like Fig. 1 (plus all gluon corrections), four-quark operators of the type considered here can mix with two-quark operators. This is especially dangerous because the two-quark operators are automatically pure $\Delta I = 1/2$ (they contain only s and d quarks) and can have the wrong chiral structure since the virtual quark in the loop violates chiral symmetry off shell (with Wilson fermions). By analogy to the determination of k_c (the critical value of the hopping constant), Bochicchio et al.[4] argue that perturbation theory will not be a reliable method of calculating the coefficients of the lower dimension operators generated. However it is still interesting to look at mixings with two-quark operators in perturbation theory; at worst we expect an indication of the order of magnitude of the coefficients. We note that on our lattice lowest order perturbation theory does predict the sign and the order of magnitude (within a factor of ~ 2) of the shift in $1/k_c$.

Upon examining the zero[th] order graph in Fig. 1, one soon sees that it in fact vanishes for all left-left operators, $O_\pm$ and Q_1 through Q_3, considered in Ref. 1. This is because all these operators have the form

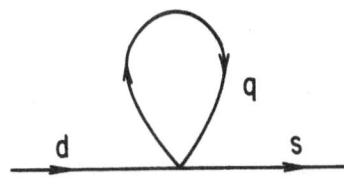

Fig. 1. The lowest (zeroth) order graph for the mixing of four-
quark operators with two-quark operators (with flavor
content $\bar{s}d$). The lable q denotes the quark field which
is contracted and may be u,d,s, or c.

$\bar{s}_a\gamma_\mu(1-\gamma_5)d_a\bar{q}_b\gamma_\mu(1-\gamma_5)q_b$ or $\bar{s}_a\gamma_\mu(1-\gamma_5)d_b\bar{q}_b\gamma_\mu(1-\gamma_5)q_a$, where q and $\bar{q}$ represent the quark fields which are contracted (u,d,s, or c), and a and b
are color indices. Thus, since there is no four-vector available to pick
up the index μ after the loop integration and spin trace is performed,
the graph must vanish. The same argument for left-left operators also
applies to any correction to Fig. 1 of the "$\langle\bar{\Psi}\Psi\rangle$ type" where the gluon
lines connect only to the fermion loop and not to the base line. This is
interesting because it implies that chiral problems which infect $\langle\bar{\Psi}\Psi\rangle$ do
not affect any of these matrix elements, at least in this direct manner.

The situation is different for the left-right operators Q_5 and Q_6.
Contractions of terms of the form $\bar{s}_a\gamma_\mu(1\mp\gamma_5)q_a\bar{q}_b\gamma_\mu(1\pm\gamma_5)d_b$ or
$\bar{s}_a\gamma_\mu(1\mp\gamma_5)q_b\bar{q}_b\gamma_\mu(1\pm\gamma_5)d_a$ do not vanish. In fact, it is quite simple to
calculate the leading contribution in 1/a, which is also the leading
chirally non-invariant contribution. For example, we find, for the cor-
rection to the operator $\bar{s}_a\gamma_\mu(1\mp\gamma_5)q_a\bar{q}_b\gamma_\mu(1\pm\gamma_5)d_b$, the result:

$$\text{Fig. 1} = (1.877 \pm .001)\,\frac{r}{2k}\,\bar{s}(1\pm\gamma_5)d\ ,$$

where r is the Wilson parameter and k is the hopping constant. For the
operator with mixed color indices, this result is simply multiplied by N,
the number of colors. By evaluating numerically the matrix elements of
the $\bar{s}d$ operator, one can then estimate the corrections to the left-right
matrix elements reported in Ref. 1. For example, for k = .162 (a typical
value in the middle of our range) we find corrections of 42% and 44% (with
positive sign) for the K-π matrix elements of Q_5 and Q_6, respectively.

Since the lowest order graph for the left-left operators vanishes,
we have looked at the next order contribution. Only two graphs, shown in
Fig. 2, contribute. The evaluation of these graphs is complicated in
general; we have again calculated only the leading contribution in 1/a,
which is proportional to $\bar{s}d$. This contribution can be found by setting
the external momentum and all the quark masses to zero. We find, for the
"color-mixed" operator $\bar{s}_a\gamma_\mu(1-\gamma_5)d_b\bar{q}_b\gamma_\mu(1-\gamma_5)q_a$ and r = 1, the result,

$$\text{Fig. 2} = ((-2.15 \pm .01) \times 10^{-2})\,\frac{g^2}{8k}\,\frac{(N^2-1)}{2N}\,\bar{s}d\ .$$

This gives a correction of 1.1% (with positive sign) to the K-π matrix
elements of Q_2 and Q_3 at k = .162. Note that corresponding correction
to the "color-unmixed" operator Q_1 vanishes due to the color trace.
Furthermore, because of the GIM cancellation, there are no such mass-

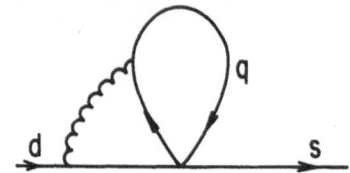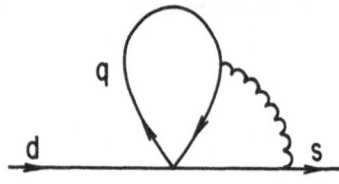

Fig. 2. The lowest non-vanishing graphs for the mixing of left-left four-quark operators with two-quark operators.

independent corrections to the operators $O_\pm$ which appear in the "charm in" approach.

Finally, we note that there is a different type of lower dimension operator, namely $\bar{s}\sigma_{\mu\nu}F^{\mu\nu}(1\pm\gamma_5)d$, which could also contribute at some level. (Operators like $\bar{s}\not{D}(1\mp\gamma_5)d$ are not new since they are proportional to $\bar{s}(1\pm\gamma_5)d$ on shell.) However, for $O_\pm$ this operator has a mass dependent coefficient (again because of GIM) and should be adequately treated by perturbation theory, where it does not appear at lowest non-trivial order.[7]

The following conclusions may therefore be drawn from these perturbative calculations. There is first of all no evidence for significant corrections to left-left operators due to mixing with lower dimension operators. This includes the operators $O_\pm$ and Q_1 through Q_3. For the left-right operators Q_5 and Q_6 there appear to be very significant corrections. Whether these corrections can be reliably taken into account with perturbation theory is unclear. In this regard it would be interesting to complete the evaluation of the other two-loop diagrams, in addition to those in Fig. 2, which make up the corrections for left-right operators. This would be an indication of how convergent the perturbation theory is.

Mixing with Two-Quark Operators: Non-Perturbative Approach

The use of CPTh gives one a handle on the mixing with two-quark operators which is independent of perturbation theory.[5] The emphasis in Ref. 5 was on operators which are chirally invariant but still unphysical, such as $\bar{s}\not{D}(1-\gamma_5)d$. Such operators must be removed since they represent renormalizations, i.e. unphysical mixings of the s and d quarks. It was shown how the evaluation of the kaon to vacuum (K → 0) matrix elements of the four-quark operators of interest can be used to remove these unwanted contributions.

Although Ref. 5 was written in continuum language, with SU(3) × SU(3) assumed throughout, the methods developed there nevertheless allow one to control certain chirally non-invariant operators which appear on the lattice. In particular, since the use of K → 0 removes the chiral operator $\tilde{\Theta}_2^{(8,1)} = \frac{8v}{f^2} \text{tr}[\Lambda\Sigma M + \Lambda(M\Sigma)^\dagger]$, it also removes any operator proportional to $\tilde{\Theta}_2^{(8,1)}$, such as $\tilde{\Theta}_2^{(8,1)}/(m_s+m_d)$, which corresponds to the chirally non-invariant operator $\bar{s}d - \frac{m_s-m_d}{m_s+m_d}\bar{s}\gamma_5 d$ (see Ref. 5). Such an operator could appear with coefficient r/a^3. Thus the statement that the K → 0 amplitude is numerically small implies that such operators are not generated with appreciable coefficients, at least with our lattice parameters. Further, the vanishing of the K → 0 amplitude when $m_s = m_d$ (which we find within our statistical accuracy), implies that the operator $\bar{s}\gamma_5 d$ (corresponding to tr[$\Lambda\Sigma - \Lambda\Sigma^\dagger$]) also is not generated with appreciable coefficient. On the other hand, because of parity invariance, K → 0 does not control the appearance of the operator $\bar{s}d$ when it appears by itself. In Ref. 1, it was implicitly assumed that the chirally non-invariant terms, generated by Wilson fermions on the lattice, are always of the form which can be controlled by K → 0. Unfortunately, this is not correct, as we have seen explicitly in perturbation theory: The operator $\bar{s}d$, by itself, is generated. (Equivalently one may say that there is a component "orthogonal" to $\hat{\Theta}_2^{(8,1)}$.) Thus K → 0 does not solve the problem of the left-right operators which was discussed above. Without invoking the full machinery of Ref. 4, or depending on perturbation theory, the only solution in this context seems to be to evaluate the Green's function made up of $\bar{s}d$ and the four-quark operator of interest, as was suggested by Sharpe.[8] For Kogut-Susskind fermions, where chiral symmetry is presumably not a problem, the evaluation of this Green's function was proposed as an alternative to K → 0. Here, it seems to be preferable, in that it would directly control the operator $\bar{s}d$ which can contaminate the K → π results for the left-right operators. Tests of this method will be made soon.[3]

One final comment on the subject of K → 0 concerns the work of Crewther.[9] He argues that K → 0 is actually of order m_M^4, not m_M^2, in CPTh, and hence not important to the order we are working. His work is not wrong, but irrelevant to the lattice calculation. His K → 0 amplitude goes like m_M^4 because he has redefined the s and d fields to cancel the s-d mixing which is generated by the four-quark operators. In other words, the vacuum has realigned. This is of course precisely what happens in the real world, but in the lattice calculation one does not know a priori how much s-d mixing is produced by the introduction of a four-quark operator into the pure QCD background. The lattice evaluation of the K → 0 amplitude just determines how much s-d mixing needs to be subtracted away. After the subtraction, but not before, the K → 0 amplitude does go like m_M^4 (i.e. it is zero to the order we are working). A similar comment on Crewther's work also appears in Sharpe's paper.[8]

Deductions from the "Figure-Eight" Graphs Alone

Contributions to the amplitudes from the "figure-eight" graphs do not suffer from any of the chiral problems discussed above. There can be no mixing with lower dimension operators because, by definition, quark lines from the operator are not allowed to contract with each other. Hence mixing is only with operators of the same dimension, with coefficients logarithmic in a, and is expected to be well determined by perturbation theory.[4] These mixings have been calculated[10,7] and put in explicity. Thus we do not expect serious lattice artifacts in the chiral behavior of the figure-eights; any violations of lowest order CPTh seen are likely to be physical.

Let us therefore assume that the CPTh expansion in powers of m_M is a good expansion, and see if there are any contradictions in the behavior of the figure-eights. It is easy to show that although CPTh predicts that the total matrix elements of any of the operators of interest vanish like m_M^2 as $m_M \to 0$, the separate figure-eight and "eye" contributions of the left-right operators go to constants as $m_M \to 0$. (For left-left operators, the figure-eight and eye contributions separately vanish as m_M.) If CPTh is a well behaved expansion, the m_M^2 terms should be much less than the constant term. This then implies that total matrix elements of the left-right operators must be much less than the figure-eight contribution alone. Thus the figure-eights (plus the assumption that CPTh works in this mass range) give a bound on the total contribution of the left-right operators, which, again assuming CPTh, can be compared to the physical rate for $K \rightarrow 2\pi$. When we do this we find that the contribution of the left-right operators is very small. (The actual bound is that the left-right contribution is much less than 50% of the physical amplitude.) While not a logical contradiction, this would be very surprising, since it disagrees with the lore that the left-right operators contribute at least a substantial fraction of the physical amplitude. Further, in the "charm in" approach, involving only left-left operators, graphs with the charm quark are found to be a factor of four greater than the total amplitude. (There is considerable cancellation between the charm and up quark graphs.) Since it is the charm quark graphs which generate the left-right operators, our bound from the figure-eights would also disagree with this result. Thus there is at least some evidence, coming only from the figure-eights, that CPTh is violated. In addition, the small size of the figure-eights (both left-right and left-left) implies that the eye graphs <u>must</u> dominate, at least if QCD is to give the explanation of the $\Delta I = 1/2$ rule. This basic conclusion of Ref. 1 therefore also follows indirectly from the evaluation of the figure-eights alone. Since the figure-eights are "safer" than the eye graphs, these are important confirmations.

CHOICE OF RENORMALIZATION POINT

The multiplicative renormalization of operators (which could include mixings with other operators of the same dimension) is what determines the relation between lattice and continuum operators (assuming that mixings with lower dimension operators have already been removed). To find this relation, one simply computes, in perturbation theory, large momentum matrix elements of the operators of interest in each scheme. On the lattice, one would get, schematically,

$$<\Theta_L> = 1 + g_L^{\,2}(a)[\ell n(p^2 a^2) + c_L] + \ldots , \qquad (1)$$

where Θ_L is some lattice operator, $g_L(a)$ is the lattice coupling as a function of lattice spacing, p is some typical external momentum, c_L is a constant, ... represents higher orders, and mixing with other operators has been ignored for notational simplicity. Similarly, in the renormalized continuum, one would get

$$<\Theta_C> = 1 + g_C^{\,2}(\mu)[\ell n(p^2/\mu^2) + c_C] + \ldots , \qquad (2)$$

where Θ_C is the corresponding continuum operator, $g_C(\mu)$ is the continuum coupling as a function of renormalization point μ, and c_C is some new constant. Formally, $g_C(\mu) = g_L(a) + \mathcal{O}(g^3)$, where g is either of the two coupling constants. One then could write

$$\Theta_C = Z(a\mu,g)\ \Theta_L , \qquad (3)$$

with Z given by

$$Z(a\mu,g) = 1 - g^2[\ell n(a^2\mu^2) + c_L - c_C] + \ldots . \qquad (4)$$

Now a is more or less fixed by practical considerations. On our lattice, with $g_L^{\,2} \simeq 1.05$, $a^{-1} \simeq 1$ GeV. One would like to choose μ so that relationship between lattice and continuum is "most convergent," i.e. so that the terms omitted in Eq. (4) are as small as possible compared to those retained. At first glance, one might guess that the proper choice is $\mu \simeq a^{-1}$. However, because of the large ratios of Λ parameters between the lattice and the continuum, this would strongly violate the assumption that $g_C = g_L + O(g^3)$. For example, using $\overline{MS}$ in the continuum one has[11] $\Lambda_C/\Lambda_L = 28.8$ for the zero flavor case. Choosing $\mu = a^{-1}$ then gives $g_C{}^2 = 2.55$, which is clearly so different from $g_L{}^2$ that the higher order terms in Eq. (4) are as significant as those kept.

The only choice that has a hope of producing a rapidly convergent relationship is therefore $\mu \simeq (\Lambda_C/\Lambda_L) a^{-1}$. This gives $g_C \simeq g_L$ by fiat. Of course, different continuum schemes would give different values of Λ_C by factors of 2 or 3, and there is no way, with a one loop calculation, that one can determine which of these is the best, i.e. most like the lattice. Such factors should make little difference at this level. In our numerical work, we use $\mu = 28.8\ a^{-1}$, but choices in the range ~ 10 to $\sim 70\ a^{-1}$ do not change things much. However, using $\mu = a^{-1}$ produces a big change: The results of the "charm in" and "charm out" approaches cease to be even qualitatively consistent.

Note that the fact that μ is proportional to a^{-1} means that μ will in general have nothing to do with typical hadronic scales as $a \to 0$. This is inevitable because matrix elements of lattice operators will always include very short distance effects as $a \to 0$, and the corresponding continuum operators must also do so. We also remark here that we have recently discovered a similar choice of continuum renormalization point in a paper by Golterman and Smit.[12]

LATTICE ALGORITHM FOR MULTI-POINT GREEN'S FUNCTIONS

An algorithm for computing three point functions on the lattice has previously been described.[13,1] We have called it "exponentiation" because it involves putting a source term into the exponent of the functional integral (i.e. into the action). Two-point functions in the presence of the source are calculated, and three point functions can then be obtained by numerically taking a (discrete) derivative with respect to the source. This technique is easy to describe in words, but in practice we use a closely related algorithm which is exact (it does not require a discrete numerical derivative) and conceptually simpler.

To explain our technique, we first need to set our notation. Quark propagators are generally calculated with a numerical algorithm, such as Gauss-Seidel or Conjugate-Gradient, which solves for v in the equation

$$Mv = b , \quad \text{or} \quad \sum_z M_{wz} v_z = b_w . \tag{5}$$

Here, M is a matrix (the quark action matrix in our case), b and v are vectors, and w and z are lattice sites. By choosing $b_w = \delta_{w,x}$, with x a fixed lattice site, one can compute the vector $v_z = (M^{-1})_{zx} \equiv G_{zx}$, which is the quark propagator from the fixed site x to any other lattice site z.

Now, for a typical three-point function, one needs the following combination of quark propagators:

$$\Gamma(x, t_y, t_z) = \sum_{\vec{z}, \vec{y}} G_{zx} G_{zy} G_{yx} . \tag{6}$$

Here t_y and t_z are the times corresponding to the four-vector lattice sites y and z. The sums over the spacial components $\vec{y}$ and $\vec{z}$ would arise from a requirement that the intermediate states carry no spatial momentum. Γ is impractical to calculate by straightforward techniques because of the factor G_{zy}. Since both $\vec{y}$ and $\vec{z}$ are summed over, one would need to apply the inverting algorithm a number of times equal to the spacial volume of the lattice. However, note that we do not actually need G_{zy} but only a particular combination of G_{zy} and G_{yx}. Calling that combination Γ', we may rewrite it as a matrix multiplication by introducing a new matrix R:

$$\Gamma'(x,t_y,z) \equiv \sum_{\vec{y}} G_{zy} \, G_{yx} = \sum_{w,u} G_{zw} \, R_{wu} \, G_{ux} ,$$

(7)

$$\text{with} \quad R_{wu} = \delta_{w,u} \, \delta_{t_w,t_y} .$$

Our algorithm can now be stated simply. With standard methods, first compute G_{ux}, the propagators from the fixed site x to all sites u. Then solve Eq. (5) again, this time choosing

$$b_w = \sum_u R_{wu} \, G_{ux} .$$

(8)

The solution v_z will be $\Gamma'(x,t_y,z)$. Then put Γ' together with G_{zx} (which is already known) to make Γ in Eq. (6). Note that we only had to solve Eq. (5) twice, instead of a number of times equal to the spacial volume. Furthermore, this method easily and conveniently generalizes to multi-point functions.

CONCLUSIONS

The main conclusions of this work appear in Ref. 1. Here, we have looked at some of the more technical aspects of the problem. We find no evidence that the mass dependence of the matrix elements of left-left operators is an artifact. Since these operators are all that are needed for the "charm in" approach, this gives us some confidence that that approach is reliable. On the other hand, the left-right operators are likely to have problems and require further study. Unfortunately, the use of the amplitude $K \to 0$ is not guaranteed to remove all of the chirally non-invariant two-quark operators which are generated with Wilson fermions. (However it is necessary to use $K \to 0$ or an equivalent amplitude, even after the chiral behavior is controlled, in order to remove chirally invariant but unphysical contributions.) We have also looked at the chiral behavior of the figure-eight graphs, whose systematic uncertainties are expected to be under better control than those of the eye graphs. Lowest order Chiral Perturbation Theory appears to be violated by these graphs as well, and their small size independently supports the conclusion that the eye graphs must dominate hadronic K decays. Finally, we have discussed the proper choice of renormalization point necessary to relate continuum operators to their lattice counterparts and have described our numerically simple algorithm for computing multi-point Green's functions on the lattice.

We wish to thank K. Barad, W. Celmaster, F. Gilman, H. D. Politzer, and M. B. Wise for helpful discussions, and G. Martinelli for useful correspondence. C. B. also would like to thank the organizers of the Wuppertal conference for making possible an extremely pleasant and informative workshop. The research of C. B., T. D., and G. H. was supported in part by the U. S. National Science Foundation, and the research of A. S. was supported in part by the U. S. Department of Energy Outstanding Junior Investigator Program. The computing was performed on the D. O. E. Livermore MFE Computing Center.

206

REFERENCES

1. C. Bernard, T. Draper, G. Hockney, A. M. Rushton, and A. Soni, Phys. Rev. Lett. 55:2770 (1985). See also C. Bernard, T. Draper, G. Hockney, and A. Soni, in "Proceedings of the Conference on Advances in Lattice Gauge Theory," Tallahassee, Florida, 10-13 April, 1985 (to be published), UCLA Report No. UCLA/85/TEP/14; T. Draper, "Lattice Evaluation of Strong Corrections to Weak Matrix Elements--$\Delta I = 1/2$ Rule," Ph.D. thesis, University of California, Los Angeles, 1984 (unpublished); C. Bernard in "Gauge Theory on a Lattice," edited by C. Zachos, et al. (Argonne National Laboratory, Argonne, Ill., 1984), p. 85.
2. J. Bijnens, H. Sonoda, and M. B. Wise, Phys. Rev. Lett. 53:2367 (1984); J. Bijnens, Phys. Lett. 152B:226 (1985).
3. C. Bernard, T. Draper, G. Hockney, and A. Soni, work in progress.
4. M. Bochicchio, L. Maiani, G. Martinelli, G. Rossi, M. Testa, "Chiral Symmetry on the Lattice for Wilson Fermions," University of Rome Report No. 452-1985, 1985.
5. C. Bernard, T. Draper, A. Soni, H. D. Politzer, and M. B. Wise, Phys. Rev. D32:2343 (1985).
6. G. Guralnik, R. Gupta, G. W. Kilcup, A. Patel, S. R. Sharpe, and T. Warnock, work in progress.
7. C. Bernard, T. Draper, and A. Soni, "Perturbative Corrections to Four-Fermion Operators on the Lattice," in preparation.
8. S. Sharpe, "On the Extraction of Kaon Decay Amplitudes from Lattice Calculations," University of Washington Report No. 40048-30 PS, 1985.
9. R. J. Crewther, "Chiral Reduction of $K \to 2\pi$ Amplitudes," Dortmund University Report No. DO-TH 85/17, 1985.
10. G. Martinelli, Phys. Lett. 141B:395 (1984).
11. A. Hasenfratz and P. Hasenfratz, Phys. Lett. 94B:165 (1980); R. Dashen and D. Gross, Phys. Rev. D23:2340 (1981).
12. M. F. L. Golterman and J. Smit, Phys. Lett. 140B:392 (1984).
13. S. Gottlieb, P. Mackenzie, H. Thacker, and D. Weingarten, Phys. Lett, 134B:346 (1984).

THE CHIRAL LIMIT IN LATTICE QCD

André Morel

Service de Physique Théorique
CEA-SACLAY
91191 Gif-sur-Yvette Cedex, France

ABSTRACT

Using the results of a quenched SU(2) simulation with staggered fermions, we discuss the small quark mass behaviour of m_π^2 and f_π^2. The importance of finite size effects and of non linearities in the zero mass extrapolation is emphasized. Evidence for scaling of m_π^2 and of a suitably determined value of f_π^2 is presented.

1. INTRODUCTORY REMARKS

This talk is based on recent results obtained in collaborations with A.Billoire, T.Jolicoeur, R.Lacaze and E.Marinari[1,2,3]. It is mainly concerned with questions related to the chiral limit of lattice QCD, namely to the behaviour of physical quantities (masses and couplings) as the quark mass m approaches zero. We are dealing with staggered fermions, for which no tuning is needed in taking this limit, other than driving m to zero. Such questions are certainly not academic ones. The physical world of the lowest lying strongly interacting particles actually corresponds, within QCD, to current quark masses of the order of a few MeV. In numerical simulations, on the other hand, the energy scale is set by the value a^{-1} of the lattice spacing, obtained by fixing some quantity, e.g. the string tension, or the ρ or proton mass, at its physical value. In practice, the value found for a^{-1} at present lies somewhere in between 1 and 2 GeV. Hence the physical world is hopefully reached for characteristic values of the bare quark mass (ma) as small as a few 10^{-3}. This is to be confronted with what is actually doable on present computers, namely (ma) $\gtrsim 10^{-2}$(Refs.2,5-8). Within the algorithms which have been used till now, this limitation is due on the one hand to the fact that the time required for computing a quark propagator with a given accuracy increases like 1/m at small m. On the other hand, even if new algorithms are developed (e.g. Fourier acceleration[4]), which overcome this difficulty, one is anyway faced with the problem of finite size effects, to be expected since the correlation length of the system increases with 1/m.

The nature of the finite size effects associated with the existence of a vanishing pion mass in the chiral limit is one of the questions we want to discuss here.

Even in the absence of finite size effects, or in cases where one is able to control their importance, the unavoidable extrapolation of actual data from (ma) $\gtrsim 10^{-2}$ down to (ma) $\sim 10^{-3}$ may lead to appreciable systematic bias, as it will be illustrated below.

Finally of course, the physics of interest in a lattice simulation crucially depends on how close one is to the continuum limit. So, an important issue concerns scaling behaviour in the low m region. Once again the question of finite size effects comes in, now because at fixed ma the correlation length (hopefully) increases with the inverse coupling squared, $\beta = 2N/g^2$ for an SU(N) gauge theory. The existence of a scaling "two-dimensional" window (in ma and β simultaneously) has to be demonstrated before conclusions are drawn on continuum QCD. We will present data on this point.

This talk is organized as follows. We first review some general aspects of those finite size effects which are associated with the existence of spontaneous symmetry breaking in the thermodynamical limit where the volume $n = L^3 \times T$ of the lattice tends to infinity (section 2). In section 3, we present the technical characteristics of quenched SU(2) QCD simulations with staggered fermions. A selection of results is commented in section 4, where we propose a method for detecting and analyzing finite size effects. Scaling behaviour for m_π^2 and f_π^2 shows up at $\beta(SU(2)) \gtrsim 2.4$, $(ma) \lesssim 0.1$, and the ratio f_K^2/f_π^2 is found to be constant for $\beta > 2.3$, and equal to $(1.24 \pm 0.05)^2$, to be compared with the experimental value $(1.21 \pm 0.01)^2$ [9] (section 5). A few conclusions are drawn in a last section.

In these notes, we only give a brief account of those numerical results which are especially relevant to the three questions discussed above, namely finite size effects, low m extrapolations and scaling. More details and a description of methods for analyzing the data can be found in Ref.2.

2. GOLDSTONE PION AND FINITE SIZE EFFECTS (FSE)

Let m_φ be some mass in physical units. The associated dimensionless correlation length is

$$\xi = (m_\varphi \, a)^{-1} \quad , \tag{1}$$

which goes to infinity either as $a \to 0$ $(\beta \to \infty)$ at fixed m_φ, or as at fixed β, $m_\varphi \to 0$ because it corresponds to the Goldstone boson of a spontaneously broken symmetry. That this seems to happen in lattice QCD with staggered fermions is well known. The regularization then preserves one axial, flavour non-singlet, subgroup $U_\epsilon(1)$ of the SU(4) $\otimes$ SU(4) chiral group of the expected continuum limit. Strong coupling arguments as well as many earlier numerical simulations indicate that this $U_\epsilon(1)$ is actually spontaneously broken in the thermodynamical limit ; the staggered fermion formalism justifies calling pion the associated Goldstone boson. Refs. to this subject can be found in 1. We want to here discuss the fate of the Goldstone pion on a finite lattice. The question is : while in the first of the two situations described above, FSE are controlled by the size of $z = \xi/L$ (L = linear lattice extension ; finite size scaling[10]), what should $\xi_\pi = 1/(m_\pi a)$ be compared with in the chiral limit ? Related discussions concerning the behaviour of order parameters for first order transitions can be found in Refs.11. We here report the results obtained in Ref.3, limiting our arguments to plausibility arguments.

Consider a lattice of size $n = L^3 \times T$, with L and T of the same order (cubic geometry). The quark condensate $\langle \bar{q}q \rangle$ can be used as an order parameter for the remnant $U_\epsilon(1)$ chiral symmetry. Spontaneous breaking in the thermodynamic limit means

$$\lim m \to 0 \ (\lim n \to \infty) \ \langle \bar{q}q \rangle_{m,n} \neq 0 \ , \tag{2}$$

while one knows from general arguments that

$$\lim m \to 0 \ \langle \bar{q}q \rangle_{m,n \text{ fixed}} = 0 \quad , \tag{3}$$

because the degenerate vacua at m = 0 are no more separated by infinite barriers. At $m \neq 0$, field configurations which differ one from the other only by $U_\epsilon(1)$ transformations yield values of the action which differ by the contribution of the explicit breaking term $m \sum_y \bar{q}q(y)$. In the limit

$m \to 0$ at n fixed, one has to integrate over the whole $U_\varepsilon(1)$ orbit of each configuration. When this is done, it is not a surprise to find that the effective parameter z which governs the finite size effects is

$$z = n(ma) \ <\bar{q}q>_{[m=0,n=\infty]} a^3 \quad , \qquad (4)$$

up to smaller terms in $1/n$ and m. From the Ward identity associated with the $U_\varepsilon(1)$ symmetry, this parameter can be rewritten

$$z = L^3 \times T \times m_\pi^2 \ f_\pi^2 \ a^4 \quad , \qquad (5)$$

showing that the characteristic length here is not $1/(m_\pi a)$, but rather the quantity $[m_\pi^2 f_\pi^2 a^4]^{-1/4}$. It is argued in Ref.3 that although one expects,

$$m_\pi^2 = sm + \text{smaller terms, at } m \to 0, \quad n \to \infty \qquad (6)$$

one should find

$$m_\pi^2 = a + b \ m^2 + \text{smaller terms, at } m \to 0, \ n \text{ fixed.} \qquad (7)$$

Such behaviours of the Goldstone pion mass will be commented upon in the forthcoming discussion of numerical results.

3. A QUENCHED SU(2) SIMULATION WITH STAGGERED FERMIONS

The data of Refs.1,2 are obtained from SU(2) simulations at $\beta = 2.3$, 2.4, 2.5. The volume of the lattices is $n = L^3 \times 24$, with $L = 6, 8$ and 12. The number of configurations used for data taking is respectively 70, 70 and 30 for the 3 lattice sizes. They are equilibrated according to the Wilson action

$$S_G = -\frac{\beta}{4} \sum_{x,\mu,\nu} \text{Tr} \ [U_P + U_P^\dagger] \qquad (8)$$

by a Metropolis Monte-Carlo method. We begin with a cold start at a first β, use 2000 sweeps for thermalization, and keep one configuration after every 100 sweeps. At the next β value, we start from the last configuration obtained and use 1000 sweeps for thermalization. The fermionic action is

$$S_F = \frac{a^3}{2} \sum_{x,\mu} \alpha_\mu(x) \ [\bar{\chi}(x)U_\mu(x)\chi(x+\mu) - \bar{\chi}(x+\mu)U_\mu^\dagger(x)\chi(x)]$$
$$+ ma^4 \sum_x \bar{\chi}(x)\chi(x) \quad , \qquad (9)$$

where $\alpha_\mu(x)$ is chosen to be

$$\alpha_\mu(x) = (-)^{x_1 + \ldots + x_{\mu-1}} \quad . \qquad (10)$$

The bare quark mass values taken for computing fermion propagators

$$G(r_o,x) = <\chi(x)\bar{\chi}(r_o)>_{\{U\}} \qquad (11)$$

for each gauge configuration are

$$ma = 0.0125, \ 0.025, \ 0.05, \ 0.1, \ 0.15 \ \text{and} \ 0.3 \quad . \qquad (12)$$

(The value 0.2 was used instead of 0.15 at $\beta = 2.3$). For each configuration, the propagators are computed for 2 origins r_o (L = 6 and 8) or 5 origins (L = 12), by the conjugate gradient method[12].

This variety of mass, lattice size and β values, together with a rather large statistical sample of gauge configurations provides us with a set of data suitable for an accurate investigation of finite size effects and scaling properties in the chiral limit.

4. DATA ANALYSIS. THE BEHAVIOUR OF $m_\pi^2(m)$

We are primarily interested in the behaviour of the lowest mass of the spectrum, which can be obtained from the large t behaviour of the following zero momentum correlation function :

$$g(t) = \sum_{\vec{x}} <\text{Tr} \ G(r_o,x) \ G^\dagger(r_o,x)> \quad . \qquad (13)$$

The average is taken over the gauge configurations and the origins r_0 used. The values of m_π follow from two-state fits of $g(t)$ to the form

$$g(t) = \sum_i R_i \left[\exp(-M_i t) + \exp(-M_i(T-t))\right] \quad , \qquad (14)$$

no indication of oscillating terms being found. The errors are obtained from the dispersion around m_π, as given by the full sample, of the values found in five subsamples. We also investigated the location of the poles in the Fourier transforms of $g(t)$ for individual configurations (see ref.2).

A first glance at possible finite size effects comes from a direct comparison of $g^L(t)$ for different spatial lattice sizes L. If $g^L(t)/g^{12}(t)$ is compatible with 1 we conclude that there is no sizeable FSE. If not, we keep the L=12 data, and perform the following analysis, inspired by the parametrizations (6) and (7). We define two variables x,y :

$$x(m) = m_\pi^2(m)/m \qquad\qquad (15)$$

$$y\left(m = \frac{m_1 + m_2}{2}\right) = \frac{m_\pi^2(m_1) - m_\pi^2(m_2)}{m_1 - m_2} \quad .$$

If Eq.(16) is applicable, one has

$$x \to s, \quad y \to s \quad \text{as} \quad m \to 0 \quad , \qquad (16)$$

while if F.S.E. are present, Eq.(7) yields

$$x \to \infty, \quad y \to 0 \quad \text{as} \quad m \to 0 \quad . \qquad (17)$$

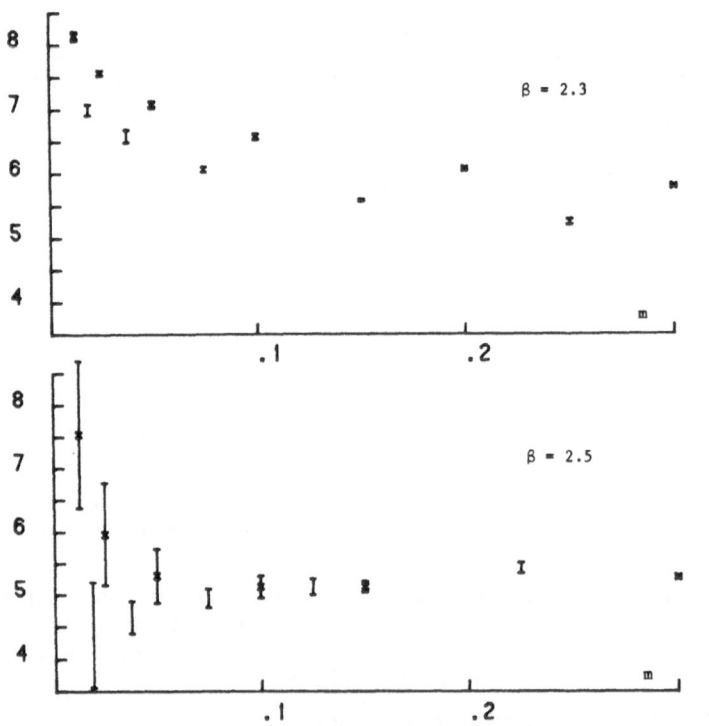

Fig. 1. The quantities x(m) [x] and y(m)[I] of Eq.(15). A negative curvature of $m_\pi^2(m)$ is seen at $\beta = 2.3$ (top) ; FSE are present at $\beta = 2.5$ (bottom). See text.

Results are displayed in Fig.1 for $\beta = 2.3$ and 2.5, L=12. At $\beta = 2.3$, where we checked directly on the correlation functions g(t) that L=8 and L=12 lead to the same results (no FSE), the data are compatible with (16). However the two quantities x and y are clearly not constants in m, *especially* at low m. They are compatible with $m_\pi^2 \to O(m)$ in the chiral limit, but with important non-linear terms, *not* due to FSE. Conversely, at β=2.5 m_π^2/m again goes up at low m, but y(m) now tends to bend down ; we conclude that this situation reveals FSE at low m, compatible with Eqs.(7) and (17). The $\beta = 2.4$ data, not reproduced here, correspond to an intermediate situation.

As a consequence, we stress that, even in the absence of FSE, reliable chiral extrapolations are hard to perform : only at very low ma is a linear behaviour of m_π^2 taking place. We believe that this result is not a peculiarity of SU(2) : application of the above criteria to the SU(3) data of Ref.8 at $\beta = 5.7$ reveals a situation close to ours at $\beta = 2.3$. The data of Ref.7 at $\beta = 6.0$ may indicate small FSE at the lowest ma value, 0.01. It may be that the observed curvature of m_π^2 at small ma signals the existence of large fluctuations associated with spontaneous symmetry breaking (logarithmic singularities[13]). However, it is not clear to us that any such effect may survive the quenched approximation, where no quark-antiquark loop (Goldstone mode propagation) is involved in g(t). This question is presently under investigation.

5. THE β DEPENDENCE OF m_π^2 AND f_π^2. THE RATIO f_K/f_π

In order to determine f_π^2, the π coupling constant squared, we decided not to use the standard PCAC relation (two flavours)

$$m_\pi^2 f_\pi^2 = \frac{m}{2} \sum_i \langle \bar{\chi}_i \chi_i \rangle \quad , \qquad (18)$$

because it is valid only in the limit $m \to 0$. Otherwise, $\langle \bar{\chi}\chi \rangle$ receives numerically important contributions from all states coupled to the same operator as the Goldstone π. We find it safer to relate f_π^2 to the residue of the correlation function at the π pole. One obtains[1] :

$$f_\pi^2 = \frac{2m^2 \, R_\pi}{m_\pi^3} \quad , \qquad (19)$$

with R_π defined as the value of R_i for the pion in a fit of g(t) to the form (14). The results of the analysis confirm the existence of finite size effects at $\beta = 2.4$ and 2.5 at the one or two lowest m values (bending down of f_π^2). Apart from that, we obtain for $f_\pi^2(m)$ a quite smooth curve, linear within errors, and much more reliable than the much steeper one given by (18) for the purpose of a chiral extrapolation.

Given the values of $m_\pi^2(m)$ and $f_\pi^2(m)$, and despite the occurrence of an unexpected behaviour of m_π^2 and/or of FSE, let us now study the β dependence. If scaling holds, and neglecting powers of β in front of exponentials in β, a quantity M with the naive dimension of a mass must behave according to

$$Ma(\beta)/Ma(\beta+0.1) = \exp\left[\frac{\pi^2}{11} \times 0.1\right] \simeq 1.3 \quad . \qquad (20)$$

In Fig.2, we plot m_π^2 and f_π^2 against m after these quantities have been scaled down to their expected values at $\beta = 2.5$. We clearly see that, especially below ma[β=2.5]$\simeq 0.1$, the data points for $\beta = 2.4$ and 2.5 nearly fall on the same curves. We even note that when they fail to do so at the lowest ma values, this is most likely to be attributed to FSE : the points $m_\pi^2(f_\pi^2)$

are slightly too high (too low) at $\beta = 2.5$. On the contrary, scaling of m_π^2 is clearly not reached at larger m, especially for $\beta = 2.3$. This non-scaling behaviour of higher masses is corroborated by the fact that if f_π^2 is (uncorrectly) determined using the relation (18), then scaling for it is worse. This we interpret as the effect of (non-scaling) high mass contributions to $\langle \chi\chi \rangle$, as mentioned above.

Encouraged by the nice behaviour of f_π^2 as a function of β and ma, we proceed to a determination of the ratio f_K/f_π. For either of the two pseudoscalars $P = \kappa$ or π we approximate m_P^2 and f_P^2 by linear functions of m and fix the lattice spacing by the physical value of f_π^2 (taken at zero quark mass). Then we have

$$f_K^2/f_\pi^2 = \frac{1 + 1.5\ \lambda\ m_K^2/f_\pi^2}{1 + 1.5\ \lambda\ m_\pi^2/f_\pi^2} \qquad , \qquad (21)$$

where λ is determined by the ratio of the slopes of m_P^2, f_P^2 versus m; the factor 1.5 accounts for the replacement of N=2 (for SU(2)) by N=3 when f_π^2 is compared to experiment. Upon insertion of the physical values of m_π^2, m_K^2 and f_π^2, Eq.(21) leads to

$$f_K/f_\pi = 1.23(4),\ 1.24(5),\ 1.24(6) \qquad\qquad (22)$$

respectively at $\beta = 2.3$, 2.4 and 2.5. The errors quoted include uncertainties

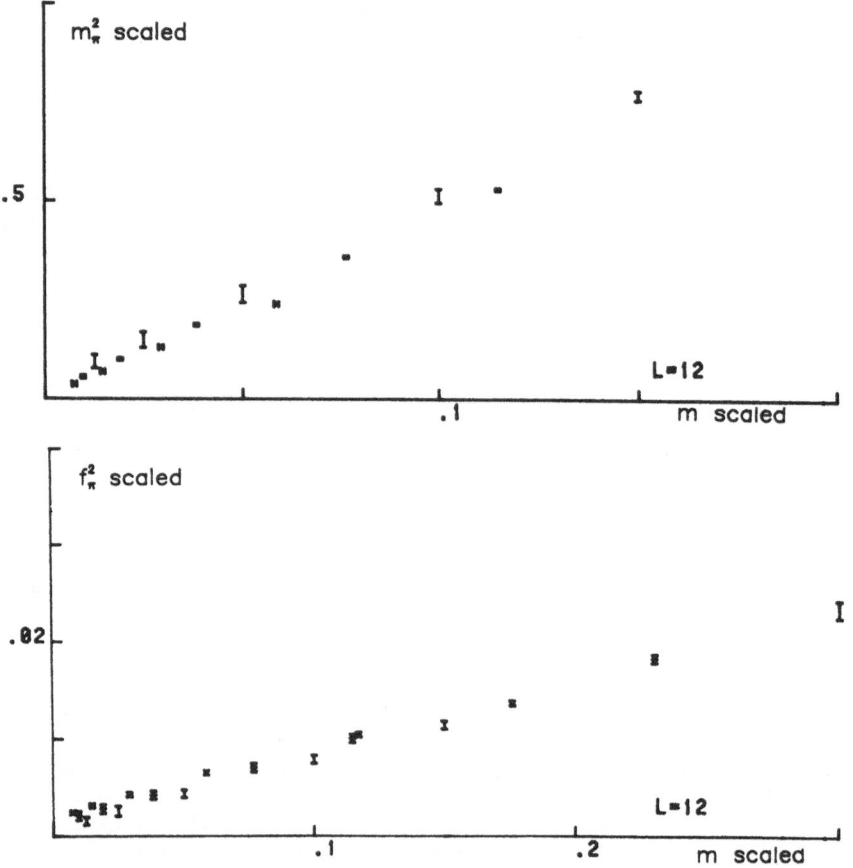

Fig. 2. The values of m_π^2 and f_π^2 for L=12, $\beta = 2.3[x]$, 2.4[∓] and 2.5[I] as functions of m. All quantities are scaled down according to their expected value at $\beta = 2.5$ (Eq.(20)).

due to the observed non linearities in m_P^2. Not only is this ratio nicely constant in β, but it happens to be very close to the experimental result 1.21 (1)[9].

6. CONCLUSIONS

We have demonstrated[1,2] that in order to be relevant for comparing lattice QCD results to the physical world, a numerical simulation requires not only good statistics but also an exploration of very low quark mass values. In this region, large lattices are needed because FSE show up very rapidly ; a method has been proposed which helps in disentangling these effects from other "true" ones, namely the occurrence of appreciable non linear terms in the m_π^2 behaviour. Although these results were obtained for SU(2), we believe them to be quite general. By the way, we note that in the quenched approximation, SU(2) is not necessarily worse than SU(3) (for mesons), while a far better insight is reached for an equal expense. We even showed that, with SU(2), very good *physical* results (f_K/f_π) could be obtained provided the analysis is carried out with much care. We emphasized in particular how important it is to determine f_π^2 from "on mass shell" data. This remark might be particularly relevant for a precise determination of the temperature where chiral symmetry is restored, and for a reliable study of its scaling properties. The intricated behaviour in m and β of the complete zero distance propagator $\langle\chi\chi\rangle$ makes it, in practice, a dangerous order parameter to be used in quantitative investigations.

It is a pleasure to thank the organizers for the excellent atmosphere and working conditions provided to us during this workshop.

REFERENCES

1. A. Billoire, R. Lacaze, E. Marinari and A. Morel, Nucl.Phys. B251 [FS13] (1985) 581.
2. A. Billoire, R. Lacaze, E. Marinari and A. Morel, "The Goldstone Pion in SU(2) lattice QCD at low quark mass : scaling and finite size effects", Saclay preprint PhT 85-180 (Nov. 1985).
3. T. Jolicoeur and A. Morel, Nucl.Phys. B262 (1985) 627.
4. G.G. Batrouni, G.R. Katz, A.S. Kronfeld, G.P. Lepage, B. Svetitsky, and K.G. Wilson, Phys.Rev. D32 (1985) 2736.
5. J.P. Gilchrist, H.Schneider, G. Shierholz and M. Teper, Phys.Lett. 136B (1984) 86 ; Nucl.Phys. B248 (1984) 29.
6. A. Billoire, E. Marinari and R. Petronzio, Nucl.Phys. B251 (1985) 141.
7. D. Barkaï, K. Moriarty and C. Rebbi, Phys.Lett. 156B (1985) 385 ; see also C. Rebbi, results presented in this workshop.
8. K.C. Bowler, D.L. Chalmers, A. Kenway, R.D. Kenway, G.S. Pawley and D.J. Wallace, Phys.Lett. 162B (1985) 354.
9. H. Leutwyler and M. Roos, Z.Phys. 25C (1984) 91.
10. M.E. Fisher, in "Critical Phenomena", Proc. 51st E.Fermi Summer School (Varenna), M.S.Green Ed. (Academic Press N.Y. 1972) ; M. Lüscher, Nucl.Phys. B219 (1983) 233 ; M. Lüscher and G. Münster, Nucl.Phys. B232 (1984) 445 ; G. Münster, Physica 124A (1984) 495.
11. V. Privman and M.E. Fisher, J.Stat.Phys. 33 (1983) 385 ; V. Privman and M.E. Fisher, "First order transitions breaking O(n) symmetry : finite-size scaling", Caltech preprint (1985) ; E. Brézin and J. Zinn-Justin, Nucl.Phys. B257 [FS14] (1985) 867 ; J.Zinn-Justin, this workshop.
12. D. Barkai, K.J.M. Moriarty and C. Rebbi, Comput.Phys.Commun. 36 (1985) 1.
13. For a review, see e.g. J. Gasser and H. Leutwyler, Phys. Reports 87C (1982) 78 ; we thank J. Zinn-Justin for pointing this possibility to our attention.

QCD SUM RULES AND SPONTANEOUS BREAKDOWN OF CHIRAL SYMMETRY [+]

E. Katznelson[*]

Physikalisches Institut
Universität Bonn
Nussallee 12
5300 Bonn 1, FRG

N.S. Craigie[*]

Institut für Theoretische Physik
Universität Heidelberg
Philosophenweg 16 und 19
6900 Heidelberg, FRG

S. Mahmood

Department of Physics
Quiad-I-Azam University
Islamabad, Pakistan

INTRODUCTION

In this talk we present some results from a Monte-Carlo simulation of a special set of four fermion correlation functions at small lattice spacing in the Susskind formalism. In a previous article [1] a number of identities were derived for meson (i.e. four fermion) correlation functions whose short distance behaviour is governed by chiral symmetry order parameters like $<\bar{\psi}\psi>$ or $<\bar{\psi}\Gamma\psi\bar{\psi}\Gamma\psi>$. In the continuum limit these relations correspond to a set of so-called odd sum rules for two point functions involving the set of currents and quark densities of the form $\bar{\psi}\psi$, $\bar{\psi}\gamma_5\psi$, $\bar{\psi}\gamma_\mu\psi$, $\bar{\psi}\gamma_5\gamma_\mu\psi$ and $\bar{\psi}\sigma_{\mu\nu}\psi$. Of particular interest are

$\psi(x)\gamma_5\gamma_\mu\psi(x)\psi(y)\gamma_5\psi(y)$ and $\psi\sigma_{\mu\nu}\psi\bar{\psi}\gamma_\mu\psi$ because the Wilson operators product expansion (OPE) tells us that in the short distance and chiral limit these are directly proportional to the order parameter $<\bar{\psi}\psi>$.

While the lattice necessarily violates the short distance behaviour of the continuum, one would expect that, as the lattice spacing goes to zero, the relation based on the OPE becomes increasingly satisfied. We argue that on the lattice it is not sufficient just to ensure a non-vanishing order parameter $<\bar{\psi}\psi>$ and vanishing pion mass m_π in order to

---

[*] Alexander von Humboldt-Fellow
[+] Presented by E. Katznelson

establish that MC is able to reproduce the continuum physics. The relation referred to above must also be reproduced.

I THE CONTINUUM PHYSICS

The correlation function we propose to test on the lattice corresponds in the continuum to the two sum rules discussed below [2].

Let us define in the chiral limit ($m_q \to 0$) the following two point function

$$q_\mu \pi_{AP}(q^2) = i \int d^4x \ <0|T\{A_\mu(x) \ P(0)\}|0> \exp(iqx) \qquad 1.1$$

where

$$A_\mu(x) = \bar{\psi}\gamma_5\gamma_\mu\psi \ , \qquad P(x) = \bar{\psi}(x)\gamma_5\psi(x)$$

and

$$(g_{\mu\alpha}q_\nu - g_{\nu\alpha}q_\mu) \ \pi_{TV}(q^2) = i \int d^4x \exp(iqx)<0|T\{T_{\mu\nu}(x)V_\alpha(0)\}|0> \qquad 1.2$$

with

$$T_{\mu\nu}(x) = \bar{\psi}(x)\sigma_{\mu\nu}\psi(x) \quad , \quad V_\alpha(x) = \bar{\psi}(x)\gamma_\alpha\psi(x)$$

The functions π_{AP} and π_{TV} satisfy the dispersion relations

$$\lim_{\substack{m \to 0 \\ q^2 \to \infty}} \pi_{AP}(q^2) = \frac{1}{\pi} \int ds \ \frac{\text{Im}\pi_{AP}(s)}{s-q^2} = \frac{<\bar{\psi}\psi>}{q^2} + O(\frac{1}{q^4}) \qquad 1.3$$

and

$$\lim_{\substack{m \to 0 \\ q^2 \to \infty}} \pi_{TV} = \frac{1}{\pi} \int ds \ \frac{\text{Im} \ \pi_{TV}(s)}{s-q^2} = \frac{\alpha(-x^2)}{\alpha(\mu^2)} \frac{<\bar{\psi}\psi>}{q^2} + O(\frac{1}{q^4}) \qquad 1.4$$

Despite being very similar the sum-rules or rather the dispersion integrals involve quite different intermediate hadronic states. In other words they are saturated in significantly different ways.

In the chiral limit only the zero mass pion state couples to the axial current, this alone saturates the sum rule (1.3). In contrast the two point function appearing in Eq. (1.4) is saturated by the whole ρ family [2], and convergence will involve cancellation between ρ,ρ' etc. in the sum-rule.

On the right hand sides of Eqs. 1.3 and 1.4 the only difference apparently is the anomalous dimension factor. At large distances ($\sqrt{q^2} < 500$ MeV) one expects a considerable difference in the behaviours of the correlation functions $\langle AP \rangle$ and $\langle TV \rangle$. They have exponential tails controlled by m_π and m_ρ, respectively.

With this brief analysis in the continuum we next examine the corresponding objects for Susskind's fermions.

2 THE LATTICE FORMULATION

The continuum relations noted above can be translated into identities of certain correlation functions in Euclidean configuration space [1]

$$\lim_{x^2 \to 0} \langle A_\mu(x) P^+(0) \rangle = X_\mu \frac{\langle \bar{\psi}\psi \rangle}{\pi^2 x^4} + \frac{N_c m}{\pi^4 x^6} \qquad 2.1$$

and

$$\lim_{x^2 \to 0} \langle T_{\mu\nu}(x) V_\alpha^+(0) \rangle = -(\delta_{\mu\alpha} x_\nu - \delta_{\nu\alpha} x_\mu)$$

$$(C_o(x^2) \frac{\langle \bar{\psi}\psi \rangle}{\pi^2 x^4} + C_m \frac{N_c m}{\pi^4 x^6}) \qquad 2.2$$

where

$$C_o(x^2) = \frac{\alpha_s(x^{-2})^{\gamma_{TV}}}{\alpha_s(\mu^2)} \qquad , \qquad C_m(x^2) = \frac{\alpha_s(x^{-2})^{\gamma_T + \gamma_m}}{\alpha_s(\mu^2)}$$

are the anomalous dimensions, m is the quark mass and N_c is the number of colours.

To find a lattice transcription of Eqs. 2.1 and 2.2 in the Kogut-Susskind formulation one should translate bilinear involving the fields $\bar{\psi}$, and ψ to lattice quantities. In the Susskind formulation the starting point is the action

$$S = \frac{1}{2} \sum_{x,\mu=1,4} (\chi(x) S_\mu(x) U_\mu(x) \chi(x+\hat{\mu})$$

$$\qquad 2.3$$

$$- \bar{\chi}(x+\hat{\mu}) S_\mu(x) U_\mu^+(x) \chi(x))$$

where $S_o(x) = 1$, $S_1(x) = (-1)^{x_2+x_3}$ $\quad S_2(x) = (-1)^{x_3+x_4}$

$$S_3(x) = (-1)^{x_4+x_1}$$

$U_\mu(x)$ are the SU(3) dynamical variables and χ are the fermion fields which carry only the colour index. To identify the Dirac fields one can split ($\chi(k)$ is χ in momentum space) into its components [3] $\psi_{\alpha_1 \alpha_2 \alpha_3 \alpha_4}(k)$ one for each $\frac{1}{16}$th of the BZ*. So $\bar\psi_{\alpha_1 \alpha_2 \alpha_3 \alpha_4}(k)$ has only a single pole in the continuum limit. The algebra of translating the BZ components of $\psi_{\alpha_1 \alpha_2 \alpha_3 \alpha_4}(k)$ provides a basis for Dirac and flavour algebras. Using the relation

$$\chi(x) = \sum_{\alpha,\mu} (-1)^{\alpha_\mu x_\mu} \int_{-\pi/2}^{\pi/2} d^4k\, e^{ik_\mu x_\mu}\, \bar\psi_{\alpha_1 \alpha_2 \alpha_3 \alpha_4}(k) \qquad 2.4$$

where $\alpha_\mu = 0, 1$. it is easy to see that $(-1)^{x_1}\chi(x)$ is equivalent to $(\sigma_1)_{\alpha_1 \alpha_1'}\, \psi_{\alpha_1 \alpha_2 \alpha_3 \alpha_4}(x)$ and that the shift $\chi(x) \to \chi(x+1)$ corresponds to

$$\psi(x) \to (\sigma_3)_{\alpha_1 \alpha_1'}\, \psi_{\alpha_1 \alpha_2 \alpha_3 \alpha_4}(x+1)^{**}$$

Thus the original action can be written in terms of the ψ fields as

$$S = \sum_x \bar\psi\, (\gamma_\mu \mathbb{1}\, \nabla_\mu \psi) \qquad 2.5$$

($\mathbb{1}$ the identity in flavour space)

where $\gamma_1 = \sigma_3 \rho_1 \tau_1$, $\gamma_2 = \rho_3 \pi_1 \tau_1$, $\gamma_3 = \sigma_1 \pi_3 \tau_1$ and $\gamma_5 = \tau_3$.

The flavour algebra can be written as [3]

$$T_1 = \sigma_3 \pi_1\,,\quad T_2 = \sigma_1 \rho_3\,,\quad T_3 = \rho_1 \pi_3 \quad \text{and} \quad T_4 = \sigma_1 \rho_1 \pi_1 \tau_3\,.$$

In addition to the above $\gamma_5 = \sigma_1 \rho_1 \pi_1 \tau_1 \sigma_3 \rho_3 \pi_3 \tau_3$ and $T_5 = \sigma_3 \rho_3 \pi_3 \tau_3$.

These give a representation of the algebra on the component fields ψ. For the fields $\chi(x)$ the algebra can be worked out by using the correspondence

$$\chi(x) \to (-1)^{x_1}\chi(x) \qquad \text{with} \qquad \psi(x) \to \sigma_1 \psi(x)$$

and
$$\qquad\qquad\qquad\qquad\qquad\qquad\qquad\qquad\qquad\qquad 2.6$$

$$\chi(x) \to \chi(x+1) \qquad \text{with} \qquad \psi(x) \to \sigma_3 \psi(x)$$

---

* BZ = Brillouin zone

** In a similar fashion one can define for $\mu=2,3,4$ a set of Pauli matrices ρ, π and τ.

We proceed by defining a set of currents needed for our M.C. using 2.6, namely

$$P^5(x) = \bar\psi(x)\gamma_5 T_5 \psi(x) = -(-1)^{x_1+x_2+x_3+x_4}\bar\chi(X)\,\chi(X)$$

$$A_4^5(x) = \bar\psi(x)\gamma_5 T_5\gamma_4\psi(x) = (-1)^{x_1+x_2+x_3+x_4}1/2\{\bar\chi(x)\,U^4(x)\chi(x+4)+hc\}$$

$$V_1^1(\chi) = \bar\psi(x)\gamma_1 T_1\psi(x) = (-1)^{x_2+x_3+x_4}\{\bar\chi(x)\,\chi(x)\}$$

and

$$T_{14}^1 = \bar\psi(x)\gamma_1\gamma_4 T_1\psi(x) = (-1)^{x_2+x_3+x_4}1/2\{\chi(x)U^4\bar\chi(x+4)+hc\} \qquad 2.7$$

We have chosen the currents which are as local as possible on the lattice. In terms of these currents (which can be called naive) the relationships 2.1 and 2.2 are both translated (in the gauge $U^4(x) = 1\!\!1$) to

$$\langle 0|R_e\,G_{i,j}(t-1)G_{i,j}(t)\,|0\rangle = \frac{\langle G(0,0)\rangle}{\pi^2 t^3} + \frac{mN_c}{\pi^4 t^5} \qquad 2.8$$

where $G_{ij}(x-y) = \chi^i(x)\chi^j(y)$ and the brackets $\langle\ \rangle$ stand for an average over gauge field configurations. For further discussion of 2.8 see [4].

3 MONTE CARLO RESULTS

A set of 40 gauge field configurations on a $8^3\times10$ lattice was generated after 1500 iterations. We used 10 hit metropolis algorithms. It was established [5] that the computer time correlation length is 50 sweeps. We selected 3 configurations at ß=5.8 and 1 configuration at ß=6.0 as a feasibility test. These configurations were separated by 1500 sweeps. The conjugate gradient method was used to calculate propagators and we use time slice implementation [6].

The results from the MC simulation for correlation function $\langle G(t-1)G^*(t)\rangle$ and the asymptotic formula (referred as SDL) on the r.h.s. of Eq. (2.8) are tabulated in Table I. for ß=5.8. A total of 40 propagators, divided into the masses m=.1a, .2a, .3a and .4a were calculated. Limited data at m=0.05a and m=.5a was obtained. The inverse lattice spacing a^{-1} for this value of ß is [7]1.2 GeV.

Table I: Results from MC simulation with Susskind fermions.
(SDL means short distance limit using measured
order $\langle\bar{\psi}\psi\rangle$.

m / T	.1 SDL	.1 MC	.2 SDL	.2 MC	.3 SDL	.3 MC	.4 SDL	.4 MC
1	.305 ± .044	.323 ± .085	.459 ± .033	.455 ± .085	.555 ± .032	.506 ± .080	.665 ± .032	.563 ± .080
2	.036 ± .005	.112 ± .035	.0517 ± .0046	.114 ± .040	.0608 ± .0043	.107 ± .032	.0716 ± .043	.105 ± .024
3	.009 ± .0012	.0168 ± .082	.0149 ± .0013	.018 ± .0072	.0175 ± .0013	.0173 ± .0072	.0205 ± .0013	.014 ± .004
4	.0043 ± .0007	.0044 ± .0002	.0062 ± .0005	.0037 ± .0014	.0073 ± .0005	.0028 ± .0011	.0085 ± .0005	.0022 ± .0006

It is evident that at these values of ß the lattice is too course
to give a satisfactory test of such identities. One needs at least a few
lattice spacing in time direction such that T<2 GeV^{-1} in order to see
the T^{-3} behaviour of the correlation function, e.g. one needs a lattice
of 24^4 at β =6.2. Nevertheless, it is interesting that some verification
of the relationship 2.8 was obtained in this MC study, limited as it is.

The results for the correlation function and their short distance
limit are summarized in the 3-dimensional plot in Figure 1 as a function
of both the mass m and lattice time t. The triangles on the plot represent the asymptotic behaviour deduced from Wilson OPE. The curves through
the MC data points are to guide the eye. For t=1 (the only spacing well
inside the range t<2.0 GeV^{-1} we have) there is substantial agreement
between the r.h.s. and the l.h.s. of Eq. (2.8). However, for small masses
there is an unexpected deviation at lattice time spacing of t=2 which
seems to indicate a large finite size effect. This effect goes away for
masses greater than m=.3. As explained above, for large t the correlation
function should have an exponential tail and SDL values should lie higher
than the MC result. Due to the effect cited above this happened only for
m >.3.

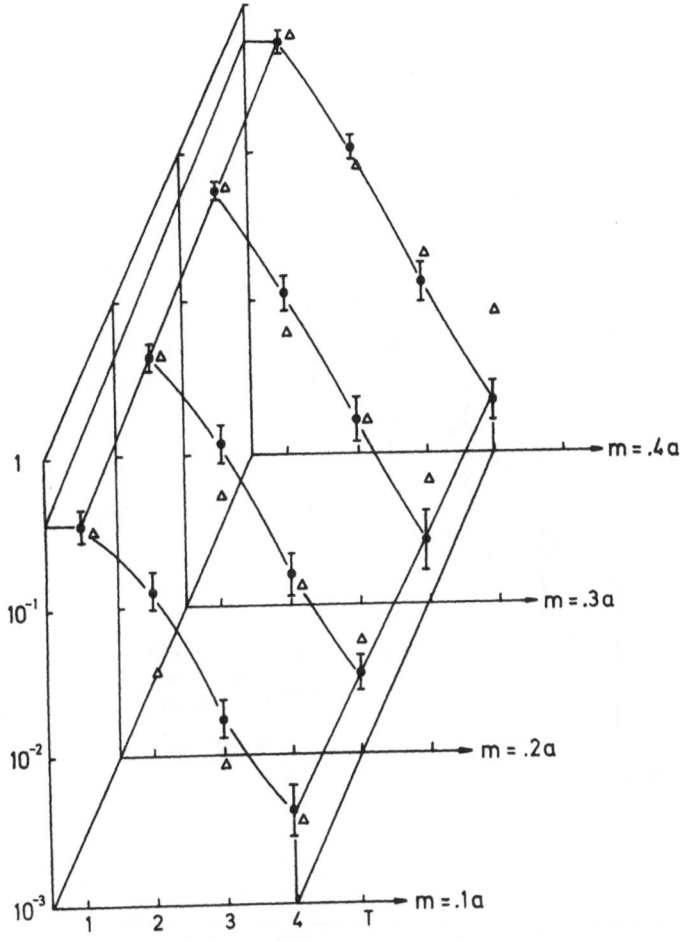

Fig. 1. Configuration-averages of <AP> for quark mass
for mass m=.1a,.2a,.3a and .4a as a function
of the time. The triangle represents the asymptotic behaviour.

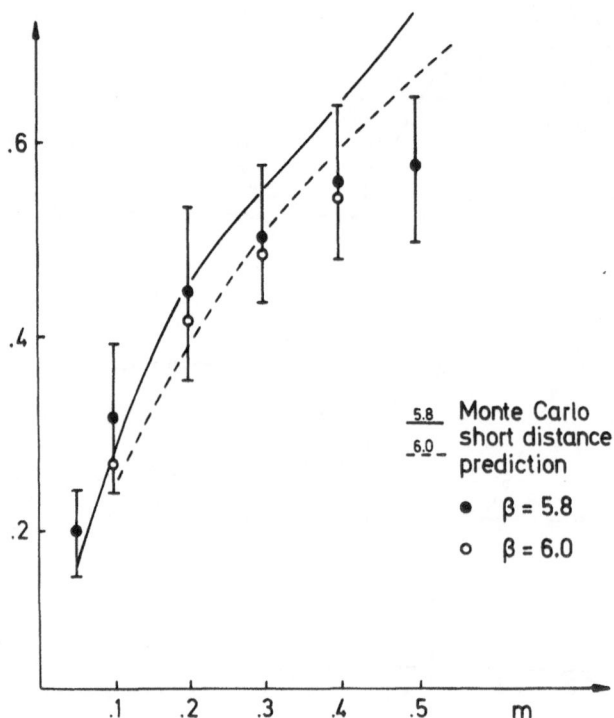

Fig. 2. The measured correlation functions
against mass. The curve is the result
of the asymptotic formula.

In Fig. 2 we have plotted the measured correlation functions against increasing mass for lattice spacing t=1 in the mass range 0.05a to 0.5a. The black dots are for ß=5.8. The agreement in the range m=.1a to .3a is reassuring and the deviation after m=0.4a is expected because of higher mass corrections. The open circles (in Fig. 2) represent the ß=6.0 points (without the statistical errors). These indicate the expected drop in the correlation function due to a drop in $<\bar{\psi}\psi>$ for smaller lattice volumes. The deviation from the SDL is for m>.4a which is the same mass as for ß=5.8 in absolute units.

To sum up, we have faced two problems in this analysis. One the finite size effect for small mass and secondly the finite mass effect for large mass. It is only for m≈.3a, .4a that both these effects are small. From this calculation it seems feasible that on larger lattices at larger ß values (e.g. 24^4 at ß=6.2), such that a $<t<2$ GeV^{-1}, the relation 2.8 can be checked to a greater accuracy.

The problem of using local currents (i.e. the need to translate the matrices to the χ field in configuration space) can probably be overcome by solving the Dirac equation in Fourier space [8]. This may create the anomalous dimensions in Eq. (2.8).

REFERENCES

[1] N.S. Craigie, E. Katznelson and C. Rebbi, Nucl.Phys. B274 (1984) 360.
[2] N.S. Craigie and F. Stern, Phys.Rev. D26 (1982) 2430.
[3] C.P. Van den Doel and J. Smit, Nucl.Phys. B228 (1983) 122.
[4] N.S. Craigie, E. Katznelson and S. Mahmood, Bonn preprint HE-85-15.
[5] E. Katznelson and A. Nobile (to appear in Comp.Phys.Comm.).
[6] D. Barkai, K.J.M. Moriarty and C. Rebbi, CERN preprint TH-4065/84.
[7] D. Barkai, K.J.M. Moriarty and C. Rebbi, Phys.Rev. D30 (1984) 1293, Ph. de Forcrand, G. Schierholz, H. Schneider and M. Teper, Phys.Lett. B152 (1985) 107.
[8] G.G. Batrouni et al., Phys.Rev. D32 (1985) 2736.

POTENTIALS

C. Michael

Physics Dept., University of Illinois
Urbana, IL 61801
and
D.A.M.T.P., University of Liverpool
Liverpool, U.K.

INTRODUCTION

The response of interacting colour fields to static colour sources is of considerable importance. This was one of the first topics to be studied using numerical simulation on a lattice and many interesting results for such potentials can be obtained. The most direct analysis is of the static interquark potential and I summarize the status of precision measurements of this quantity. As well as the usual static potential in which the gluonic field is in the lowest energy configuration, there is also a spectrum of excited gluonic states which are relevant to hybrid meson spectroscopy. Recent results are reported on this topic. The lowest energy state of the gluonic field between static colour sources has fluctuations in its electric and magnetic fields and these can be studied on a lattice and related to the spin-spin and spin-orbit potentials of heavy quark spectroscopy. Recent progress in this area is reviewed. The remaining topic to be discussed is the response to static adjoint colour sources. Here an isolated source is possible - it forms a colour singlet "glue-lump" state by binding with the gluon field. The spectrum on a lattice is discussed - it has theoretical interest and possible experimental relevance should gluino states exist. The potential between two such sources - the adjoint potential - is then presented.

Although the numerical evaluations to be discussed here all involve pure gauge colour fields - thus without dynamical quarks - the results should provide a close guide to those of full QCD. In this field numerous successful comparisons with experiment are possible and one can obtain some guidance in areas as yet unexplored experimentally.

STATIC INTERQUARK POTENTIAL

Extracting $V_o(R)$. The most direct determination of the potential $V_o(R)$ for static fundamental-colour sources at separation R is from a ratio of averages of R × T and R × T-a rectangular Wilson loops since

$$\underset{T \to \infty}{\text{Limit}} \quad \frac{W(R,T)}{W(R,T-a)} = \lambda_o(R) = e^{-aV_o(R)} \, .$$

where $\lambda_o(R)$ is the eigenvalue of the transfer matrix (i.e. the attenuation in taking one step of length a in the time direction) in the presence of the static sources at separation R. The first determinations of $V_o(R)$ were done using T ≈ R rather than T → ∞. Recent precise lattice evaluations show that this is inaccurate - for example one needs T > 7a at R = 3a to attain a stable value of the ratio $\lambda_o(R)$.[1] Extraction of the largest eigenvalue $\lambda_o(R)$ can be improved by modelling the spectrum of eigenvalues as a discrete set.[2,3] In particular, variational methods can be used effectively.[2]

Rotational Invariance. Another requirement on a lattice is that R ≫ a. Again this has often been interpreted as R ⩾ a. A test is to consider the potential between static sources separated off the lattice axis and to study whether rotational invariance of the potential $V_o(R)$ is exactly restored. It is known[4,5] that rotational invariance is approximately restored for β > 2.2 in SU(2) and for β > 5.6 in SU(3). Recent precise determinations[6,1] show small discrepancies for R ⩽ 3a at larger β-values too. A simple prescription to understand the size of these discrepancies is to replace the Coulomb term in the continuum static potential by a lattice propagator corresponding to one gluon exchange.[4,1] This gives excellent agreement for R ⩾ 2a, see fig. 1, so that one may use this prescription to correct the accurately determined lattice results from R = 2a to 3a in extracting the continuum potential.

The String Tension. The potential $V_o(R)$ is accurately measured in lattice evaluations and different methods agree well. To extract a value for the string tension K is more subjective since it involves fitting the shape of $V_o(R)$ to a form such as $-eR^{-1} + c + KR$. The influence of points at small R is relatively large and relatively big changes in the value of K result from modifications to the form of the fit such as using the lattice propagator instead of R^{-1}. Furthermore some groups fix e = π/12 from

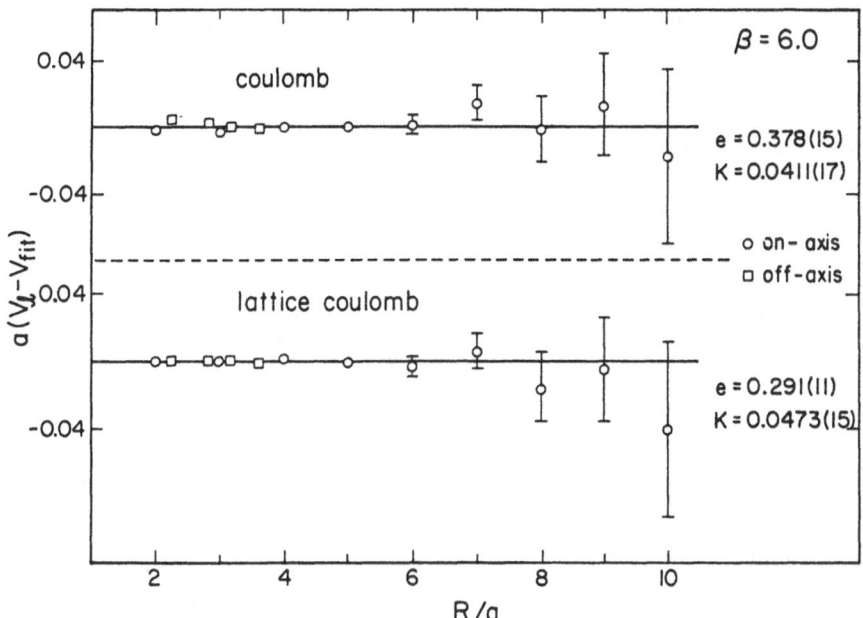

Fig. 1   The lattice potential $V_\ell(R)$ compared with a fit of the form $-eR^{-1} +$
c + KR. The evaluation is for SU(3) colour at β = 6.0 from ref. 9
and includes off-axis potential values. When the R^{-1} term in the
fit is replaced by a lattice one gluon exchange propagator, the
fit is improved as shown in the lower comparison. The string tension value K is also substantially different in the two fits.

Lüscher's fluctuating string analysis while others fit only at larger R but
with e = 0. Some control is provided by fitting to on-axis and off-axis R-
values simultaneously as emphasized above. Some recent determinations are
collected in the Table. A satisfactory feature is that results for the
string tension K obtained using endless (Polyakov) loops[7] are now in
reasonable agreement with these obtained from the static potential. The
finite-temperature corrections to the former[3] and the Coulomb-like corrections to the latter are now both seemingly understood. Recent results for
SU(2) colour are contained in refs. 1, 11, 12.

Scaling. The values of the string tension extracted from lattice
evaluations tend to decrease with increasing β compared to the two-loop
perturbative expectation. This departure from asymptotic scaling is
comparable to that seen in other observables, as well as from MCRG results.
Thus scaling (rather than asymptotic scaling) is tenable for β > 5.7 in

SU(3) and hence continuum results for ratios of physical quantities can be extracted from such lattices.

The SU(3) String Tension K from recent lattice evaluations.

(Values of $\sqrt{K}/\Lambda_L$ are displayed)

β	ON + OFF-AXIS	RECT. LOOPS		POL. LOOPS
5.7	$119 \pm 10^{(9)}$			$112 \pm 2^{(7)}$
5.8	$108 \pm 10^{(6)}$			
6.0	$93 \pm 5^{(9)}$	$94 \pm 9^{(8)}$	$90 \pm 2^{(10)}$	$87 \pm 3^{(7)}$
6.3		$90 \pm 12^{(8)}$	$79 \pm 4^{(10)}$	

Conclusion. A consistent picture emerges of a static potential V(R) which shows confinement and which is in broad agreement with the phenomenological potentials needed for heavy quark spectroscopy. At larger R one can compare the picture with a string model. The expected universal $\pi R^{-1}/12$ term is obscured by the Coulomb component, however, although in 2 + 1 dimensions, there is evidence for the string fluctuation term.[13] Lattice studies also yield information directly about the width of the gluon flux distribution between static quarks.[14]

EXCITED GLUONIC POTENTIALS

Classification. The gluonic field between static colour sources at separation R (taken along the z axis) can be classified by the group of rotations about this axis together with end-to-end exchanges and space inversions. Similar classifications arise for the electronic field in diatomic molecules. The ground state has a cigar-like symmetry with $J_z = 0$ about the separation axis. An example of an excited state is one with $J_z = \pm 1$ about this axis. It is worthwhile to study the potential energy of these $J_z = \pm 1$ states for different values of separation R. Then, for heavy quarks, the Born-Oppenheimer approximation will be valid and the meson bound states will be given by solving the Schröedinger equation in this potential, taking proper account of the intrinisic angular momentum in the gluonic field. Such meson states can achieve additional J^{PC} values to those allowed for $q\bar{q}$ states. Thus one experimental signature of such hybrid mesons will be exotic J^{PC} values.

Lattice evaluation. Using the variational method to extract potentials between static quarks, we have studied the excited gluonic states on a lattice -- where they are classified by the discrete group D_{4h} if the separation is along a lattice axis. We find discrete energy levels at fixed separation R. The lowest excitation is found to be in the E_u representation which corresponds to $J_z = \pm 1$ or equivalently to a transverse colour magnetic field. Results have been presented for SU(2) colour[2,1] and for the more physical case of SU(3) colour.[15,9] Our recent results are illustrated in fig.2. Also shown is the string model expectation of an excitation energy of πR^{-1} compared to the ground state (A_{1g}) gluonic field.

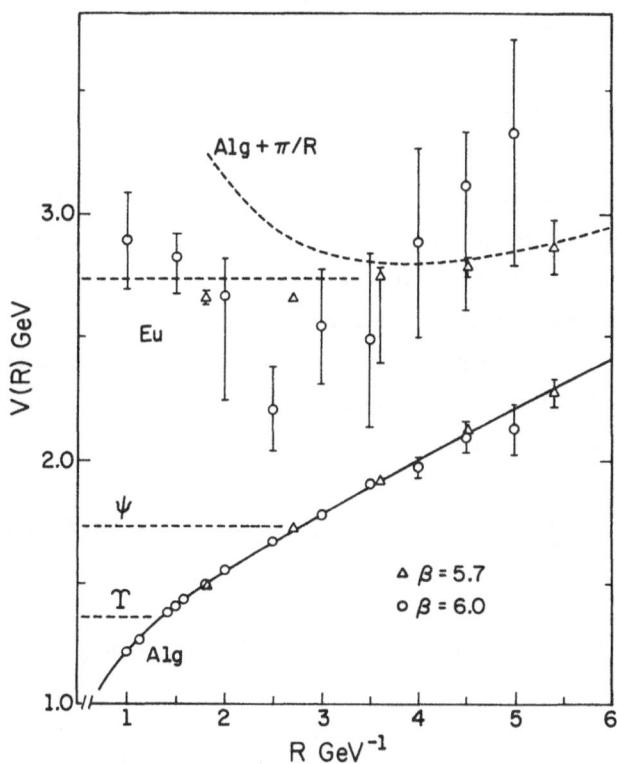

Fig. 2   The lattice potential with the ground state gluon field (A_{1g}) and with the gluon field excited to the E_u representation.[9] The points at ß = 5.7 and 6.0 in SU(3) are combined using a(6.0) = 0.554 a(5.7) = 0.5 GeV^{-1}. As a guide, the lowest $c\bar{c}$ and $b\bar{b}$ states are shown for the A_{1g} potential (the ψ and T respectively) as well as an esti-mate for the lowest bound state in the E_u potential. The differ-ence gives the excitation energy for the lowest hybrid meson - i.e. for $c\bar{c}$ the lowest hybrid state lies around 1 GeV above the ψ mass.

This is seen to be in reasonable agreement at larger R where the string picture is more likely to apply.

Implications for spectroscopy. The potential we find is rather flat in R so that it is easy to estimate the mass of the lightest hybrid meson states. They lie above the $D\bar{D}$ threshold for $c\bar{c}$ systems and above the $B\bar{B}$ threshold for $b\bar{b}$ systems likewise. There is no mechanism suppressing decay of these hybrid mesons by quark antiquark pair production to two conventional mesons ($B\bar{B}$, etc). Thus they should be quite broad resonances. Moreover the rather flat potential implies that radial and orbital excitations in this potential will be very closely spaced in energy. Hence the spectrum will be dense. The flat potential also implies that the wavefunction at the origin will be small so that coupling to e^-e^+ will be small. This dense spectrum of broad resonances will be hard to disentangle experimentally – although there will be exotic J^{PC} values present.

SPIN-DEPENDENT POTENTIALS

Formulation. A heavy quark of mass m acts as if it has a colour magnetic moment of strength gm^{-1}. Thus to order m^{-2}, the spin-spin potential will depend on the correlation of the colour magnetic fields at $\underline{r}_1$ and $\underline{r}_2 \equiv \underline{r}_1 - \underline{R}$. A quark moving in an electric field experiences a magnetic field and this yields an expression for the spin-orbit potential in terms of correlations of one electric and one magnetic field. See Gromes [16] for a careful review of the derivation of these relations:

$$V = V_o(R) + \frac{1}{2m^2 R}\left[(\underline{R} \times \underline{P}_1)\left\{\underline{S}_1\big(V_o'(R) + 2V_1'(R)\big) + \underline{S}_2\, 2V_2'(R)\right\} \right.$$

$$\left. - (\underline{R} \times \underline{P}_2)\left\{\underline{S}_2\big(V_o'(R) + 2V_1'(R)\big) + \underline{S}_1\, 2V_2'(R)\right\} \right]$$

$$+ \frac{\underline{S}_1 \cdot \underline{S}_2}{3m^2} V_4(R) + \frac{1}{m^2}\left(\frac{\underline{R} \cdot \underline{S}_1\, \underline{R} \cdot \underline{S}_2}{R^2} - \frac{\underline{S}_1 \cdot \underline{S}_2}{3} \right) V_3(R)$$

$$\frac{R^k}{R} V_A'(R) = -g^2\, \frac{\epsilon^{ijk}}{2T} \int\!\!\int_{-T/2}^{T/2} (t'-t) < B^i (\underline{r}_A, t)\, E^j(\underline{r}_1, t') > dt\, dt'$$

$$\left(\frac{R^i R^j}{R^2} - \frac{\delta^{ij}}{3}\right) V_3(R) + \frac{\delta^{ij}}{3} V_4(R) = \frac{g^2}{T} \int\!\!\int < B^i (\underline{r}_1, t)\, B^j (\underline{r}_2, t') > dt\, dt'$$

where A = 1 or 2. For orientation, one gluon exchange contributes

$$V_0 = \frac{-e}{R} \qquad V_1' = 0 \qquad V_2' = \frac{e}{R^2} \qquad V_3 = \frac{3e}{R^3} \qquad V_4 = 8\pi e \delta^3(\underline{R})$$

with $e = (4/3) \, g^2/4\pi$ for SU(3) colour; while Lorentz scalar exchange has $V_0' \neq 0$, $V_1' = -V_0'$, all others zero.

Lattice evaluation. The electric and magnetic fields can be evaluated by using a × a loop insertions in the R × T rectangular Wilson loop used to evaluate the static potential $V_0(R)$. Then if U is the path ordered product around such an insertion in the ij plane: $U - U^+ \approx 2iga^2 F_{ij}$ which enables B or E to be evaluated at $\underline{r}_1$ or $\underline{r}_2$. See fig. 3a. The pioneering work of the Liverpool group[17] lead to a determination of the spin-dependent potentials V_3 and V_4 which was consistent with one gluon exchange, see fig. 4. Subsequent work[18] has confirmed this and has shown that the spin-orbit potential V_2' is similarly close to one gluon exchange. A study of the spin-orbit potentials V_1' and V_2' is more delicate on a lattice since the integral over $|t' - t|$ is weighted by $|t' - t|$ which makes it converge rather slowly. Since large $|t' - t|$ values involve large Wilson loops with small expectation values and hence large statistical errors, this gives a big uncertainty. This can be circumvented by noting that the large $|t' - t|$ behaviour of the correlation is controlled by the lowest excited gluonic mode (see previous section) which has been accurately determined.[1,9] With this advance, the spin-orbit potentials were evaluated[19] and the result is shown in fig. 3b. Here a substantial contribution to V_1' is seen and this receives no contribution from one gluon exchange. The R-dependence of V_1' suggests that it is associated with the confining component (linear potential $V_0 \approx KR$). This long range spin-orbit component is consistent with coming from a Lorentz scalar confining potential.

Normalization. Both the components like one gluon exchange in V_2', V_3 and V_4 and the confining mechanism component in V_1' when extracted by lattice evaluation as discussed above have a strength some 35% of the contribution of these components to the lattice static potential $V_0(R)$. The lattice measurement of the E and B fields involves a size scale (the lattice spacing a) in the measuring insertion which is different from the physical size scale of $\sim m^{-1}$ over which these fields are sampled. The appropriate anomalous dimension to take account of this has been calculated perturbatively.[17] This provides the scaling behaviour to be expected for these spin-dependent potentials. To make a quantitative comparison of the normalization, however, would involve extending the two loop lattice calculation of Billoire[20] to the spin-dependent gluon exchange component and this has not been done.

233

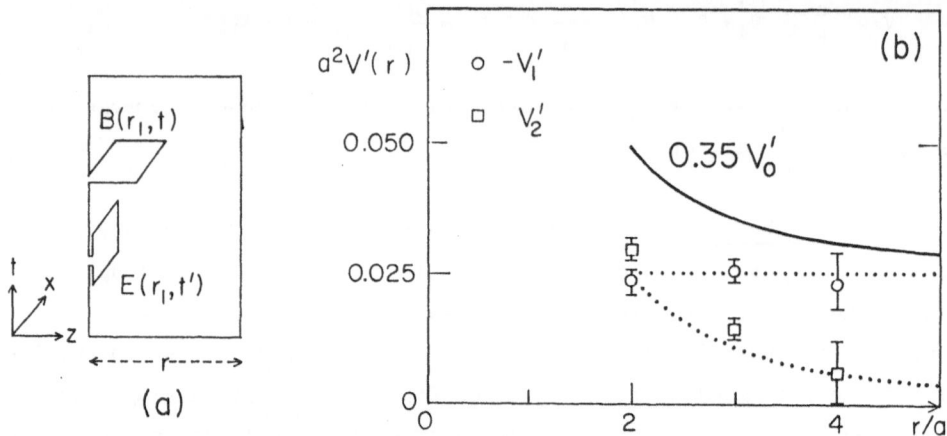

Fig. 3   (a) An illustration of the Wilson loop evaluated to determine the
"same-side" spin-orbit potential V_1' (r). (b) Results for the
spin-orbit potentials V_1' and V_2' from ref. 19. Here V_0 is the
spin-averaged static potential and the relationship of V_2' to the
coloumbic component of V_0 and of V_1' to the confining component
of V_0 is illustrated by the dotted lines.

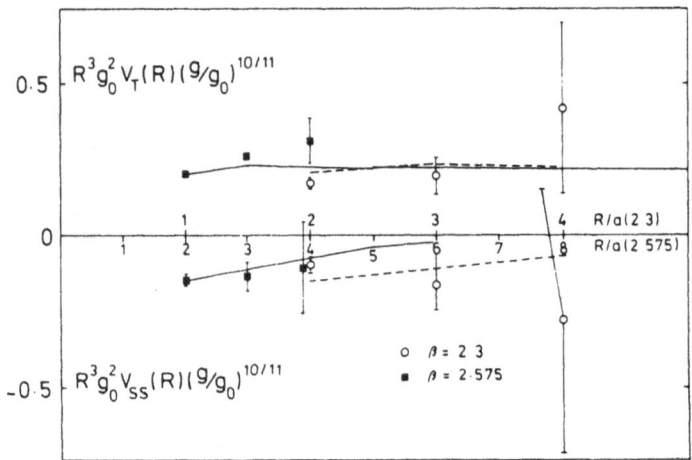

Fig. 4   The spin-spin potentials V_3 and V_4 (where $V_3 \equiv g_0^2 V_T$ and $V_4 \equiv g_0^2 V_{SS}$)
from lattice evaluation.[17] The solid and dashed curves are given
by one lattice gluon exchange. One sees that the tensor potential
V_T is consistent with a scaling behaviour and an R^{-3} dependence as
expected. The scalar potential V_{SS} is also consistent with one
lattice gluon exchange and this will approach a δ-function at R = 0
as β is increased.

    Phenomenology. The data on charmonium χ-states   now enables the
phenomenological interquark potential to be closely determined.[21] This
fixes the strength of the long-range spin-orbit potential and the value

obtained is consistent with Lorentz scalar exchange -- the result given by lattice evaluation of QCD.[19]

ADJOINT POTENTIALS

One adjoint source. A static source in the colour adjoint representation has several features of interest. Such an isolated source will form a colour singlet state by binding a gluonic field around it. The resulting state, the "glue-lump", has much in common with a glueball although it seems to be easier to study on a lattice. It is like a glueball with one gluon "nailed down". Should a massive gluino exist, than it will act like a static colour adjoint source and the glue-lump states will be of relevance to the gluino-gluon spectrum: the glueballino. On a lattice the energy levels of these states can be determined.[22,23,24] The $J^{PC} = 1^{\pm-}$ levels lie lowest with only a small energy separation ($\sim$ 50 MeV) with the 1^{+-} level as the ground state.

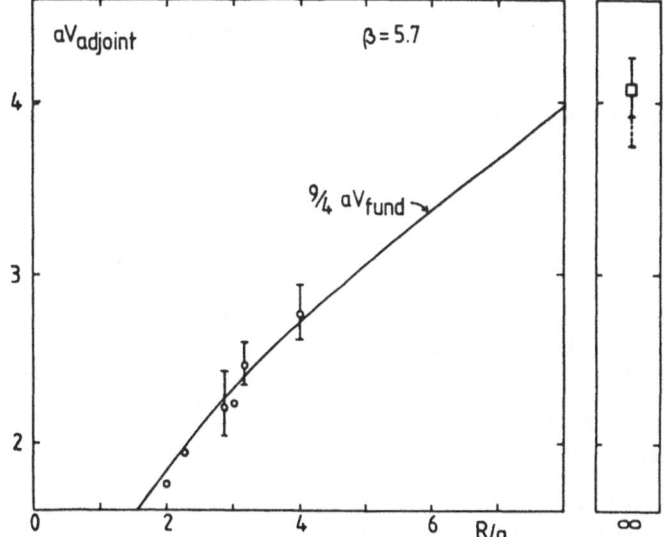

Fig. 5   The potential between static sources in the adjoint representation of SU(3) from ref. (24). Both on-axis and off-axis points are shown. The point at R = ∞ represents twice the energy of an isolated static adjoint source – the "glue-lump". This provides an upper limit for the adjoint potential. Also illustrated is the Casimir ratio 9/4 times the fundamental source potential – this agrees very well in the R-range determined.

<u>Two adjoint sources</u>. The potential energy between two adjoint sources at separation R is expected to show three regimes: (i) at small R one gluon exchange should dominate with a strength related by the adjoint representation Casimir to that in the fundamental source potential, (namely 8/3 for SU(2) colour and 9/4 for SU(3) colour), (ii) a region with a linearly rising component ($K_A R$) corresponding to a gluon flux connecting the two sources with adjoint string tension K_A and (iii) a large R region where the potential energy is constant at a value given by twice the isolated adjoint source energy discussed above. The lattice results in SU(2)[23,1] and SU(3)[24] support this picture. The SU(3) result is shown in fig. 5. There is a region ($R \lesssim 7$ GeV^{-1}) which can be approximated by a Coulomb plus linear form for the potential so that an adjoint string tension can be determined. Recent results give $K_A/K_F = 2.5 \pm 0.25$ for SU(2)[1] and 2.2 ± 0.4 for SU(3).[24] These ratios are consistent with the Casimir ratio as first noted by Ambjorn et al.[25] They presented an explanation based on dimensional reduction - the string-like flux configurations are essentially like a two dimensional field theory and hence the results are like QCD in 2 dimensions where the string-tension is proportional to the Casimir ratio.

SUMMARY

Much progress has been made in a quantitative study of the interactions of static colour sources using lattice QCD. This improves our understanding of these interactions and provides a theoretical basis for phenomenological potential models.

<u>References</u>

1. A. Huntley and C. Michael, Nucl. Phys. (to be published); Liverpool preprint LTH 133.
2. L. A. Griffiths, C. Michael and P. Rakow, Phys. Lett. 129B, (1983) 351.
3. N. A. Campbell, P. Rakow and C. Michael, Phys. Lett. 139B, (1984) 288.
4. C. Lang and C Rebbi, Phys. Lett. 115B, (1982) 137.
5. A. Hasenfratz, P. Hasenfratz, U. Keller, and F. Karsch, Z. Phys C25, (1984) 191.
6. R. Sommer and K. Schilling, Z. Phys. C29, (1985) 95.
7. P. de Forcrand, G. Schierholz, H. Schneider and M. Teper, Phys. Lett. 160B, (1985) 137.
8. K. C. Bowler et. al. Phys Lett. 163B, (1985) 367.
9. N. A. Campbell, A. Huntley and C. Michael, (in preparation).
10. P. de Forcrand and J. Stack (private communication).
11. F. Gutbrod, DESY preprint 85-092, (1985).
12. B. Berg and A. Billoire, DESY preprint 85-082, (1985).
13. J. Ambjorn, P. Olesen and C. Petersen, Nucl. Phys. B244, (1984) 262.
14. J. W. Flower and S. W. Otto, preprint CALT-68-1272, (1985).
15. N. A. Campbell, et al., Phys. Lett. 142B, (1984) 291.
16. D. Gromes, Z. Phys C26, (1984) 401.
17. C. Michael and P. Rakow, Nucl. Phys. B256, (1985) 640.

18. P. de Forcrand and J. Stack, Phys. Rev. Lett. 55, (1985) 1254.
19. C. Michael, University of Illinois at Urbana-Champaign preprint P/85/12/1985 (1985).
20. A. Billoire, Phys Lett. 104B, (1981) 472.
21. J. L. Rosner, Proc. Kyoto Conference 1985 (to be published); University of Chicago preprint EFI 85-63, and references therein.
22. L. A. Griffiths, C. Michael and P. Rakow, Phys. Lett. 150B, (1985) 196.
23. C. Michael, Nucl. Phys. B259, (1985) 58.
24. N. A. Campbell, I. H. Jorysz and C. Michael, Phys. Lett. B (to be published); Liverpool preprint LTH134.
25. J. Ambjorn, P. Olesen and C. Petersen, Nucl. Phys. B240 [FS12] (1984) 189.

QUENCHED HADRON MASSES USING A 16^4 LATTICE

R.D. Kenway

Physics Department
University of Edinburgh
Edinburgh, Scotland

INTRODUCTION

In this paper I report preliminary results of hadron mass calculations in quenched QCD performed by K.C. Bowler, C.B. Chalmers, G.S. Pawley, D. Roweth, D.J. Wallace and myself using the Edinburgh DAPs. The memory limitation on these machines has restricted us to 16^4 spacetime lattices for the quark propagator calculations using the even/odd partitioned conjugate gradient algorithm [1], having transformed the gauge configurations to temporal gauge. Anticipating that the time extent may be insufficient to observe asymptotic decay of the hadron propagators for $\beta \gtrsim 5.7$, we have imposed Dirichlet boundary conditions on the quark propagators in the time direction (antiperiodic in space) with the source on timeslice 5. This permits their subsequent extension to a $16^3 \times 24$ lattice via the distant source method [2] and periodic extension of the gauge field configurations. The data I present are from 80 propagators at two β-values and five quark masses, using Kogut-Susskind fermions. These took approximately 600 hours of DAP time to produce, 75% of which was wasted on I/O. We are exploring more efficient ways of partitioning the fermion matrix so as to reduce this stretch factor [3].

RESULTS AT $\beta = 5.7$

We have analysed 8 configurations separated by 448 pseudo-heatbath sweeps following 560 sweeps from the last configuration at $\beta = 6.0$ used in [4]. On each we have calculated propagators at quark masses $m_q = 0.01$, 0.04, 0.09, 0.16 and 0.5. These required, respectively, 700 + 500, 300, 150, 120 and 40 conjugate gradient iterations using 32-bit arithmetic for convergence to a squared residual of between 10^{-8} for the lightest and 10^{-16} for the heaviest quark masses. These values were arrived at by requiring agreement between all hadron propators on one configuration and a random gauge transformation of it to at least 3 significant figures. At the lightest quark mass it is essential to restart the conjugate gradient after about 700 iterations to prevent the accumulation of rounding errors.

We have constructed propagators for the four local meson operators used in [5]. Both parity states propagate in each of these channels, except the pseudoscalar, necessitating fits to the sum of an exponential

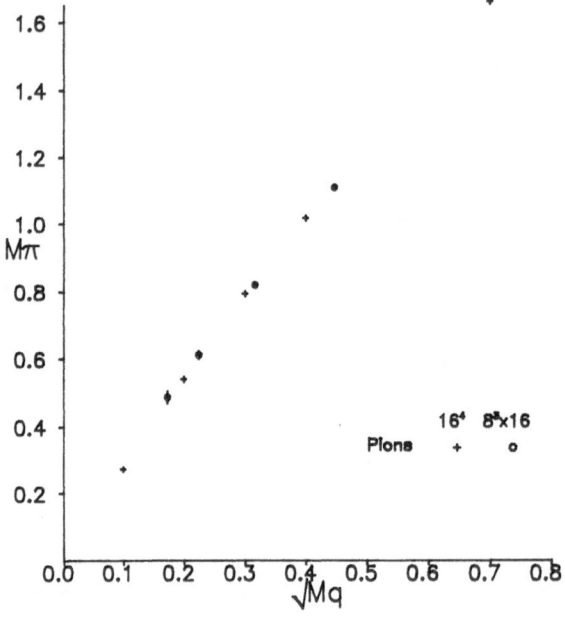

Fig. 1. The pion mass in lattice units at β = 5.7.

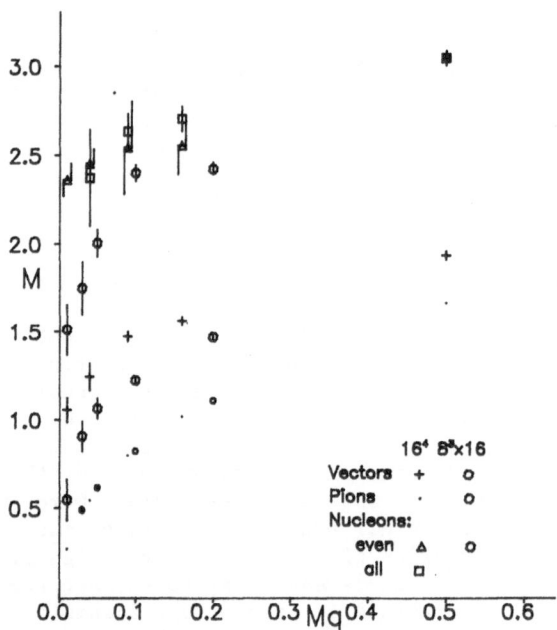

Fig. 2. Hadron masses in lattice units at β = 5.7.

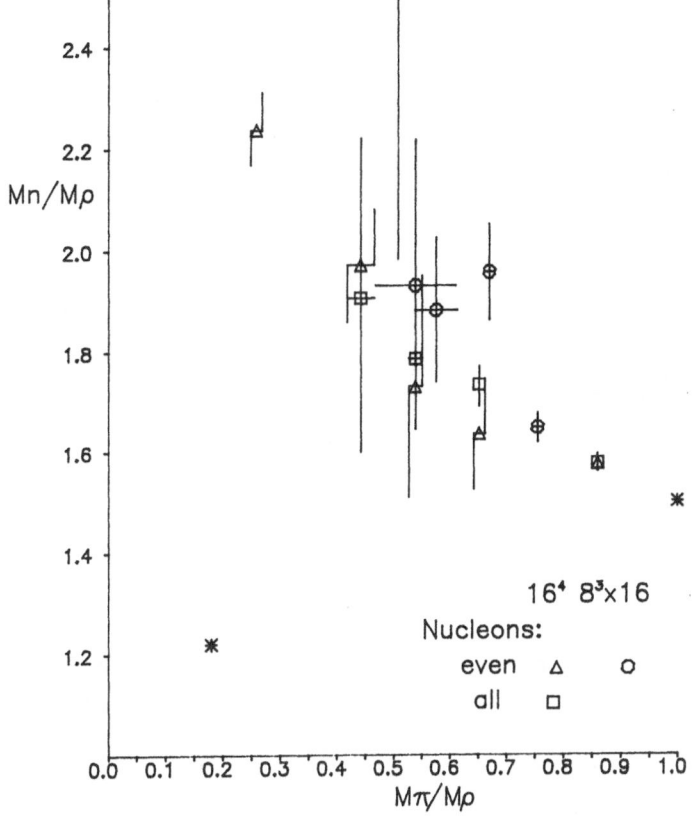

Fig. 3. Mass ratios at β = 5.7; the starred points are the experimental and infinitely-heavy quark values.

and an oscillating exponential. Our data is not of sufficiently high quality to support more than a 4-parameter fit, so we attempt to remove excitations at short times by successively dropping data points near the source until the resulting masses stabilize. For the pion we are able to include a radial excitation in the fit. We construct two nucleon time-slice propagators: one involves the sum over all spatial sites and the other the sum over a spatial lattice of spacing 2 which includes the source. For free fermions with antiperiodic boundary conditions in space, these nucleon propagators differ, contrary to [6]. However, in the inter-acting theory this difference goes away in the average over gauge config-urations and we may regard any discrepancy between the two mass estimates as a measure of the error in our calculation.

At β = 5.7 we can compare our results with those on an $8^3 \times 16$ lattice. Fig. 1 shows the pion mass plotted against $\sqrt{m_q}$. Any spatial finite size effect in the data itself is small, but the   linear extrapolation of the smaller lattice data is consistent with a larger non-zero intercept at $m_q = 0$ [7]. The full data for pion, rho and $\frac{1}{2}^+$ nucleon is shown in fig.2 together with our $8^3 \times 16$ results. There appears to be a spatial finite size effect tending to decrease the rho and nucleon masses at small quark masses, although the errors are large. Note that with only 8 propagators on the 16^4 lattice our error estimation is crude (the spread in the measurements from two consecutive blocks of 4) and the actual error may be larger than that shown. Where both yield successful fits, the two nucleon propagators give consistent mass estimates. In fig. 3 we plot the

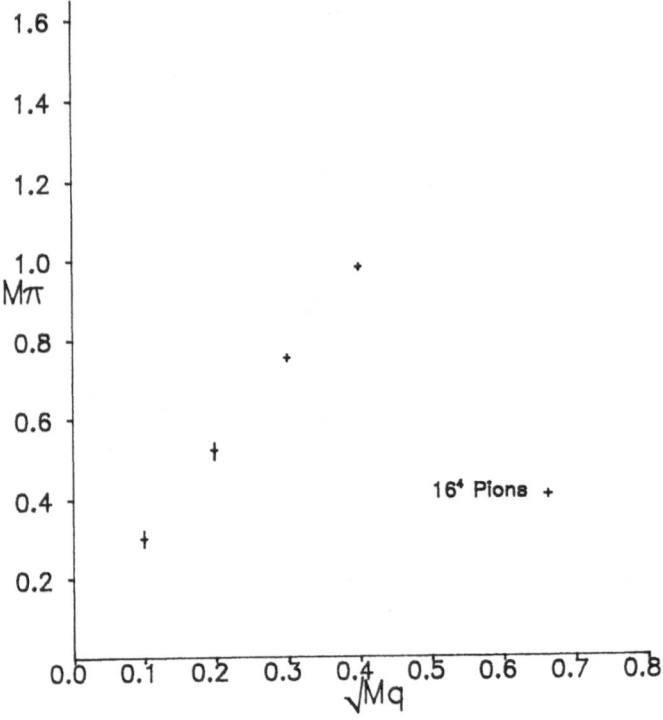

Fig. 4. The pion mass in lattice units at β = 6.0.

nucleon/rho mass ratio versus the pion/rho mass ratio [8]. The spatial finite size effects approximately cancel in the ratio and both spatial lattice sizes support the conclusion that m_n/m_ρ increases as m_π/m_ρ decreases and that there is no crossover to the light quark regime necessary for agreement with the experimental values.

RESULTS AT β = 6.0

Here we have analysed 8 of the 16^4 configurations used in [4], separated by 896 pseudo-heatbath sweeps. The same quark masses and convergence criteria as at β = 5.7 were used, although there is faster convergence of conjugate gradient.

The pion mass is plotted against $\sqrt{m_q}$ in fig. 4. We see a clear signal for an excited state in the pseudoscalar channel and this has been included in our fits. Our pion mass deviates from the estimates of [5] as the quark mass decreases, being almost 20% higher at m_q = 0.01. A linear fit through our data supports a larger non-zero intercept at m_q = 0 than on the 16^4 lattice at β = 5.7 [7]. We also see a signal for the pion in the scalar channel with a mass in rough agreement with fig. 4 for all but the lightest quark mass, suggesting flavour symmetry restoration at the higher quark masses.

The other hadron masses are shown in fig. 5. We do not attach much

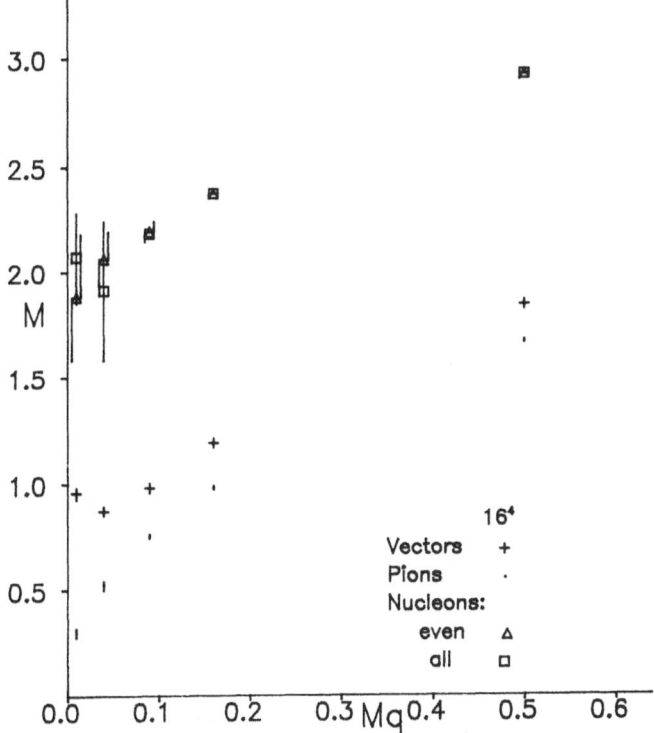

Fig. 5. Hadron masses in lattice units at β = 6.0.

significance to the up-turn in the rho mass because of our low statistics. There is a signal for the second rho in the pseudovector channel, but it is systematically higher in mass than the first. Both our rhos and nucleons are higher in mass than those of [5]. However, there is no evidence of asymptotic decay in our nucleon propagators and so it is probable that their masses are overestimated.

Hence, plotting the mass ratios as in fig. 6 is almost certainly premature. We anticipate that the $16^3 \times 24$ calculations currently in progress will result in a drop in the nucleon mass. We may therefore take encouragement from the flattening out of the mass ratio plot at $m_n/m_\rho \sim 2$ for small m_π/m_ρ and hope that the larger lattice will expose the crossover from the heavy to light quark regimes. However, it seems unlikely that m_n/m_ρ will drop sufficiently to agree with experiment. So, to the extent that β = 6.0 is in the scaling region, we may be on the verge of showing that the quenched approximation gives the wrong hadron masses.

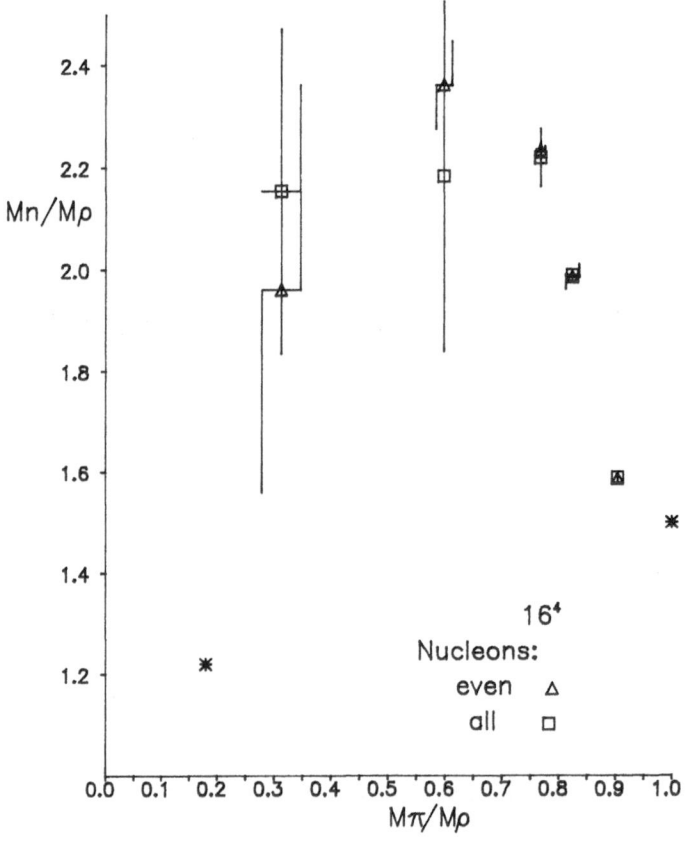

Fig. 6. As fig. 3 at β = 6.0.

REFERENCES

[1] K.C. Bowler, D.L. Chalmers, A. Kenway, R.D. Kenway, G.S. Pawley and
    D.J. Wallace, Nucl. Phys. B240 [FS12] (1984) 213
[2] R.D. Kenway, Phys. Lett. 153B (1985) 327
[3] C.B. Chalmers, R.D. Kenway and D. Roweth, Algorithms for inverting
    the fermion matrix on large lattices, in preparation
[4] K.C. Bowler, A. Hasenfratz, P. Hasenfratz, U. Heller, F. Karsch,
    R.D. Kenway, H. Meyer-Ortmanns, I. Montvay, G.S. Pawley and
    D.J. Wallace, Nucl. Phys. B257 [FS14] (1985) 155
[5] D. Barkai, K.J.M. Moriarty and C. Rebbi, Phys. Lett. 156B (1985) 385
[6] A. Morel and J.P. Rodrigues, Nucl. Phys. B247 (1984) 44
[7] A. Billoire, R. Lacaze, E. Marinari and A. Morel, The Goldstone pion
    in SU(2) lattice QCD at low quark mass: scaling and finite size
    effects, Saclay preprint PhT 85-180, November 1985
[8] K.C. Bowler, D.L. Chalmers, A. Kenway, R.D. Kenway, G.S. Pawley and
    D.J. Wallace, Phys. Lett. 162B (1985) 354

A CORNUCOPIA OF LATTICES

Gregory Kilcup

Lyman Laboratory of Physics, Harvard Univesity
Cambridge, MA 02138

Viewed from a sufficiently broad perspective, the present state of computational QCD is rather pleasing. After a few years and a few teraflops of numerical experiments we are resolving a spectrum which mimics reasonably well the spectrum obtained by real experiments at considerably more expense and effort over a few decades. Even for those who find quark chemistry for its own sake uninteresting, this should be exciting news. We can now have some measure of confidence when we try to recoup the investment in the lattice, and extract information which is not yet available in the wallet cards. Just as the calculation $(g_e - 2)$ is more than a confirmation of QED, the computation of various matrix elements of 3 and 4 quark operators will provide a unique window to short distance physics. Soon we will be able to see, for example, whether or not the standard model gives an adequate account of CP violation, and whether certain extensions of the standard model can be made consistent with low energy phenomenology. Of course, to have complete confidence we will probably have to spend a few more teraflops making contact with the continuum (possibly taking a short cut via improved actions), even before introducing dynamical fermions on realistic lattices.

In this talk I would like to give a progress report on a lattice project at Los Alamos which is working in these directions. The people involved are Rajan Gupta and Gerry Guralnik of Los Alamos, Apoorva Patel of San Diego, and Steve Sharpe and me at Harvard, with technical advice provided by Tony Warnock of Cray Research. The projects are basically of two sorts. There is sneaking up on the continuum:

- Determination of MCRG flows under the $\sqrt{3}$ blocking transformation [1]
- β-function along Wilson and improved action lines

And there are the things to do once we reach the continuum:

- Hadron spectrum
- Coupling constants
- Matrix elements (chief among these: $\Delta I = \frac{1}{2}$ rule)

To attack these problems we are fortunate enough to have access to CRAY's at Los Alamos National Laboratory and at the Magnetic Fusion Energy Computer Center at Livermore. The (quenched) lattices we are using comprise:

- $18^3 \times 42$ on the Wilson axis at $\beta = 6.2$ [configurations and propagators generated on LANL CRAY-XMP/48's]

- $6^3 \times 14$ obtained from the above by two $\sqrt{3}$ blockings (equivalent Wilson axis $\beta_{eff} \approx 5.48$) [propagators on MFE CRAY-XMP/22]

- $8^3 \times 16$ ($\beta_{eff} \approx 5.57$) obtained in the same way from Phillipe de Forcrand's $24^3 \times 48$'s at $\beta = 6.3$ [propagators on MFE CRAY-XMP/22]

- $8^3 \times 16$ on the Wilson axis at $\beta = 5.6$ [configurations and propagators on MFE CRAY-XMP/22]

- $12^3 \times 30$ with a four parameter improved action as determined by MCRG studies ($\beta_{eff} \approx 5.84$) [configurations and propagators on MFE CRAY-2]

Quantum Numbers

Since our ultimate goal is to determine matrix elements for which chiral symmetry is very relevant, we eschew Wilson fermions, and choose the formalism whose chiral properties we find easier to understand, i.e. staggered fermions. The price one pays, of course, consists in having to deal with a more complicated spin/flavor structure. The course of disentangling spin and flavor has led us through some rather fun group theory, which I'd like to spend a few minutes sketching. See ref. [2] for details.

In the Wilson formalism flavor symmetry is unbroken, while the Poincaré group is reduced to the semidirect product $SW_4 \boxtimes (Z)^4$, where SW_4 ("Spezial Würfel") denotes the hypercubic group, whose representations are well known [3], and $(Z)^4$ are the discrete translations. It is then trivial to decompose continuum states into their lattice multiplets. In the staggered formalism, on the other hand, the relevant symmetry group is the semidirect product of translations with a certain 12288 element finite group which mixes up spacetime and $SU(4)$ flavor symmetries in a nontrivial way. To classify the states one proceeds as in Wigner's little group method for the Poincaré group—but on the lattice a massive particle can have many inequivalent sorts of momenta. For a particle at rest, for example, the little group turns out to be isomorphic to $(SW_4 \otimes D_4)/Z_2$ where D_4 is the dihedral group of the square.[†] The Z_2 factor generated by $-1 \otimes -1$ means that not all representations of $SW_4 \otimes D_4$ are allowed—the two factors must agree on how to represent the element -1. We then find three baryonic representations, consisting of the the three SW_4 representations $(\frac{1}{2}, \frac{1}{2})$, $\overline{(\frac{1}{2}, \frac{1}{2})}$, $(\frac{3}{2}, \frac{1}{2})$ which send $-1 \mapsto -I$, times the two dimensional representation of D_4. The mesons sit inside the other $10 \times 4 = 40$ allowed representations. To

---

[†] I am ignoring parity and charge conjugation here. Including them enlarges D_4 to $Z_2 \otimes \Gamma_4$ (Γ_4 is the 32 element group of Euclidean gamma matrices), but adds nothing essential.

decompose the continuum representations one simply follows the chain of groups

$$SU(2)_S \otimes SU(4)_F \longrightarrow SU(2)_S \otimes SO(4)$$
$$= SU(2) \otimes [SU(2) \otimes SU(2)] = [SU(2) \otimes SU(2)] \otimes SU(2)$$
$$\longrightarrow SW_4 \otimes D_4.$$

For example, the 15-plet of vector mesons decomposes as

$$(1, 15) \longrightarrow (1, 1, 1) \oplus (1, 1, 0) \oplus (1, 0, 1)$$
$$\longrightarrow 6_\mu \oplus \boxminus_\mu \oplus (1, 0)_i$$

where μ runs over the 4 one dimensional reps of D_4, and i runs over only the 3 nontrivial ones. This agrees with the work of Golterman and Smit [4], who obtained the $\vec{p} = 0$ representations previously in a slightly more cumbersome fashion.

In a similar way one can go on to consider $\vec{p} \neq 0$ representations. For example, the symmetries of the $\vec{p} = (0, 0, p_z)$ sector form a certain 256 element group, which has 56 nonfaithful mesonic represenations of dimensions one and two (the transverse and longitudinal parts of vector particles need not be degenerate), and 2 eight dimensional baryonic representations. Looking at these states can be a useful in several ways: to test the restoration of rotational symmetry, to determine the momentum dependence of certain matrix elements, and, as a practical matter, to improve the signal for determining particle masses.

NLT operators

The practical complication of staggered fermions shows itself when one tries to find operators to project out particular quantum numbers: it is in principle impossible to construct operators of definite parity which are local (or even quasi-local) in time. The appearance of particles of both parities in the correlators then makes the fitting a trickier business. In practice we have found a way to help alleviate this headache. If one considers correlators of local operators with operators split by one unit in time (e.g. $\bar{\chi}(n)U_t(n)\chi(n + \hat{t})$), one has the same set of intermediate states as in the local case, but we find in practice that only one of the opposite parity partners makes a measurable contribution. One can understand this by appealing to the continuum limit. In the 16 component formalism the introduction of a timelike link corresponds to the addition of an extra γ_0. Since ρ's and π's of a given flavor can be created by two gamma matrix structures differing by a γ_0, they can contribute to the local-nonlocal correlator, while A_1's and B's cannot. This trick for enhancing the signal allows us to extract the ρ mass from lattices where the local-local data looks none to good. And best of all, there are no extra propagator costs.

Ward Identities

Another tool we have found useful is the technique of deriving lattice Ward identities via a hopping parameter expansion. The prototypical example of such an identity is the one everyone uses as a consistency check on the propagator:

$$m_q \sum_n |G(n; 0)|^2 = tr[G(0; 0)] \equiv \langle \bar{\chi}\chi \rangle$$

the saturation of which leads to the lattice version of the Gell-Mann–Oakes–Renner relation. Equally trivially one can consider taking the derivative of this identity to find a relation for the slope of the condensate which can be concisely written

$$Int_1(m_q) \equiv (1 - m_q \frac{\partial}{\partial m_q}) \langle \overline{\chi}\chi(m_q) \rangle = 2m_q \sum_{n \text{ even}} |G(n;0)|^2.$$

So again one can obtain useful information with no extra work. In a similar fashion one can go on to derive lattice analogs of the usual current algebra identities. For example, one can relate $\langle \pi | \overline{\chi}\chi | \pi \rangle$ to $\langle 0 | O_\pi | \pi \rangle$, which tells us that measuring the matrix element is merely roundabout way of determining the same old f_π. More usefully, one can derive relations for weak interaction matrix elements which show which lattice operators have the correct chiral behavior[2].

A Small ($6^3 \times 14$) Playground [5]

The ensemble for which we have the best statistics consists of the coarsest lattices. We have 40 configurations distributed according to some random improved gauge action obtained by two applications of the $\sqrt{3}$ blocking transformation to $18^3 \times 42$ lattices generated at $\beta = 6.2$. If we could use some improved fermion action as well, we would then be probing indirectly the physics of $\beta = 6.3$ at some fraction of the computer cost for doing it directly. This has been considered by our conference hosts[6] for Wilson fermions—we hope they will soon shed some light on the staggered case as well. In the meantime we have computed propagators for the standard *unimproved* staggered fermion action. We are therefore investigating some region of coupling constant space which is *a priori* no closer and no further away from continuum physics than the corresponding point on the Wilson axis, but which may be interesting in its own right.

We computed quark propagators with antiperiodic boundary conditions for $m_q \in \{.25, .1, .06, .045, .03, .025\}$, the lightest mass being the limit beyond which the pion correlation length becomes too large for the small lattice. Determining f_π either from the π correlator or from $\langle \overline{\chi}\chi \rangle$ and m_π^2/m_q we find the lattice scale: $a^{-1} \sim .47 GeV$. Alternatively, from m_ρ we get the estimate $a^{-1} \sim .52 GeV$, consistent within errors. The nucleon and $\Lambda(\frac{1}{2}^-)$ masses are less certain, but are also consistent with this scale. Our lightest pion is then $m_\pi \sim 220 MeV$. These features of the spectrum are encouraging, but not to be taken too seriously, since we are definitely far from the continuum—the non-Goldstone pion, for example, shows no signs of being degenerate with the true pion. We merely take this unreasonably good spectrum as an indication that the small lattices make a reasonable testing ground for the numerical experiments we will make on bigger lattices.

For $m_q \in \{.25, .1, .06, .045, .03\}$ we also computed quark propagators in the presence of a pion source (for two values of the source time), enabling us to look at various three point functions. The cleanest of these is the $\epsilon\pi\pi$ correlator. Fitting the data to a form which includes the allowed physical processes for the different time orderings of the operators, we can determine the $\epsilon\pi\pi$ coupling constant, m_π, m_ϵ and the matrix element $\langle \pi | \overline{\chi}\chi | \pi \rangle$. This reveals a remarkable feature: while for the heavier quark masses m_ϵ agrees with its two point determination, for the lighter masses it comes out consistently higher. The π mass, on the other hand, is always in precise agreement with its the two point value.

Our interpretation is simple: in the two point functions one is not necessarily measuring the true m_ϵ. Instead the ϵ can decay into two π's, which dominate the channel if they are light enough. Indeed, when $m_\epsilon^{2pt} \neq m_\epsilon^{3pt}$, we find $m_\epsilon^{2pt} \approx 2m_\pi$. One might be inclined to disbelieve this explanation on the grounds that in the quenched approximation it demands that a flavor nonsinglet pion annihilate into gluons. One must keep in mind, though, that the symmetry suppressing this process is a lattice shift symmetry, which is only realized in an infinite ensemble of gauge configurations. Further, we have reason to expect large breaking of shift symmetries in our ensemble. If we measure Wilson loops on the blocked lattices we find the larger loops ($2 \times 5, 3 \times 3$ and 3×4) correspond to a Wilson axis β of about 5.48, which is consistent with previous MCRG results. The smallest loops, on the other hand, correspond to $\beta \sim 5.2$. Thus we have much more short distance fluctuation than we would have for a lattice of the same size on the Wilson axis. To say something more about the effect of such fluctuations we have also examined ...

A Matching Pair ($8^3 \times 16$)

Our second ensemble of blocked configurations was obtained from Phillipe de Forcrand's $24^3 \times 48$'s ($\beta = 6.3$), twice blocked down to $8^3 \times 16$. Again matching large Wilson loops gives an equivalent $\beta_{eff} \sim 5.57$, while smaller loops show more fluctuation. Computing propagators for $m_q \in \{.1, .05, .03, .02, .015\}$, we again find that for the lighter masses $m_\epsilon \approx 2m_\pi$. To make a direct comparison, we also generated 29 configurations with a standard Wilson action at $\beta = 5.6$, computing propagators for $m_q \in \{.1, .05, .02\}$. We find an amusing agreement for m_π^2/m_q between the two lattices, despite the difference in regularization schemes. We find an encouraging agreement between the two for the ρ mass—we would have been upset not to. But in the Wilson axis ensemble we find very clearly that $m_\epsilon > 2m_\pi$ for small m_q. We take this as support for our claim that short distance fluctuations are responsible for the two pion contamination in the ϵ channel.

A Definite Improvement ($12^3 \times 30$)

As part of the MCRG program we are also generating gauge configurations with a certain four parameter gauge action which includes the plaquette in the fundamental, sextet and octet representations, as well as the 1×2 planar loop in the fundamental. The ratios of the coefficients were determined by a two lattice Wilson loop matching method [7]. Computing propagators for $m_q \in \{.05, .03, .02, .01\}$, we find preliminary results from 34 configurations on a $12^3 \times 30$ lattice give a spectrum almost unreasonably consistent. From a multihit average of Wilson loops we extract a string tension corresponding to $a^{-1} \sim 1.2 GeV$. Likewise from f_π we find $a^{-1} \sim 1.2 GeV$. From the ρ mass we get $a^{-1} \sim 1.12 GeV$. From the nucleon and its negative parity partner we get the much less reliable estimates $a^{-1} \sim 1.04 GeV$ and $a^{-1} \sim 1.18 GeV$ respectively. We see no sign of flavor restoration—neither the two flavors of local pion nor the two local ρ are degenerate. But I stress that these results are preliminary. In particular, thanks to a compiler bug on the new CRAY-2 at MFE we have not yet been able to obtain the local-nonlocal correlators which greatly improve the mass determinations. We are currently continuing to explore the same action at weaker coupling.

A Realistic Chunk of Spacetime? $(18^3 \times 42)$

The final set of lattices in our horn of plenty is an ensemble of $18^3 \times 42$'s on the Wilson axis at $\beta = 6.2$, which, if we believe the MCRG results, may be close to the asymptotic scaling region. At this point we have analyzed only 13 lattices, with $m_q \in \{.1, .03, .01\}$. We are slowly accumulating data for $m_q = .005$ as well. Although we can give reliable results only for the masses of the local ρ's and π's we already find a very heartening result: in contrast to the coarser lattices, here the two ρ's are degenerate, *and so are the two π's*. Taking the scale from m_ρ, we get $a^{-1} \sim 1.8 GeV$, implying that for $m_q = .01$, $m_\pi \sim 290 MeV$. The fact that the π's remain degenerate even for light quark masses we take as strong evidence that flavor symmetry is being restored at $\beta = 6.2$. This greatly simplifies our analysis of weak interaction matrix elements. We have looked at mesons at spatial momenta $\vec{p} \in \{(0,0,0), (0,0,1), (0,1,1), (1,1,1), (0,0,2)\}$, and confirm the naive (continuum) dispersion relation, the $|\vec{p}|^4$ terms being too small to emerge from the noise. We are continuing to accumulate statistics (at a cost of ~ 5 CRAY-XMP hours per consort of propagators), and expect to have an ensemble of 30 lattices by Christmas. If we continue to observe flavor restoration for $m_q = .005$ (corresponding to $m_\pi \sim 200 MeV$), we can feel rather confident that the other numbers we extract may have something to do with the real world. Indeed they may even have something to do with short distance physics.

References

[1] R. Cordery, R. Gupta and M. Novotny, Phys. Lett 128B (1983) 425 ;
    A. Patel and R. Gupta, Nucl. Phys. B251 [FS13] (1985) 789.

[2] G. Kilcup and S. R. Sharpe, "A Toolkit for Staggered Fermions," Harvard preprint HUTP/A086 (1985).

[3] M. Baake, B. Gemünden and R. Oedingen, J. Math. Phys. 23 (1982) 91; J. Math. Phys. 24 (1983) 1021 ;
    J. Mandula, G. Zweig and J. Govaerts, Nucl. Phys. B228 (1983) 91.

[4] M. Golterman and J. Smit, Nucl. Phys. B225 (1985) 328 ;
    M. Golterman, "Staggered Mesons," Amsterdam preprint (1985).

[5] G. Kilcup et al., "ϵ Beyond the Naive Mass Spectrum," Harvard preprint HUTP/A069 (1985).

[6] K.-H. Mütter and K. Schilling, Nucl. Phys. B250 [FS10] (1984) 275.

[7] R. Gupta et al., Phys. Lett. 161B (1985) 352.

THE H PARTICLE ON THE LATTICE[*]

Paul B. Mackenzie

Institute for Advanced Study
Princeton, NJ 08540, USA

H. B. Thacker

Fermi National Accelerator Laboratory
Batavia, IL 60150, USA

ABSTRACT

Jaffe's doubly strange spin zero dibaryon (the H particle) is one of the few hadrons made of light quarks which could conceivably exist and yet not have been observed up to the present, making it an interesting test of the standard methods of lattice gauge theory. We have investigated this state and found that the mass of the dibaryon is above the two Λ threshold, making it unstable to strong decay.

About ten years ago, Jaffe used the bag model to consider to consider various exotic multiquark states to determine which had some likelihood of being stable. The most probable candidate he found was the spin zero doubly strange dibaryon, which he named the H particle,[1] made of six light quarks. Using bag model parameters which fit the low lying hadrons quite well, he predicted the mass of this state to lie about 80 MeV below the mass of two Λ's, its lightest strong decay channel, making it stable against strong decay. At first glance, it is not clear how such a light state (mass of a little over 2 GeV) could have escaped detection up to now. However, the only serious search for the particle, a 1977 Princeton-Brookhaven experiment[2] looking at the process PP→H K⁺K⁺, was not very sensitive, especially in the mass range $M_H > 2.0$ GeV. There are two nuclear emulsion events which seem to show nuclei containing a pair of Λ's decaying as two sequential normal Λ decays,[3] arguing against the possibility of the reaction $\Lambda\Lambda \to$ H, which one would expect if $M_H < 2M_\Lambda$. However, the interpretation of these events is controversial. New experiments have been proposed which promise to be much more definitive.[4]

The H particle has recently come back into the news as a possible source for anomalous cosmic ray muon events recently reported from the X-ray source Cygnus

---

[*] Talk given by Paul Mackenzie

X-3.[5] If these events are real, they must have been caused by neutral particles (in order that galactic magnetic fields not have altered their direction) and yet they could not have been caused by any known neutral particle. Neutrons would have decayed before traversing the distance from Cygnus X-3, the photon spectrum is thought to be understood and is not large enough to account for the observed flux, and the azimuthal angle distribution of neutrino events should be different from the one that is observed. If the H were light enough, it would be a possible source for these events. (Its mass would actually have to be rather close to the PP threshold.[6]) The scenario for the doubly strange H to be produced at Cygnus X-3 requires that the star be composed entirely of strange quark matter, right up to the surface. It has previously been suggested that the existence or nonexistence of the H would shed light on the likelihood for the existence of stable strange matter.[7]

In addition to the obvious physics importance of such a particle, its possible existence is of interest to those of us who are in the process of developing reliable algorithms for lattice QCD calculations. Understanding the reliability of these calculations requires complicated and detailed error analysis which is not easy for specialists to agree on and which is difficult for nonspecialists to evaluate. It would be a striking test of the emerging calculational methods to predict accurately (to within stated error bars) a physical quantity before its experimental determination. Most of the things which can be easily calculated by crude Monte Carlo methods (the low lying hadron masses, for example) have been well known experimentally for 20 or 30 years. The mass of the H particle (if it exists) is almost unique in being almost as easy to calculate as the proton mass, and yet not having been measured so far.

The H has been considered in a wide variety of phenomenological models and has always been found to be either stable or almost so. In addition to Jaffe's original estimate of 80 MeV, more detailed bag model calculations have given binding energies of 230 [8] to -10 MeV.[9] Dibaryons have recently attracted attention in chiral models, where they have an interesting interpretation as solitons associated with an SO(3) subgroup of flavor SU(3). Several estimates of the H mass have been done in chiral models, which are somewhat less successful than quark models in parameterizing the known hadron masses. These range from 1.03 GeV to 2.10 GeV [10], compared with the ΛΛ threshold at 2.23 GeV.

In bag and quark models, spin splittings are given by the one gluon exchange hyperfine operator. In the flavor SU(3) symmetry limit, we have for the hyperfine splitting

$$\Delta E \propto - \sum_{\substack{i<j \\ quark\ pairs}} (\lambda^i \sigma^i) \cdot (\lambda^j \sigma^j), \tag{1}$$

where the λ^i and σ^i are the Gell-Mann and Pauli matrices. Jaffe showed how to compute this quantity without considering the details of the six quark wave functions in terms of the casimir operator of the combined colorspin SU(6) group, and obtained for the H particle a minimum eigenvalue of -24 from the [490] representation of color-spin SU(6). He obtained a binding energy estimate of 80 MeV in the bag model. The stability of the H in quark models is simplest to see in a naive quark model calculation with hyperfine interactions. The hadron masses are fit to the form

$$M = \sum_{\substack{i \\ quarks}} m_i - \sum_{\substack{i<j \\ quark\ pairs}} \frac{(\lambda^i \sigma^i) \cdot (\lambda^j \sigma^j)}{m_i m_j} W, \tag{2}$$

where W parameterizes the wave function at the origin and the QCD coupling constant, and m_i are the masses of the quarks. Using the color-spin-flavor wave function of the H, which we derive later, the factor $-24/m^2$ from the eigenvalue of eqn. (1)

becomes $-5/m_u^2 - 22/m_u m_s + 3/m_s^2$ when flavor SU(3) symmetry breaking is introduced. Using the parameters of reference 11, which fit the baryon spectrum quite well, and making the unjustified assumption that W is the same for baryons and dibaryons, we obtain a hyperfine splitting for the H of -348 MeV for the second term in eqn. (2). This yields a mass of 2.18 GeV and a binding energy of 50 MeV.

The foregoing has ignored mixing between colorspin SU(6) representations induced by the flavor SU(3) symmetry breaking. Very recently, Rosner[12] has presented an improved quark model calculation which includes this refinement. He finds a small admixture of the [189] representation of SU(6) giving a binding energy for the H of 53 MeV.

The results of our lattice calculation of the mass of the H were first reported in reference 13. To examine the H in lattice QCD, we analyzed the quark propagator data accumulated for the calculation of hadronic coupling constants performed with Gottlieb and Weingarten.[14] For a detailed description of the definitions, methods and results in that calculation, as well as a list of standard references for lattice gauge theory, see reference 14. We use a coupling constant $\beta \equiv 6/g^2 = 5.7$, which corresponds to a lattice spacing of roughly .9 inverse GeV, depending on the quantity used to set the mass scale. We worked on a 6^2x12x18 lattice, with 18 taken as the Euclidean time direction. The transverse size of 6 is roughly the size of a single hadron. A total of 20 gauge configurations was analyzed. Each configuration was separated by 500 Metropolis sweeps, after equilibrating for 1000 sweeps. Quark propagators were calculated for hopping parameters K=.325, .34, and .355, which correspond to pions of mass around 900, 750, and 600 MeV respectively. The hadron masses obtained for these quark masses must be extrapolated to the correct physical limit of M_π =138 MeV. We will analyze a scaled mass splitting for the H which is not very sensitive to the extrapolation. The valence approximation was used, ignoring the effects of internal quark loops.

Statistical errors were estimated from the fluctuations of analyses performed on data sets with one lattice at a time removed (the 'jackknife method'), as described in reference 14. This is a much more stable and reliable procedure than the alternative of analyzing subensembles containing relatively small numbers of lattices. Removal of small sets of contiguous lattices allows testing for the presence of correlations between gauge configurations. Negligible correlation was found for masses with 500 sweeps separating the configurations.

A rough guide to the reliability of the calculations may obtained by comparing the spectrum results of reference 14 with the known hadron masses. The mass splittings for the six baryons not used as inputs agreed with experiment to well within the statistical errors of 20-30%. The $\pi\rho$ splitting was too low by 30%; the KK* splitting was too low by over a factor of two. The Nρ mass ratio was too large by 25%. In no case did a mass splitting come out with the wrong sign.

An explicit expression for the quark model wave function of the H is required for the lattice calculation. This is much more complicated than the wave functions for the two and three quark hadrons. The color and spin part of the H wave function is the color singlet, spin singlet part of the [490] representation of the combined colorspin SU(6) group.[1] The [490] is represented by the Young tableau with two rows of three boxes each. The color and spin wave function may be obtained by first symmetrizing the color and spin indices of two trios of quarks, yielding spin 1/2 color octets and spin 3/2 color decuplets. Only the two octets may be combined to make an overall color singlet. When the color and spin indices of the two spin 1/2 color octets are combined into a color singlet and spin singlet, three pairs of indices from the two octets are antisymmetrized as required by the [490] symmetry. Flavor indices are then arranged to obtain Fermi symmetry, yielding an SU(3) flavor singlet. This results in a huge number of explicit terms. The correctness of this expression was checked by applying the colorspin operator of eqn. (1) in explicit form to it to obtain the correct eigenvalue, -24. The number of terms in the wave function is squared in the two point function calculation, making the analysis program with this expression much too time consuming. A more tractable form of the wave function may be obtained by reexpressing it in terms of pairs of quarks of the same flavor. The quarks in each pair must be antisymmetric in overall color and spin, and so must be in either spin 0 color sextets or spin 1 color 3*s. The allowed combinations are three sextets, three 3*s, or two 3*s and a sextet. There are 138 nonzero terms in all (out of a possible 15^3). The 15x15 flavor pair propagators may by constructed relatively quickly at each lattice site from the quark propagators. These are combined with the flavor pair wave function into the full wave function. The calculation of the H two point function from the quark propagators performed in this way required two weeks of VAX 11/780 CPU time, compared with a few days for all the rest of the spectrum and coupling constant analysis combined. The flavor pair wave function may also be obtained directly by starting from an arbitrary combination of trios of flavor pairs in color and spin singlets and using isospin raising operators to find the combination which gives a flavor singlet.[15] The [490] symmetry then follows from Fermi statistics. Because of the complexity of the wave function, the correctness of the derivation and programming was checked by deriving and programming the wave function along the two completely separate routes: the flavor pair basis and the quark basis. The results were checked and agreed at an intermediate step and in the output of the analysis programs.

If the H exists as a stable particle, it is thought to be a tightly bound six quark state, with a radius possibly not much larger than the radii of the ordinary hadrons. On the lattice in the infinite volume limit, we should find the pole for the H in its proper place. In a finite volume, if the H really is mostly a tightly bound six quark state, finite volume errors should be comparable to those for the ordinary hadrons. If the H is unstable, the dominant singularity in the H two point function will be that of the lightest physical state to which it couples, at $2M_\Lambda$. If the H prefers to exist as a pair of independent Λ's, finite volume effects may be very large when the two Λ's are squeezed into a lattice barely big enough to fit a single hadron, making the dominant singularity in the H propagator appear to be above $2M_\Lambda$.

Our results are shown in Figure 1. Defining the effective mass as the logarithm of the ratio of propagators at adjacent time slices, we show the relative mass splitting $(M_H - 2M_\Lambda)/M_\Lambda \cdot 1115 MeV$, extrapolated to the physical quark mass limit. At short times, the splitting is very small, but always positive. A large positive splitting develops asymptotically. The qualitative effect is very insensitive to quark mass. Very similar graphs are obtained for all quark masses used in the calculation, as well as for the extrapolated results. As a further check, we extrapolated all three flavors of quarks to the chiral symmetry limit and obtained almost identical results for the splitting. This is in contrast to the calculations in chiral models which are extremely sensitive to the details of chiral symmetry breaking.[16]

The fact that the singularity in the H two point function appears above rather than at $2M_\Lambda$ is a finite volume effect, so there are clearly large finite volume errors in this calculation. However, finite volume errors on a tightly bound H would not necessarily act to decrease its splitting. In a study of finite volume effects in the Sine-Gordon model with periodic boundary conditions,[17] the binding was increased as

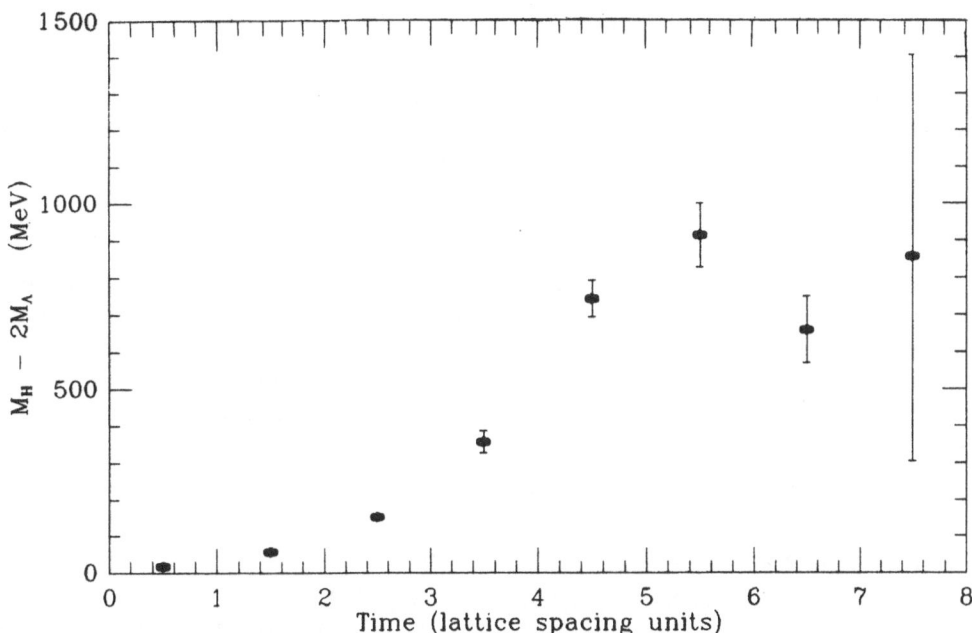

Figure 1. The mass splitting between the H and two Λ's extrapolated to the physical quark masses. Vertical lines are statistical errors. The splitting is positive for all Euclidean time separations and for all quark masses used in the calculation.

the volume decreased and the particles were pushed deeper into their potential well. Furthermore, the same sign of the splitting is observed in our data at short times before the quarks have spread out enough to feel the effects of the finite volume. On the other hand, the H wave function must contain some component of ΛΛ state. The closer the H is to the ΛΛ threshold, the larger this component is expected to be, becoming the full wave function if the H is above threshold. It is reasonable to guess that finite volume effects associated with this part of the wave function tend to push the mass up. For a very slightly bound H, it is conceivable that finite volume effects could make the H appear unbound on our lattices.

Another source of uncertainty is the question of whether the lattice is long enough in the time direction that we are seeing the asymptotic form of the splitting. The last four data points in Figure 1 are consistent with being flat, but because of the large statistical error of the last point the last three are also consistent with a falling splitting as time increases. This raises the question of whether the very large time behavior might be different than the short and intermediate time behavior. This type of behavior might be expected if the H existed as a stable, lightly bound deuteron-like object. (This is not a very likely possibility, based on meson exchange potential potential model calculations.[18]) A potentially serious way that this calculation could go wrong is in underestimating the hyperfine splitting between the H and the center of the dibaryon multiplet. The H is stable in quark model estimates precisely because of a very large spin splitting of this sort (about -350 MeV). In the spectroscopy data described above, most of the spin splittings for the known hadrons agreed well with experiment, but two of them were low by up to a factor of two. A worst case scenario for this calculation to go wrong might be that the H is somewhat bound in real life, a bad misestimate of the spin splitting makes it somewhat unbound on the lattice with our approximations, and finite volume errors magnify that effect into the very large splitting seen in our data. We do not consider this likely, but it cannot be excluded.

To sum up, we do not see a negative splitting between the H and two Λ's at any combination of quark masses or at any separation of the hadron operators. Much better lattice calculations for the H will be possible in the near future, and very likely much better experiments to test the calculations.

Acknowledgments

We thank William A. Bardeen for collaboration in obtaining a convenient form for the wave function, and Jonathan Rosner and R. L. Jaffe for discussions. We thank Steven Gottlieb and Don Weingarten for collaboration in the work of reference 14. PBM thanks the Fermilab theory group for hospitality while this work was begun.

The work of PBM was supported by the U.S. Department of Energy under Grant No. DE-AC02-76ER02220.

References

1) R. L. Jaffe, Phys. Rev. Lett. **38**, 195 (1977).
2) A. S. Carroll et al., Phys. Rev. Lett. **41**, 777 (1978).
3) D. J. Prowse, Phys. Rev. Lett. **17**, 782 (1966); M. Danysz, Nuc. Phys. **49**, 121 (1963).
4) V. Fitch, private communication; P. D. Barnes, Proc. of the Second LAMPFF II Workshop; and P. D. Barnes and J. Franklin, Brookhaven AGS proposal (1985), unpublished.
5) G. Baym, E. W. Kolb, L. McLerran, T. P. Walker, and R. L. Jaffe, Phys. Lett. **160B**, 181 (1985).
6) I. B. Khriplovich and E. V. Shuryak, Novosibirsk preprint 85-117 (1985).
7) E. Witten, Phys. Rev. **D30**, 272 (1984); E. Farhi and R. L. Jaffe, Phys. Rev. **D30**, 2379 (1984).
8) M. Soldate, unpublished.
9) K. F. Liu and C. W. Wong, Phys. Lett. **113B**, 1 (1982).
10) A. P. Balachandran, F. Lizzi, and V. G. J. Rodgers, Nuc. Phys. **B256**, 525 (1985); S. A. Yost and C. R. Nappi, Phys. Rev. **D32**, 816 (1985); R. L. Jaffe and C. L. Korpa, MIT preprint (1985).
11) J. Rosner, Quark Models, 1980 St. Croix Summer School lectures, published in Virgin Islands Inst. 1980, 1.
12) J. Rosner, Enrico Fermi Institute preprint EFI 85-84.
13) Paul B. Mackenzie and H. B. Thacker, Phys. Rev. Lett. **55**, 2539 (1985).
14) S. Gottlieb, P. B. Mackenzie, H. B. Thacker, and D. Weingarten, Fermilab preprint 84/98-T (1985), submitted to Nuclear Physics B. This reference contains a list of standard lattice gauge theory references.
15) We thank W. A. Bardeen for showing us this method.
16) We thank Chiara Nappi for a conversation on this point.
17) D. Hochberg and H. B. Thacker, Nuc. Phys **B257**, 729 (1985); H. B. Thacker, Fermilab preprint conf-85/83-T (1985).
18) A. T. M. Aerts and C. B. Dover, Phys. Rev. **D28**, 450 (1983).

A NEW METHOD FOR INVERTING FERMIONIC MATRIX

J. Wosiek

Institute of Computer Science
Jagellonian University, Krakow

ABSTRACT

The Alternate Directions Implicit method is generalized and used to compute quark propagators in quenched QCD. Preliminary results on the chiral properties of the theory and on the spectrum of hadrons agree with those obtained with other approaches.

INTRODUCTION

A lot of CPU time is spent nowadays on the computation of quark propagators.[1,2,3] Therefore it might be useful to explore the alternative, to commmonly used conjugate gradients, techniques of inversion of the large sparse matrices. I would like to present here preliminary results of the application of the Alternate Directions Implicit (ADI) method to quenched QCD in four dimensions. This approach offers some physical insight into the very process of inversion. It turns out also that the ADI algorithm converges slightly faster than the conjugate gradients.

A METHOD

We are looking for the solution of a system of linear equations $M_F X = B$, with the right hand side describing a unit source. Assume that the matrix M_F can be decomposed into a sum $M_F = \Sigma_1 + \Sigma_2$. Then one can rewrite above equations in the form

$$(r+\Sigma_1)X = (r-\Sigma_2)X + B,$$
$$(r+\Sigma_2)X = (r-\Sigma_1)X + B, \qquad (2.1)$$

with the auxiliary relaxation parameter r. Define the iterative procedure, $X^{(0)} = B$,

$$X^{(N+1/2)} = (r + \Sigma_1)^{-1} [(r - \Sigma_2)X^{(N)} + B],$$
$$X^{(N+1)} = (r + \Sigma_2)^{-1}[(r - \Sigma_1)X^{(N+1/2)} + B]. \qquad (2.2)$$

For suitable r Eqs. (2.2) converge to the solution $X = M_F^{-1}B$.

We shall work with Kogut-Susskind fermions in four dimensions.[8] In this case the most symmetric (and optimal) splitting of M_F is given by

$$2\Sigma_1 = im_q a - S_x - S_y, \quad 2\Sigma_2 = im_q a - S_z - S_t,$$

$$(S_\mu)_{mn} = \alpha_\mu(n) \ U_\mu(n) \ \delta_{n+\mu,m} + \alpha_\mu(m) \ U_\mu^\dagger(m) \ \delta_{n-\mu,m} \ . \quad (2.3)$$

Matrices $r \mp \Sigma_1$ $(r \pm \Sigma_1)$ describe two-dimensional propagation of a fermion in a background field. They correspond to the set of decoupled two-dimensional theories. We will refer to this property as the "dimensional reduction": full ADI iteration is basically composed of two inversions of the fermionic matrix restricted to the two (alternating in their orientation) planes. Of course every two-dimensional inversion must be performed for all parallel planes. These are completely decoupled procedures, and can be easily vectorized. Also a simple and rather flexible segmentation of the algorithm is possible. In the most extreme case only one plane of the link and fermionic variables must reside in the fast memory at each stage of the calculation.

The rate of convergence of (2.2) depends on the value of r. We have fixed r by minimizing the spectral radius of the error propagating matrix

$$F = (r + \Sigma_2)^{-1} (r - \Sigma_1) (r + \Sigma_1)^{-1} (r - \Sigma_2) \quad (2.4)$$

for free fermions. In this case $r \pm \Sigma_1$ and $r \pm \Sigma_2$ commute and the problem is reduced to finding the minimum of

$$\max_\sigma \ (\frac{r - \sigma}{r + \sigma})^2 \quad (2.5)$$

σ being the eigenvalue of Σ_i. One obtains $r_o = i \sqrt{m^2 a^2 + \lambda_{max}^2}$ with $\lambda_{max} = 2\sqrt{2}$. It turns out that the optimal value of r for the interacting case is smaller than that by ca 30%. This reduction may come from the effective decrease of the range of eigenvalues resulting from the disordering effect of the gauge variables.

Other choices or r are of course possible. In particular, it would be very interesting to extend the Jordan-Wachspress theory of optimal parameters to our case. It is known that for simpler cases, iteration dependent J-W parameters may speed up the convergence remarkably.

We have tested the method on a small $(4^3 x8)$ lattice. Figure 1 shows the error Δ, defined as

$$\Delta = ||M_F X^{(n)} - B||/\sqrt{N_F} \ , \quad N_F = 3 \ L_T L_S^3 \ , \quad (2.6)$$

as the function of the number of iterations n. Two observations are evident from this graph. 1) the algorithm converges for small quark masses though the critical slowing down is substantial, 2) it is more effective to reach the small quark mass through the series of converging iterations at higher (progressively decreasing) values of $m_q a$ (solid line) than by performing many iterations starting from $X^{(0)} = B$ (dashed line) at small $m_q a$. In the first approach one generates results for intermediate masses and prepares better starting point for the next $m_q a$. The spectral radius extracted from Fig. 1 agrees nicely with that predicted by Eq. (2.5) with $\lambda_{max} = 1.7$ instead of the free value 2.8.

As is readily seen from Eq. (2.2) the ADI approach is useful only when one can easily invert matrices $r \pm \Sigma_i$. They are sparse matrices corresponding to the nearest neighbours interactions on the two-dimensional torus (periodic boundary conditions are used). For an L^4 lattice $r \pm \Sigma_i$ are built from $3L \times 3L$ blocks describing one-dimensional

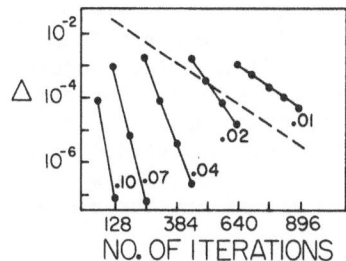

Fig. 1. The error Δ (cf. 2.6) of the approximate solution $\chi^{(n)}$ as a function of a number of iterations. Solid lines --$m_q a$ changing successively from .10 to .01 as indicated; dashed line - iteration independent $m_q a = .01$.

propagation of a fermion. In terms of these blocks, $r \pm \Sigma_i$ are a modified tri-diagonal matrices - modification coming from the periodic boundary conditions. We have generalized recursive Thomas algorithm for solving tri-diagonal systems and used it to invert $r \pm \Sigma_i$ exactly. Of course other methods could also be utilized for this purpose. Choice of the particular procedure depends on the available resources.

RESULTS

We have used the ADI technique to compute the spectrum of lattice QCD in the quenched approximation. Preliminary results of this calculation were reported elsewhere.[4] The simulation was done on the 16×8^3 lattice at $\beta = 6.0$ using the Bochum Cyber 205 computer. The SU(3) algebra was performed using the 8-bit coding technique of Bunk et al.[5] This is a very effective, space saving method which requires only one CDC word to store a SU(3) matrix. Twelve uncorrelated gauge configurations were used for averaging fermionic observables. The configurations were prepared starting from twelve larger (28×16^3), well thermalized at $\beta = 5.8$, configurations generated by A. Konig for other purposes.[7] The small configurations were cut out from large lattice and then, imposing periodic boundary conditions on 16×8^3 lattice, we have run 400 sweeps in order to thermalize the local heating at the boundaries, and achieve the equilibrium at $\beta = 6.0$. Thermalization was performed with the aid of the RAUPE program written in Wuppertal by Schilling et al.[6]

In principle one could invert the fermionic matrix for arbitrary small quark mass $m_q a \neq 0$. Spectral radius as given by Eq. (2.5) is smaller than 1 for $m_q a > 0$. Aside from the critical slowing down, it is not recommended to go to the too small $m_q a$ on a finite lattice, since the finite size effects become more and more important. This bias is stronger in the Goldstone boson channel. The integrated pion propagator becomes flatter in time with decreasing quark mass and the influence of the finite boundaries grows accordingly. Therefore, we have restricted quark masses to $m_q a = .16, .09, .04$ and $.0225$. However, during the tests the method performed satisfactorily also for smaller values of $m_q a$ ($=.01$). For larger lattices (hence smaller $m_q a$), critical slowing down becomes a problem for ADI as well as for other approaches. In this situation, proposed in Ref. [3], pre-conditioning of the M_F is very promising. The idea of Ref. [3] is rather general and it would be very interesting to apply it to the ADI case. According to the discussion in Sect. II, we were decreasing quark mass successively starting from the largest value of 0.16. For every $m_q a$ we did enough iterations to reach the accuracy of 10^{-6} (cf. Eq. (2.6)) which was sufficient for present purposes. This required 64 iterations for $m_q a = .16$ changed gradually to 200 at $m_q a = 0.225$.

Masses of the low lying hadrons are extracted in a standard, by now, way.[8,9,10] The quark propagator[+] in a given background is

$$G_{ab}(n,0) = \{ M_F^{-1}[U] \, B^{(b,0)} \}_{(a,n)} \tag{3.1}$$

where a and b define colour and $n(=n^\mu)$ specifies a lattice site. From (3.1) one constructs integrated over the space hadronic propagators averaged over the gauge configurations. In four available[++] mesonic channels they read

$$T_P(n_t) = \sum_{\vec{n},a,b} \langle |G_{ab}(n,0)|^2 \rangle_U \tag{3.2a}$$

$$T_{VT}(n_t) = \sum_{\vec{n},a,b} [(-)^{n_x} + (-)^{n_y} + (-)^{n_z}] \langle |G_{ab}(n,0)|^2 \rangle_U \tag{3.2b}$$

$$T_{PV}(n_t) = \sum_{\vec{n},a,b} [(-)^{n_x+n_y} + (-)^{n_x+n_z} + (-)^{n_y+n_z}] \langle |G_{ab}(n,0)|^2 \rangle_U \tag{3.2c}$$

$$T_S(n_t) = \sum_{\vec{n},a,b} (-)^{n_x+n_y+n_z} \langle |G_{ab}(n,0)|^2 \rangle_U \tag{3.2d}$$

and for a spin 1/2 baryons

$$B(n_t) = \sum_{\substack{\vec{n} \text{ even} \\ a,b,c \\ a',b',c'}} \varepsilon_{abc}\varepsilon_{a'b'c'} \langle G_{aa'}(n,0) G_{bb'}(n,0) G_{cc'}(n,0) \rangle_U \tag{3.2e}$$

In Kogut-Susskind formalism both parities contribute to a given channel, therefore we expect the following physical states to saturate the large n_t limit of T's

---

[+] More precisely this is the propagator of a χ field [8].

[++] We restrict the analysis to the local operators.

$$T_P = R_\pi e^{-m_\pi a n_t} + (n_t \to 16 - n_t) \, , \tag{3.3a}$$

$$T_{VT} = R_\rho e^{-m_\rho a n_t} + (-)^{n_t} R_B e^{-m_B a n_t} + (n_t \to 16 - n_t) \, , \tag{3.3b}$$

$$T_{PV} = R_{A_1} e^{-m_{A_1} a n_t} + (-)^{n_t} R_\rho e^{-m_\rho a n_t} + (n_t \to 16 - n_t) \, , \tag{3.3c}$$

$$T_S = R_S e^{-m_S a n_t} + (-)^{n_t} R_\pi e^{-m_\pi a n_t} + (n_t \to 16 - n_t) \, , \tag{3.3d}$$

$$B = R_P \left[(-)^{n_t} e^{-m_p a n_t} + (-)^{n_t} e^{-m_p a(16 - n_t)} \right]$$

$$+ R_{N*} \left[(-)^{n_t} e^{-m_{N*} a n_t} + e^{-m_{N*} a/16 - n_t)} \right] \, . \tag{3.3e}$$

In principle higher recurrences should also be included in Eqs. (3.3) This however is limited by the statistics and the number of n_t points available. At present only the excited state in the pion channel can be seen.

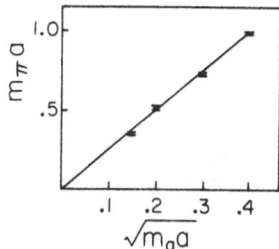

Fig. 2.   Pion mass as a function of $\sqrt{m_q a}$. Solid line shows the linear fit (3.4).

Our analysis of the quark propagators is similar to that of Ref. [10]. We have fitted the MC data with the formulas (3.3) changing number of states and the minimal distance from the source. As usual, increasing the number of states destabilizes the fit. Too few parameters result in the poor fit and in an overestimation of the masses. Apart from the reasonable χ^2 we have checked for stability of the results when one, two or three, closest to the source, points were removed from the data. In general two states (four parameters) fits were stable against the change of n_t^{min}, but this depended on the particular channel and on the value of n_t^{min}.

The pion mass is extracted from the two states fit to T_p (n_t). Second state is required in the direct[+] channel. The pseudoscalar propagator does not show any oscillatory behavior, hence there is no evidence for the opposite parity partner. This is in accord with the non relativistic quark model. No exotic 0^{--} (e.g. hybrid) state is seen. The fit gives consistent results for n_t^{min} = 1 and 2. Figure 2 shows the dependence of the $m_\pi a$ on the square root of the quark mass. Pion mass extrapolates smoothly towards 0 in the chiral limit as expected for the Goldstone boson associated with the spontaneously broken chiral symmetry. The linear fit of the form

$$m_\pi a = \omega\sqrt{m_q a} + b \tag{3.4}$$

gives ω = 2.46 ± .05 in close agreement with the result of refs. [9,10]. The linear relation (3.4) holds for larger range of the quark masses than can be expected from the Gell-Mann-Ochs-Renner relation

$$m_\pi^2 f_\pi^2 = 2\, m_q \langle\bar\psi\psi\rangle \tag{3.5}$$

This is evident from the Fig. 3 where our results for the quark condensate

$$a^3\langle\bar\psi\psi\rangle = 1/2 \sum_a \langle G_{aa}(0,0)\rangle_U \tag{3.6}$$

are shown. In the same range of quark mass $\langle\bar\psi\psi\rangle$ varies by a factor of two. All this change must be therefore balanced by the dependence of the f_π on m_q in order to produce a straight line in Fig. 2.[++]

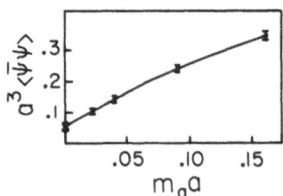

Fig. 3.   Quark condensate as measured from Eq. (3.6) for four
          values of $m_q a$. Quadratic fit and its extrapolation
          to $m_q a$ = 0 are also shown.

[+] In the terminology of Ref. [10].

[++] It was shown in Ref. [9] that finite size effects, which are crucial to $\langle\bar\psi\psi\rangle$ at small $m_q a$, set in for yet smaller quark masses than considered here.

A π-like state contributes also to the scalar amplitude T_s as an opposite parity partner of the $S(0^{++})$ resonance. Indeed we see the π signal with much smaller than S residuum. Two states fit to T_s is stable for $m_q a = .04$ and $t_{min}=2,3$. For $m_q a = 0.225$ the signal is vanishing with errors growing accordingly. The fit gives pion mass similar to that obtained from the pseudoscalar channel, but with necessarily bigger errors.

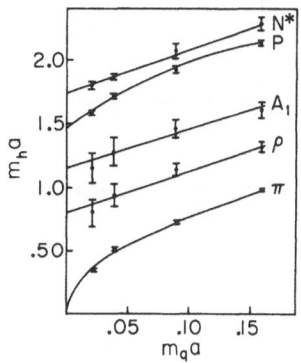

Fig. 4.   Masses of lowest lying hadrons versus the quark mass.
Solid lines result from the linear or quadratic (proton only) fits. The square root curve for a pion is taken from Fig. 2.

Vector-tensor propagator exhibits oscillatory behavior, hence both parity partners ρ and B couple to this channel, cf. (3.3b). The ρ state is also seen in the pseudovector channel together with the A_1 resonance. Two states fits were stable, but fitted parameters had much larger than m_π errors. To our accuracy, values of m_ρ obtained from both channels were consistent. Figure 4 summarizes our results on the spectrum of low lying hadrons. Solid lines show linear/quadratic fits of the $m_q a$ dependence.

In the three quark channel we measured only the correlations between the local operators, which are dominated by a proton and an N^* resonance. Two state fit was stable for n_t^{min} = 2,3. The ratio of the proton to rho masses is shown in Figure 5. It is weakly dependent on the quark mass, what may give more credibility to the extrapolation procedure. As usual, this number is much higher than the experimental value 1.2. Since our result is consistent with the number obtained by Edinburg Group (8^3x16 lattice) and is much bigger than that quoted by Barkai et al. (16^3x32 lattice), we feel that large part of the discrepancy with the experiment can be caused by finite size effects.

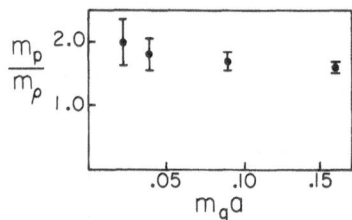

Fig. 5. Ratio of a proton to rho masses as a function of $m_q a$.

Using pion and rho masses as an input one obtains for the quark mass and for the lattice constant a

$$m_q = 1/2(m_u + m_d) = 3.5 \text{ MeV},$$ (3.7)

$$a^{-1} = .93 \text{ GeV}.$$

From the intercept in Fig. 3 and from (3.7) one obtains for the condensate (experimental values are quoted in the parentheses)

$$\langle \bar{\psi}\psi \rangle = 2(215 \text{ MeV})^3 \qquad (2(225 \pm 25 \text{ MeV})^3) .$$ (3.8)

For non-strange particles we obtain

$$m_{A_1} = 1070 \pm 100 \text{ MeV} \qquad (1270 \text{ MeV}),$$

$$m_B = 930 \pm 100 \text{ MeV} \qquad (1230 \text{ MeV}),$$

$$m_S = 740 \pm 60 \text{ MeV} \qquad (980 \text{ MeV}),$$

$$m_P = 1360 \pm 60 \text{ MeV} \qquad (940 \text{ MeV}).$$
(3.9)

From m_K mass one gets the mass of a strange quark

$$m_s = 90 \text{ MeV}$$ (3.10)

Then the m_{K*} and m_ϕ are predicted to be

$$m_{K*} = 890 \text{ MeV} \qquad (892 \text{ MeV}) \tag{3.11}$$

$$m_\phi = 1050 \text{ MeV} \qquad (1020 \text{ MeV})$$

Very good agreement of these two numbers with the experiment is partly due to the fact that the vector propagator was earlier used as an input to fix a lattice constant. Therefore we conclude only that the dependence of the vector mass on $m_q a$ is correctly reproduced by Monte Carlo simulations.

To summarize, we have proposed and tested a new technique for computing quark propagators in quenched QCD. The method works for the quark mass $m_q a \neq 0$ and is a useful alternative to other commonly applied approaches, mainly thanks to the feature of dimensional reduction. Our results on the chiral properties of QCD and on the spectrum of low lying hadrons are consistent with those of other authors. Encouraged by this overall agreement we plan to perform in the future more detailed study of the hadronic spectrum on larger lattices.

This work was done in Wuppertal during the fellowship of the Alexander Humboldt Foundation. I would like to thank the Theory Group in Wuppertal for their hospitality. I would like to thank also Drs. J. P. Wefel and T. G. Guzik for their substantial help in the preparation of this typescript.

REFERENCES

1. D. Weingarten, Nucl. Phys. B257 [FS14] (1985) 629.
2. D. Barkai, K. J. M. Moriarty and C. Rebbi, CERN preprint, CERN-TH 4065/84, November 1984.
3. G. G. Batrouni, G. R. Katz, A. S. Kronfeld, G. P. Lepage, B. Svetitsky and K. G. Wilson, Phys. Rev. D32 (1985) 2736.
4. J. Wosiek, Jagellonian University preprint, TPJU 18-85, November 1985.
5. B. Bunk and R. Sommer, Wuppertal preprint NU-B-85-8 (1985).
6. K. H. Mutter and K. Schilling, Nucl. Phys. B230 [FS10] (1984) 275.
7. A. Konig, K. H. Mutter and K. Schilling, Phys. Lett. 147B (1984) 145.
8. H. Kluberg-Stern, A. Morel, O. Napoly and B. Petersson, Nucl. Phys. B220 [FS8] (1983) 447; A. Morel and J. P. Rodriques, Nucl. Phys. B247 (1984) 44.
9. J. P. Gilchrist, G. Schierholz, M. Schneider and M. Teper, Nucl. Phys. B248 (1984) 29; K. C. Bowler, D. L. Chalmers, A. Kenway, R. D. Kenway, G. S. Pawley and D. J. Wallace, Nucl. Phys. B240 [FS12] (1984) 213.
10. D. Barkai, K. J. M. Moriarty and L. Rebbi, "Hadron Masses in Quenched QCD," preprint, February 1985.

(USES OF) AN ORDER PARAMETER FOR

LATTICE GAUGE THEORIES WITH MATTER FIELDS

Mihail Marcu *

Fakultät für Physik
Universität Freiburg
Freiburg, West-Germany

ABSTRACT

The ideas around an order parameter that tests the existence of charged states in gauge theories with matter fields are reviewed. The order parameter is used to show that the phase diagram of the Z_2 Higgs-gauge model in 4 dimensions has a line of second order phase transitions with mean field critical exponents.

1. INTRODUCTION

Although gauge theories with matter fields have been extensibely studied in the last fifteen years, several outstanding theoretical questions are still unsolved. In QCD confinement has yet to be proven. In the standard model for weak and electromagnetic interactions the challenge lies in a better understanding of the Higgs particles. In QED the efforts to construct the theory rigorously have not been successful.

In pure matter theories, local order parameters are a powerful tool in finding out whether charged states exist. Settling this point often means understanding the phase diagram of the theory. In gauge theories there are no local order parameters. In the absence of matter fields however, the Wilson loop criterion [1] distinguishes between the confined and the nonconfined phases.

In gauge theories with matter fields there are no local order parameters and the Wilson loop obeys a perimeter law throughout the phase diagram (strictly speaking this is only true for theories with matter fields in the fundamental representation of the gauge group and without additional global symmetries that are spontaneously broken only in part of the phase diagram). Nevertheless, the question of the existence of charged states plays the same outstanding role as in the case of a pure matter theory.

A candidate for a charged state is obtained by separating two localized observables: the first one has the quantum numbers we decided to call "charge" and stays at a given space location; the second, with anticharge quantum

---

* Address after April 1986: II Institut für Theoretische Physik, Universität Hamburg.

numbers, is sent to infinity. The charge-anticharge "dipole" state has to fulfill two requirements:

- (i) it is created by acting with a gauge invariant operator on the vacuum;
- (ii) its energy stays bounded as the separation is increased.

It is by no means certain that the candidate for the charged state is indeed charged. The typical phase diagram [2] of a gauge theory with matter fields in the fundamental representation has (Fig.1):

- (i) a free charge phase, where charged states exist (example: QED);
- (ii) a screening region, with Debye-type screening of the charge (example: Higgs mechanism);
- (iii) a confinement region, with charge screening through fragmentation (example: QCD).

If there is no additional global symmetry, the confinement and screening regions are not separated by phase transition lines [2,3].

If the candidate for the charged state is indeed charged, its vacuum overlap ρ (i.e. its scalar product with the vacuum) is zero. If ρ is nonzero, a cluster argument (see section 4) shows that the charge is screened; the state produced by charge separation could have been also obtained by acting with some local operator on the vacuum. Thus ρ is an order parameter distinguishing between the free charge and the screening-confinement phases.

An order parameter defined along these lines was proposed by the author in joint work with Klaus Fredenhagen [4-7]. It will be discussed in detail in sections 3 and 4, after the appropriate notation will have been introduced in section 2. The status of rigorous results concering this order parameter and related problems will be briefly reviewed in section 5.

In section 6 the behaviour of the dipole states at finite charge-anticharge separation is discussed. The main result is a qualitative distinction

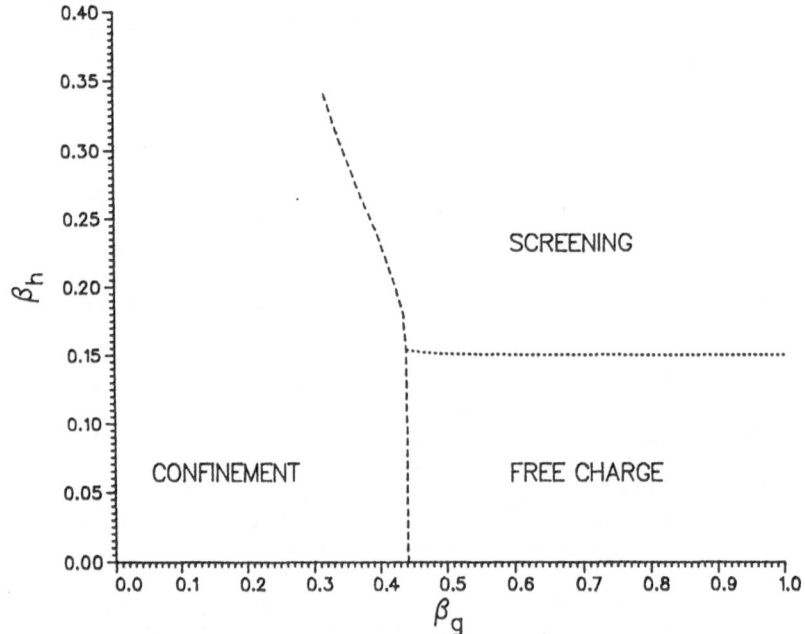

Fig.1. Phase diagram of a lattice gauge theory with matter fields in the fundamental representation on the example of the Z_2 theory in 4 dimentions. The gauge coupling is $\beta_g = 1/g^2$. The matter coupling is β_h (hopping parameter). The dotted line extends up to $\beta_g = \infty$.

between the screening and the confinement region. A new way to compute some masses is pointed out.

The order parameter can be expressed in terms of rations of simple expectation values in the Euclidean path integral formulation of the theory under consideration. Thus it can be computed in a Monte Carlo simulation. This conputation is not very difficult in the free charge phase and in the screening region. In the confinement region however, almost insurmountable problems arise.

In section 7 the results of a simulation of the Z_2 gauge theory with Z_2 matter fields [8], away from the critical regions, are briefly dicussed. In the confinement region the computation of the order parameter turns out to be many orders of magnitude away from the reach of today's most powerful computers. A way to circumvent this problem is presented in section 8. In the screening-confinement region the order parameter is equal to a parameter proposed by Bricmont and Fröhlich [9] (to be referred to as the BF parameter), which can be measured. A theorem in ref. [5] shows that the BF parameter tests the existence of a bound state between a source and a dynamical charge (in the free charge phase this is a "hydrogen atom").

In the literature there are several attempts to define order parameters in lattice gauge theories with matter fields [9-11]. These quantities fail to reproduce the phase diagram of Fig.1. (the autors of ref [9] agree with the interpretation of their criterion given in section 8 [12]). The order parameter proposed by Kennedy and King [13] works well only in the case of non-compact lattice QED [14].

The Monte Carlo investigation of the vacuum overlap order parameter in the Z_2 theory was done jointly with Thomas Filk and Klaus Fredenhagen. After gaining experience away from the critical region, we investigated the screening-free charge transition. Here the order parameter turned out to be extremely useful numerically. In section 9 results are shown which very strongly suggest that the whole transition line (the dotted line of Fig.1) is second order with mean field exponents.

Section 10 contains conclusions and a brief enummeration of outstanding problems.

2. NOTATION

Let us consider a lattice gauge theory with scalar matter fields in the fundamental representation of the gauge group G (scalars are assumed here for. simplicity; in ref. [6] for example, the discussion is in terms of fermions). The action S is

$$S = -\beta_g \sum_P \mathrm{Tr}\, U(\partial p) - \beta_h \sum_l (\varphi^+ U \varphi)(1) - \sum_x V(\varphi(x)) \qquad (2.1)$$

U(1) is the element of G at the link 1, $\varphi(x)$ is the matter field at the site x, $U(\partial p)$ is the ordered product of link variables around the plaquette p, $\varphi^+ U \varphi(1)$ is the gauge-matter interaction, β_g and β_h are the couplings, and $V(\varphi)$ is a usual $\varphi 4$ potential.

It is helpful to introduce a pictorial notation for certain expectation values. Let L be a path from x to y (i.e. an ordered set of connected links with x and y as boundary). U(L) is the path ordered product

$$U(L) = P \prod_{l \in L} U(1) \qquad (2.2)$$

and we denote

$$\left\langle \overset{L}{\underset{x \quad y}{\frown}} \right\rangle = \frac{1}{Z} \sum_{U,\varphi} \left(\sum_{a,b} \varphi_a(x) U_{ab}(L) \varphi_b^\dagger(y) \right) \exp(-S(U,\varphi)) \qquad (2.3)$$

Z is the partition function, a and b are indices in the fundamental representation of G, and the path drawn inside the expectation value brackets is L. Similarly, for a closed path we denote

$$\left\langle \overset{L}{\underset{}{\bigcirc}} \right\rangle = \frac{1}{Z} \sum_{U,\varphi} (\mathrm{Tr}\ U(L)) \exp(-S(U,\varphi)) \qquad (2.4)$$

For a lattice gauge theory, the "canonically quantized" formulation is obtained [3,15,16] by going to the temporal gauge and deriving the transfer matrix $T = \exp(-H)$ (H = Hamiltonian) such that, up to a constant,

$$Z = \mathrm{Tr}\ T^{n_\tau} \qquad (2.5)$$

n_τ is the number of sites in Euclidean time direction. T is an operator in a Hilbert space whose vectors will be denoted by $|...\rangle$. $|0\rangle$ is the vacuum. In deriving T, the usual convention is that the configurations in a time slice form the Hilbert space basis. In this basis, the diagonal time-zero matter and gauge field operators will be denoted by $\hat{\varphi}(x)$ and $\hat{U}(1)$, x and 1 being a spatial point and a spatial link (i.e. in the time-zero hyperplane). The path ordered product $\hat{U}(L)$ is defined in the same way as (2.2) with hats on the U's.

3. THE FINITE ENERGY DIPOLE STATES

The simplest way to define a gauge invariant state with a charge at the spatial point x and an anticharge at the spatial point y is described by the formula:

$$|\underline{x},\underline{y},\underline{L}\rangle = \sum_{a,b} \hat{\varphi}_a(\underline{x})\ \hat{\varphi}_b^\dagger(\underline{y}) \hat{U}_{ab}(\underline{L}) |0\rangle \qquad (3.1)$$

L being some spatial path from x to y. The charge-anticharge pair is connected by a flux tube that ensures gauge invariance. It is well known that the energy of a long thin flux tube is proportional to its length [1]. Thus the energy of (3.1) will diverge as $y \to \infty$. In order to obtain a finite energy candidate for a charged state, one has to consider more general dipole states. Every dipole state however will contain some sort of flux tube. If the thickness of this flux tube stays finite as $y \to \infty$, we still expect the energy to diverge.

In refs. [4-7] a method for regularizing the energy of the dipole state (3.1) was proposed. The idea is to translate $\hat{U}(L)$ into Euclidean time. Let us define

$$\hat{U}^{(n)}(\underline{L}) = T^n \hat{U}(\underline{L}) T^{-n} \qquad (3.2)$$

For simplicity, L will be from now on the straight line connecting x with y. The divergent contribution to the energy of (3.1) comes from $\hat{U}(L)$. Now we replace $\hat{U}(L)$ by $\hat{U}^{(n)}(L)$. The higher energy components of $\hat{U}^{(n)}(\underline{L})|0\rangle$ will be more suppressed as n is increased and the state

$$|\underline{x},\underline{y},n\rangle = \sum_{a,b} \hat{\varphi}_a(\underline{x}) \varphi_b^\dagger(\underline{y}) \hat{U}_{ab}^{(n)}(\underline{L}) |0\rangle \qquad (3.3)$$

might have a chance of ending up with a finite energy. In terms of time-zero operators, (3.2) is a dipole state with a flux tube of thickness 2n (each conjugation with the transfer matrix increases the thickness by one lattice spacing). Thus n has to become infinite as $y \to \infty$.

Let us discuss two bounds on the energy of $\hat{U}^{(n)}(\underline{L})|0\rangle$. Since we are on the lattice, it is easier to consider expectation values of $\exp(\pm H)$ rather than H itself.

The first bound essentially proves that the energy of (3.3) diverges as the distance $|\underline{x}-\underline{y}|$ becomes infinite and n has a fixed value. The energy of $\hat{U}^{(n)}(\underline{L})|0\rangle$ is roughly minus the logarithm of

$$\frac{\langle 0| \hat{U}^{(n)\dagger}(\underline{L})\ T\ \hat{U}^{(n)}(\underline{L})\ |0\rangle}{\langle 0| \hat{U}^{(n)\dagger}(\underline{L})\ \hat{U}^{(n)}(\underline{L})\ |0\rangle} = \frac{\langle \overset{|\underline{x}-\underline{y}|}{\boxed{\qquad\qquad}}\ 2n+1\rangle}{\langle \underset{|\underline{x}-\underline{y}|}{\boxed{\qquad\qquad}}\ 2n\rangle} \tag{3.4}$$

On the r.h.s. the lengths of the rectangle sides are written inside the expectation value symbol. We shall use this convention for quantities of the type (2.3) – (2.4) whenever possible. By interchanging space and Euclidean time, we see that for large values of $|\underline{x}-\underline{y}|$ the r.h.s. is $\exp\left(|\underline{x}-\underline{y}|(E_{2n}-E_{2n+1})\right)$ E_m being the energy of a source-antisource pair at distance m (this is the usual interpretation of the Wilson loop). Since the source-antisource potential is attractive, it follows that the r.h.s. goes to zero as $\underline{y}\to\infty$. Thus the energy of (3.3) becomes infinite if $\underline{y}\to\infty$ while n is kept fixed.

The second bound will show that the energy of (3.3) stays finite if $n\to\infty$ at least linearly with $|\underline{x}-\underline{y}|$. Since we need an upper bound on the energy of $\hat{U}^{(n)}(\underline{L})|0\rangle$, we consider the quantity

$$\frac{\langle 0| \hat{U}^{(n)\dagger}(\underline{L})\ T^{-1}\ \hat{U}^{(n)}(\underline{L})\ |0\rangle}{\langle 0| \hat{U}^{(n)\dagger}(\underline{L})\ \hat{U}^{(n)}(\underline{L})\ |0\rangle} = \frac{\langle \overset{|\underline{x}-\underline{y}|}{\boxed{\qquad\qquad}}\ 2n-1\rangle}{\langle \underset{|\underline{x}-\underline{y}|}{\boxed{\qquad\qquad}}\ 2n\rangle} \tag{3.5}$$

Using only the Schwarz inequality and the perimeter law for Wilson loops, the r.h.s. may be estimated to be smaller than a constant times

$$\exp 2\alpha\left(\frac{|\underline{x}-\underline{y}|}{n}+2\right) \tag{3.6}$$

α being the exponent in the perimeter law. The proof for this result is essentially identical with the proof of Proposition 6.1 in ref. [5] (see also [7]). The energy of (3.3) is thus uniformly bounded if $n\to\infty$ at least linearly with $|\underline{x}-\underline{y}|$

4. THE VACUUM OVERLAP ORDER PARAMETER

As discussed in the introduction, the crucial dynamical question is: is the limit

$$|\underline{x}\rangle = \lim_{\substack{\underline{y}\to\infty \\ n\gtrsim |\underline{x}-\underline{y}|}} \frac{|\underline{x},\underline{y},n\rangle}{\||\underline{x},\underline{y},n\rangle\|} \tag{4.1}$$

a charged state? The precise mathematical meaning of this limit will not be discussed here (see refs. [5,7]). We will rather concentrate on the physical arguments.

Let us denote by $\tilde{\rho}_n(|\underline{x}-\underline{y}|)$ the vacuum overlap of $|\underline{x},\underline{y},n\rangle$:*

$$\tilde{\rho}_n(|\underline{x}-\underline{y}|) = \frac{\langle 0|\underline{x},\underline{y},n\rangle}{\||\underline{x},\underline{y},n\rangle\|} \tag{4.2}$$

If $|\underline{x}\rangle$ is a charged state, it will be orthogonal to the whole vacuum Hilbert

_____

* In ref. [6] $\tilde{\rho}$ was erroneously defined to be the square of (4.2).

space $\mathcal{H}_o$ ($\mathcal{H}_o$ is the completion of the space obtained by acting with local operators on the vacuum). In particular, its scalar product $\widetilde{\hat{\rho}}_{\infty}(\infty)$ with the vacuum will be zero. If on the other hand the charge is screened, $|\underline{x}\rangle$ is in $\mathcal{H}_o$ and there exists some local gauge invariant operator such that

$$\underset{\underline{y}\to\infty,\; n\gtrsim|\underline{x}-\underline{y}|}{Lim} \quad \frac{\langle 0|A|\underline{x},\underline{y},n\rangle}{\|\,|\underline{x},\underline{y},n\rangle\,\|} \;\neq\; 0 \tag{4.3}$$

As $\underline{y}\to\infty$ the anticharge will get outside the localization region of A. If further the separation $|\underline{x}-\underline{y}|$ exceeds the screening (or fragmentation) length, we expect the l.h.s. to cluster:

$$\frac{\langle 0|A|\underline{x},\underline{y},n\rangle}{\|\,|\underline{x},\underline{y},n\rangle\,\|} \quad \underset{\substack{\underline{y}\to\infty \\ n\gtrsim|\underline{x}-\underline{y}|}}{\widetilde{}} \quad \langle 0|A|\underline{x}\rangle\,\langle 0|\underline{y}\rangle \tag{4.4}$$

Since thr limit was assumed to be nonzero, it follows that $\langle 0|y\rangle = \widehat{\hat{\rho}}_{\infty}(\infty)$ is nonzero. Summing up, we conclude that $\widehat{\hat{\rho}}_{\infty}(\infty)$ may be used as an order parameter:

$$\widehat{\hat{\rho}}_{\infty}(\infty) = \begin{cases} 0 & \text{free charge} \\ \neq 0 & \text{screening-confinement} \end{cases} \tag{4.5}$$

In ref. [7] we propose to call $\widetilde{\hat{\rho}}$ the "vacuum overlap order parameter".

One of the appealing features of this order parameter is its simple expression in terms of expectation values of the type (2.3) – (2.4):

$$\widetilde{\hat{\rho}}_n(|\underline{x}-\underline{y}|) = \frac{\langle\;\raisebox{0.3ex}{\tiny$\overset{|\underline{x}-\underline{y}|}{\longrightarrow}$}\;n\;\rangle}{\langle\;\substack{n \\ n}\;\rangle^{\frac{1}{2}}} \tag{4.6}$$

Still, if one is interested in questions as the continuum limit, the products of time-zero operators at the same point occuring in the denominator might be a problem (the singularities coming from the corners should cancel between the numerator and the denominator). The quantity $\rho_n(|\underline{x}-\underline{y}|)$:

$$\rho_n(|\underline{x}-\underline{y}|) = \frac{\langle\;\raisebox{0.3ex}{\tiny$\overset{|\underline{x}-\underline{y}|}{\longrightarrow}$}\;n\;\rangle}{\langle\;\;2n\;\rangle^{\frac{1}{2}}} \tag{4.7}$$

avoids this problem. Since roughly

$$\rho_n(|\underline{x}-\underline{y}|) \;\sim\; \langle\varphi^{+}\varphi\rangle\;\widetilde{\hat{\rho}}_n(|\underline{x}-\underline{y}|) \tag{4.8}$$

and $\langle\varphi^{+}\varphi\rangle > 0$, $\rho_{\infty}(\infty)$ also behaves as (4.5) and it may be used alternatively as an order parameter.

$\rho_n(|\underline{x}-\underline{y}|)$ also has a direct physical interpretation [7]. It plays a role similar to that of $\widetilde{\hat{\rho}}$ if we consider finite energy dipole states with sources rather than dynamical matter fields at $\underline{x}$ and $\underline{y}$.

5. COMMENTS ON RIGOROUS RESULTS

In the screening-confinement region the powerful tool of convergent expansions is available [3]. In the convergence region, a general proof that $\rho_{\infty}(\infty)$ is nonzero should be feasible. For the Abelian case, proofs are available for $G=Z_2$ [5] and $G=U(1)$ [17]. In the nonabelian case, the technical difficulties are expected to be greater.

For Abelian gauge groups, Griffith inequalities can be used to show that

$\rho_\infty(\infty)$ is zero in a sizable part of the free charge phase [5]. The numerator of (4.7), computed at (β_g, β_h), is smaller than the two-point function $\langle 0 | \varphi(\underline{x}) \; \varphi^+(\underline{y}) | 0 \rangle$ computed in the pure matter theory $(\beta_g = \infty)$, at coupling β_h. the denominator is larger than the same Wilson loop computed in the pure gauge theory ($\beta_h = 00$, at coupling β_g. The result follows using the exponential decay to zero for the two-point function in the pure matter theory and the perimeter law in the pure gauge theory. For G-U(1) a similar argument was given in ref. [18].

The behaviour (4.5) of the vacuum overlap order parameter* is just an indication for the (non)existence of charged states. For $G=Z_2$, a completely rigorous treatment of the charged state problem was given in ref. [5]. It is likely that the arguments presented there for the screening-confinement region can be generalized whenever there is a convergent expansion. This has not yet been done however. The proof of the existence of charged states in the free charge phase turned out to be more difficult even in the Z_2 case [5]. The main problem was our inability to regularize the charge operator. In theories with an additive charge (e.g. G=U(1)), charge regularization might still turn out to be feasible. However, only in the case of a discrete gauge group there is a convergent expansion in the free charge phase.

A problem worth investing a lot of work into is the continuum limit [6] of the vacuum overlap order parameter. It is interesting that the behaviour (4.5) was recently proven to be correct in the continuum Schwinger model [19].

6. THE DIPOLE STATES AT FINITE SEPARATION

Knowledge of the behaviour of $\rho_h(r)$ at finite values of r and n is important both in proving properties of the order parameter and in numerical computations.

Let us denote by m_q the energy of a source with charge quantum numbers, by m_c the mass of the charged state, by m_s the inverse screening length in the screening region, by m_0 the bare mass of the matter field and by σ the pure gauge theory string tension computed at the same value of β_g as that of the full theory. All these quantities are in dimensionless lattice units.

In a first approximation the numerator of (4.7) behaves as ($r = |\underline{x} - \underline{y}|$, c_i are constants):

$$\exp(-m_c r - m_q(r+2n)) \qquad \text{free charge phase}$$
$$\rho_\infty(\infty)\,(1 + c_1 \exp(-m_s r))\,\exp(-m_q(r+2n)) \qquad \text{screening region} \qquad (6.1)$$
$$\rho_\infty(\infty)\,\exp(-m_q(r+2n)) + c_2 \exp(-\sigma r n - m_0 r) \quad \text{confinement region}$$

and the denominator behaves as:

$$\exp(-m_q(r+2n)) \qquad \text{free charge phase}$$
$$\exp(-m_q(r+2n)) \qquad \text{screening region} \qquad (6.2)$$
$$\exp(-m_q(r+2n)) + c_3 \exp(-\sigma r n) \qquad \text{confinement region}$$

In the limit $r \to \infty$, $n \to \infty$ these formulas reproduce (4.5). For a finite value of r (6.1) and (6.2) will be a good approximation only if n exceeds a certain

---

* An additional argument for this behaviour emerged during a controversy with Phys.Rev.Lett. which delayed the publication of ref. [6] for ten months. One of the referees claimed that $\rho_\infty(\infty) = 0$ in all phases. The other essentially stated that $\rho_\infty(\infty) \neq 0$ in all phases. Eq. (4.5) would be a compromise between the two.

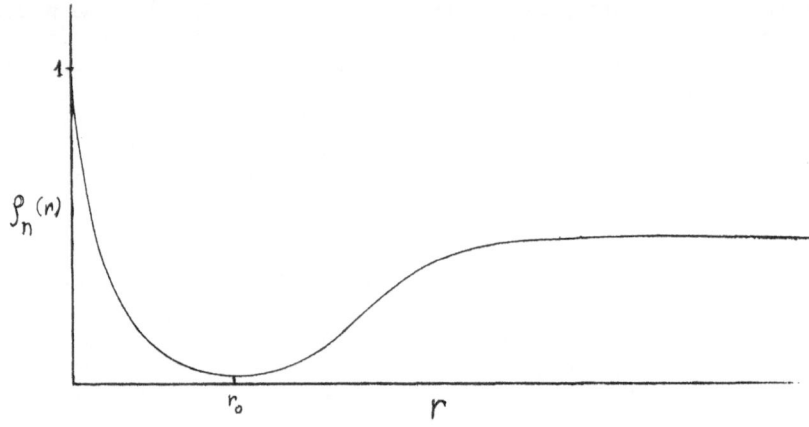

Fig. 2. The qualitative behaviour of $\rho_n(r)$ in the confinement region.
n is assumed to be large enough.

r-dependent value. If this is the case, then $\rho_n(r)$ will decay exponentially
to zero in the free charge phase and to a constant in the screening region.
This decay may be used to determine the charged state mass m_c and the inverse
correlation length m_s. m_s is in fact a very interesting quantity. If we are
at couplings where the Higgs particle is stable, m_s is the Higgs mass.

In the confinement region $\rho_n(r)$ has completely different short and long
distance properties. For $r < r_o = 2m_q/\sigma$ and n large enough, the surface term in
(6.1)-(6.2) becomes dominant and $\rho_n(r)$ will decay like $\exp(-m_c r)$. This situation is similar to the free charge phase. Thus at short distances, the charge
and the anticharge behave as if they were free. We call this property "asymptotic charge localizability" [7]. For $r > r_o$ (n large enough!) the perimeter
term in (6.1)-(6.2) becomes dominant and $\rho_n(r)$ approaches its true asymptotic
value. The qualitative behaviour of $\rho_n(r)$ is depicted in Fig. 2.. The dramatic change at $r \approx r_o$ is due to the onset of fragmentation at this scale.
r_o is also the scale where the quark-antiquark potential changes from a
linear function to a constant.

The gauge invariant two-point function $\rho_\infty(r)$ has all the properties
described in this section and the advantage that we need not worry about n
any more. In practical calculations however (e.g. Monte Carlo), the maximal
values of both n and r may be restricted to relatively small values. If we
try to keep n as small as possible (i.e. not larger than linear in r), deviations from (6.1)-(6.2) could occur for small values of n. This is due to the
high energy components of (3.3) not being sufficiently suppressed. Typically
this will happen when n is not larger than the correlation length for two-
point functions of local gauge invariant operators.

Note that in the confinement region $\rho_n(r)$ is n-dependent even if the
high energy components of (3.3) are well suppressed and (6.1)-(6.2) is a good
approximation.

7. MONTE CARLO SIMULATION AWAY FROM THE CRITICAL REGIONS

In order to gain numerical experience with the vacuum overlap order
parameter, a Monte Carlo simulation was done for the Z_2 lattice gauge
theory with Z_2 matter fields in d=3 and 4 dimensions. This was in joint
work with Thomas Filk and Klaus Fredenhagen [8].

$\rho_n(r)$ turned out to be easy to measure in the free charge phase and in
the screening region. We chose r=n and r=2n with $r \leq 10$ and $n \leq 5$ on a lattice of

size 22^d. As long as the couplings were not close to lines of second order phase transitions, the behaviour (6.1)-(6.2) was very well obeyed. To a high degree of accuracy $\rho_n(r)$ and $\rho_{r\Lambda}(r)$ were equal. The main reason for the easy numerics is the smallness of m_q.

In the confinement region both the numerator and the denominator of (4.7) become extremely small as r is increased. The statistical errors however are roughly r- and n-independent and comparable to their value in the rest of the phase diagram. In order to reduce the errors we used a variance improvement technique [20]. Although we gained many orders of magnitude in precision, we were never able to see more than the exponential decay for $r < r_0$. With our present methods, it can be estimated that mapping out the dip in Fig. 2. is way beyond the possibilities of today's most powerful computers. Thus a direct computation of the order parameter $\rho_\infty(\infty)$ in a Monte Carlo simulation is not feasible in the confinement region.

8. THE HYDROGEN ATOM AND THE BRICMONT-FRÖHLICH PARAMETER

The state $|\underline{x}^{(e)}\rangle$ defined by

$$|\underline{x}^{(e)}\rangle = \hat{\varphi}(\underline{x})|0\rangle \tag{8.1}$$

is not gauge invariant; it contains a source with charge quantum numbers (or "external charge") at the spatial point $\underline{x}$. Since $|\underline{x}^{(e)}\rangle$ is a local excitation of the vacuum, its total dynamical charge as measured through Gauss's law is that of the vacuum. Therefore this state contains an anticharge in addition to the external charge. Now the question arises: is the lowest energy state in the Hilbert space sector containing $|x^{(e)}\rangle$ a bound state? (by sector the set of states with a given dynamical and external charge is meant). If this is the case, then in the free charge phase this bound state is similar to the hydrogen atom. In the confinement region it would be a meson composed of a very heavy and a light quark.

A projection on the lowest energy state $|\underline{x}_0^{(e)}\rangle$ may be achieved using the transfer matrix:

$$|\underline{x}_0^{(e)}\rangle = \lim_{n \to \infty} \frac{T^n |\underline{x}^{(e)}\rangle}{\|T^n |\underline{x}^{(e)}\rangle\|} \tag{8.2}$$

$|\underline{x}_0^{(e)}\rangle$ is a bound state if "its wavefunction is quadratically integrable", or, in more formal terms, if it is an eigenvector of T. If $|\underline{x}_0^{(e)}\rangle$ is a bound state, then it contains an anticharge localized around the source at $\underline{x}$. If on the other hand it is not a bound state, the anticharge will escape to infinity and we are left with a charge localized around $\underline{x}$ that can be measured through Gauss's law. This is possible only if charged states exists, i.e. in (part of) the free charge phase.

In ref. [5] (Proposition 5.5) we proved a result reducing the existence problem for the bound state to the investigation of the scalar product $\langle \underline{x}^{(e)}| \underline{x}_0^{(e)}\rangle$: $|\underline{x}_0^{(e)}\rangle$ is an eigenvector of T in the Hilbert space sector with an external charge at $\underline{x}$ and total dynamical charge like the vacuum if and only if $\langle \underline{x}^{(e)}| \underline{x}_0^{(e)}\rangle \neq 0$. Actually if this is not the case, then $|\underline{x}_0^{(e)}\rangle$ is the ground state in the Hilbert space sector with an external charge at $\underline{x}$ and a total dynamical charge like that of the charged state $|\underline{x}\rangle$ [5,7].

If a bound state exists, the quantity $|\langle \underline{x}^{(e)}| \underline{x}_0^{(e)}\rangle|^2$ is the probability of the anticharge to be at $\underline{x}$ (wavefunction squared at the origin).

In terms of expectation values of the type (2.3), the criterion for existence of bound states reads:

$$\omega_n := \frac{\langle \uparrow n \rangle^2}{\langle \uparrow 2n \rangle} \xrightarrow[n\to\infty]{} \begin{cases} \neq 0 & \quad \text{if bound state exists} \\[6pt] = 0 & \quad \text{no bound state} \end{cases} \tag{8.3}$$

ω_∞ is the parameter of Bricmont and Fröhlich [9]. Since a situation where both the charged state and the hydrogen atom exist is possible, the BF parameter cannot be used as a criterion for confinement.

It was argued that external charge – dynamical anticharge bound states always exist in the screening-confinement phase since the anticharge cannot escape to infinity because there are no free charges in this phase. In ref. [5] (Theorem 5.6) we derived a result that, taken together with (8.3), proves this assertion using convergent expansion techniques. This result is:

$$\omega_\infty = \rho_\infty(\infty) \qquad \text{in the screening-confinement phase} \tag{8.4}$$

Intuitively, this is true when cluster properties are strong enough so that:

$$\frac{\langle \overset{|\underline{x}-\underline{y}|}{\boxed{}} n \rangle}{\langle \boxed{} 2n \rangle^{1/2}} \underset{\substack{\underline{y}\to\infty \\ n \gtrsim |\underline{x}-\underline{y}|}}{\approx} \frac{\langle \overset{\underline{x}}{\uparrow} n \rangle}{\langle \uparrow 2n \rangle^{1/2}} \cdot \frac{\langle \overset{\underline{y}}{\uparrow} n \rangle}{\langle \uparrow 2n \rangle^{1/2}} \tag{8.5}$$

Actually this was proven only in the convergent expansion regions of the Z_2 theory. Aside from technical problems, the proofs should be similar for any compact gauge group. The physical argument presented above makes it very likely that (8.4) is true also outside the convergent expansion regions.

The BF parameter is accessible to Monte Carlo simulations since, in a first approximation,

$$\langle \uparrow n \rangle = \rho_\infty(\infty) \exp(-m_q n) \quad \text{in the screening-confinement phase} \tag{8.6}$$

In the confinement region in particular, there are no surface effects like in (6.1)-(6.2) and ω_n approaches its asymptotic value much faster than $\rho_n(r)$. We thus have an indirect method of measuring the vacuum overlap order parameter.

During the simulation of the Z_2 theory we also computed ω_n [8]. As expected, in the screening region it turned out that $\omega_\infty = \rho_\infty(\infty)$ even for couplings where a convergent expansion is not available. Moreover, ω_∞ has a jump across the first order transition line separating the screening from the confinement region (the line with an isolated endpoint).

In the free charge phase, the BF parameter is very difficult to obtain in a Monte Carlo simulation.

9. LINE OF SECOND ORDER PHASE TRANSITIONS IN THE 4-DIMENSIONAL Z_2 THEORY

In the 4-dimensional Z_2 theory the order of the screening – free charge transition has been an open problem. A brief review of the strong evidence that the whole line is second order with mean field exponents [21] will be given here. The transition line (dotted line in Fig.1) will be denoted by $\beta_{h,c}(\beta_g)$ (this defines a curve in the coupling constant plane).

In the screening phase, to a good approximation (n is large enough!):

$$\rho_n(|\underline{x}-\underline{y}|) = \sum_M \left[\tanh(\beta_h) \, \exp(-m_q(\beta_g, \beta_h)) \right]^{|M|} \tag{9.1}$$

M being a path from $\underline{x}$ to $\underline{y}$ containing no link more than once. In the Ising

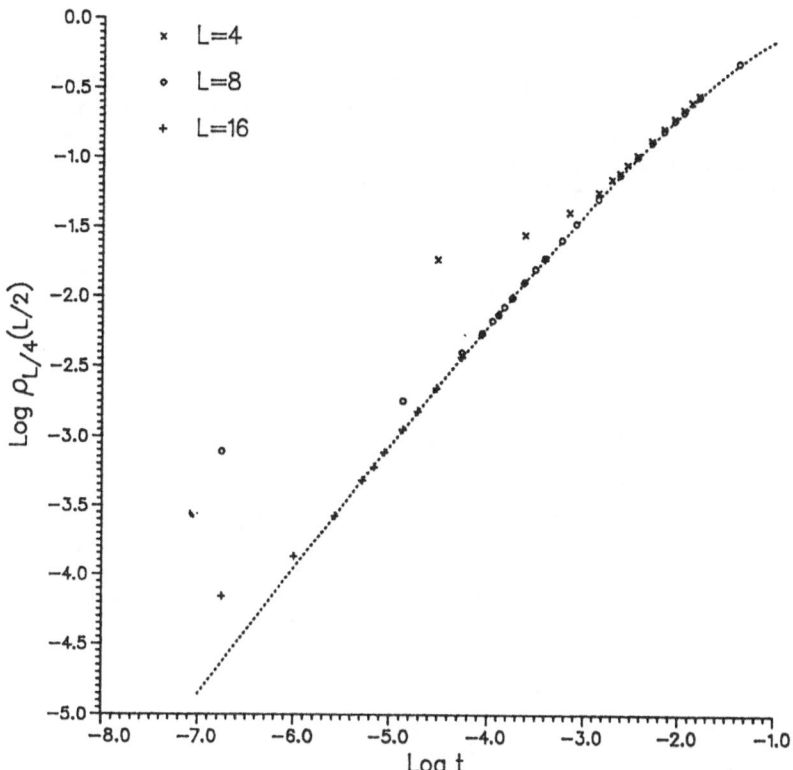

Fig. 3.   Scaling plot for the order parameter.

limit $(\beta_g = \infty$, $m_q = 0)$ a common approximation to the spin-spin correlations at high temperature is recovered. For β_g fixed, (9.1) breaks down if the square bracket attains a certain β_g-independent value. This gives an approximate equation for the transition line:

$$\tanh(\beta_{h,c}(\beta_g)) \ \exp(-m_q(\beta_g, \beta_{h,c}(\beta_g))) = \tanh(\beta_{h,c}(\infty)) \qquad (9.2)$$

m_q can be easily measured in a simulation via (6.2). Using our Monte Carlo data for m_q and the available value for the Ising critical coupling, we obtained [21] $\beta_{h,c}(.5) = .15081 \pm .00003$.

The above argument suggests that for fixed β_g there is an Ising-like transition with shifted critical coupling. In a high precision simulation we computed $\rho_{r/2}(r)$ and $\rho_r(r)$ for various lattice sizes L. Then we fitted the Ising scaling law for the magnetization square [22] to our data:

$$\rho_{\infty}(\infty) = \text{const } t \ \left| \ln(t) \right|^{2/3} \quad , \quad t = 1 - \beta_{h,c}(.5)/\beta_h \qquad (9.3)$$

The results of Fig.3 show that (9.3) is very well obeyed, the best fitted value for $\beta_{h,c}(.5)$ being $.15082 \pm .00003$ [21].

Close to the critical point finite size effects occur. Following [23], we made the following finite size scaling Ansatz:

$$\rho_{L/4}(L/2) = \frac{g(t)}{L^2} \ f(tL^2 g(t)), \quad g(t) = \left| \ln|t| \right|^{1/3} \qquad (9.4)$$

A nice universal finite size scaling curve f was obtained for $\beta_{h,c}(.5)$ between .15081 and .15085 [21].

Estimating $\beta_{h,c}(.5)$ from other quantities led to a result an order of magnitude less accurate than using the order parameter.

10. CONCLUSIONS AND OUTLOOK

The vacuum overlap order parameter is relevant in understanding charged states, confinement and Higgs mechanism in lattice gauge theories, and may also be successfully used in numerical studies. The outstanding problem at present is to understand its continuum limit. To this aim both theoretical and numerical work is necessary. Studies of Z_n, $U(1)$ and $SU(2)$ models might be a good start.

In ref. [5] two other order parameters have been introduced, which should be also investigated in more detail in the future.

REFERENCES

1 K.Wilson, Phys.Rev.D10 (1974) 2445.

2 E.Fradkin and S Shenker, Phys.Rev.D19 (1979) 3682.

3 K.Oosterwalder and E.Seiler, Ann.Phys.110 (1978) 440.

4 K.Fredenhagen, Freiburg preprint THEP 82/9 (talk presented at the Colloquim in honour of Prof. R. Haag's 60th birthday, Hamburg, Nov.1982)

5 K.Fredenhagen and M.Marcu, Commun.Math.Phys.92 (1983) 81.

6 K.Fredenhagen and M.Marcu, Phys.Rev.Lett.56 (1986)223.

7 K.Fredenhagen and M.Marcu, "Order parameters for lattice gauge theories with matter fields", to be published.

8 T.Filk, K.Fredenhagen and M.Marcu, "Numerical investigation of the vacuum overlap order parameter", to be published.

9 J.Bricmont and J.Fröhlich, Phys.Lett.122B (1983) 73.

10 G.Mack and H.Meyer, Nucl.Phys.B200 (1982) 249.
    H.Meyer-Ortmanns, Nucl.Phys.B230 (1984) 31 and B235 (1984) 115.

11 J.Bricmont and J.Fröhlich, Nucl.Phys.B230 (1984) 407.

12 J.Bricmont and J.Fröhlich, private communication.

13 T.Kennedy and C.King, Phys.Rev.Lett.55 (1985) 776 and Princeton preprint (1985).

14 C.Borgs and F.Nill, Max-Planck-Inst. preprint MPI-PEE/PTh 59/85 (1985).

15 M.Lüscher, Commun.Math.Phys.54 (1977) 283.

16 J.Kogut, Rev.Mod.Phys.51 (1979) 659.

17 J.Barata, private communication.

18 K.Kondo, Prog.Theor.Phys.74 (1985) 152.

19 J.L.Alonso and A.Tarancón, Phys.Lett. 165B (1985) 167.

20 R.L.Dobrushin, Teory.Prob.Appl.13 (1969) 197.
    O.E.Lanford and D.Ruelle, Commun.Math.Phys.13 (1969) 194.
    G.Parisi, R.Petronzio and F.Rapuano, Phys.Lett.128B (1983) 418.

21 T.Filk, K.Fredenhagen and M.Marcu, DESY preprint 86-003, to appear in Phys.Lett.B.

22 M.A.Moore, Phys.Rev.B1 (1970) 2238.
    E.Brézin, J.C. Le Guillou and J.Zinn-Justin, Phys.Rev.D8 (1973) 2418.
    F.J.Wegner and E.K.Riedel, Phys.Rev.B7 (1973) 248.
    O.G.Mouritsen and S.J. Knak Jensen, Phys.Rev.B19 (1979) 3663.

23 E.Brézin and J.Zinn-Justin, Nucl.Phys.B257 (1985) 867.

THE GF11 SUPERCOMPUTER [*]

J. Beetem, M. Denneau, and D. Weingarten

IBM T. J. Watson Research Center
Yorktown Heights, NY 10598

ABSTRACT

GF11 is a parallel computer currently under construction at the IBM Yorktown Research Center. The machine incorporates 576 floating- point processors arranged in a modified SIMD architecture. Each has space for 2 Mbytes of memory and is capable of 20 Mflops, giving the total machine a peak of 1.125 Gbytes of memory and 11.52 Gflops. The floating-point processors are interconnected by a dynamically reconfigurable non-blocking switching network. At each machine cycle any of 1024 pre-selected permutations of data can be realized among the processors. The main intended application of GF11 is a class of calculations arising from quantum chromodynamics.

A detailed treatment appears in the IEEE Proceedings of the 12th Annual International Symposium on Computer Architecture, Boston, June, 1985, and in the proceedings of the Frontiers of Quantum Monte Carlo conference to be published in the Journal of Statistical Physics in 1986.

*Presented by D. Weingarten

PARALLEL COMPUTING IN LATTICE THEORY

Henrik Bohr

The Technical University of Denmark
Building 423, DK-2800 Lyngby Denmark

Abstract

Some parallel processor systems are presented and applied to lattice theories in physics. The first type of system is an analogue parallel computer which simulates dynamical models instantaneously. The second type is a digital computer system with a parallel Non-Von-Neumann architecture that is especially suited for simulating lattice models. It does that as fast as present days supercomputers, such as CRAY and CYPER 205, but with a cost/benefit ratio which is 0,5 % to that of CRAY or CYBER.

Introduction

In this review we shall present some computer systems which are parallel processors (ref. 1) and which a group of us have assembled at the Technical University of Denmark. They represent examples of the fifth generation computers which aim at increasing the computing power by a factor of 10 - 100. It is expected that a parallel processing set-up can achieve this goal if a set of microprocessors in an efficient way can share the computations to be done among each other and process them in parallel. Yet there is no fully developed software to support such a parallel processing in an efficient and general way and also no standard parallel processor hardware is yet available to the extend of being really at the fifth generation level. On the other hand there are many experimental attempts going on in building parallel processor systems such as ours.

The advantage of parallel processing is well-known. Take, for example (ref. 2), a simulation of a large 2 dimensional array of atomic spins, say 100x100, that each should be updated throughout a series of several hundred operations. An ordinary serial computer would spend most of its time repeatedly retrieving and storing a vast amount of data successively for all the 10.000 spins with relatively little time devoted to actual computation. By contrast, a parallel processor system with, say hundred micro-processors working in parallel on each 100 spins, could take several pieces of data and perform a series of operations on them in parallel before returning them to a distributed memory. Then most of the time would be spend on processing which takes typically shorter time than accessing a memory and the entire processing should in principle go hundred times faster than the update of the whole system managed by a single processor.

Since there is yet no general software available for parallel computers it is important to choose a field in which parallel programming is straight-forwardly applicable. For such a set of problems a fairly synchronous and uniform data-flow between eventual sub-divisions of each problem is desirable. Most problems with simulation of dynamical systems are of that kind. By dynamical systems we think of systems with mutually coupled elements that can be described by a set of coupled differential equations. They are solved numerically on computers by the finite element method and in terms of parallel processing a subdivision of the finite set of element can be handled by each microprocessor. Parallel processing is straight-forward if the coupling between the elements of such a dynamical system is local and therefore the dataflow between the subdivisions becomes local too and homogeneous. Most dynamical models in physics are of that kind and especially lattice gauge models. Thus the advantage of parallel processing should be obvious in these cases.

We shall in the succeeding chapters discuss our attempts in parallel processing along those lines.

2. Analogue parallel processors

We shall first describe and demonstrate an analogue/digital computer MOSES (Modular Symbolic Simulator) recently constructed in Denmark. Its superiority to other conventional computers is due to the speed with which it can simulate even complicated systems of non-linear differential equations. It does that instantaneously regardless the size the system has. We shall demonstrate it on lattice models in theoretical physics. Since the computer (see fig. 1) is basically built up of a large number of independent but connectable processor modules that are manually movable on a board-wire system, it is natural to associate the computer with af lattice, where the lattice links are represented by the modules. Therefore the size of lattice one can handle is proportional to the number of processor modules and the size of the board system. At the moment the computer can only handle small lattices in a realistic physical model (e.g. Gauge theory) but an improved version of MOSES, capable of handling more lattice points is under consideration.

The speed of the computer can be described in terms of how long a time it takes to perform an iteration of a given lattice of size, say 10^4, i.e. to update all the field variables of the lattice links so as to produce an average action result for a given set of initial values. MOSES can produce 1000 such iterations in a millisecond and this number will be the same for any lattice size. For a 10^4 lattice this is roughly 10^3 times faster than a Cray-1. For studying small lattice systems the advantage of MOSES is that it can produce a large number of iterations quickly, e.g. 10^5 iterations in a second and thus improve the data in the critical region of a phase trasition.

The obvious drawback of the MOSES computer is that the cost of the machine grows linearly with the lattice size for 2-dimensional lattices and for 4-dimensional lattices the cost grows quadratically with the lattice size (not taking planned improvements into account).

In the next section we shall present some technical details about MOSES.

3. The computer system MOSES

MOSES - the modular symbolic simulator (ref. 1) - is a versatile, graphical simulation system specifically designed for simulation of dynamic and stochastic systems. MOSES was designed and built in the Tecnical University of Denmark by Kaj Jensen. The computing capacity of MOSES is embedded in small brick-like processor modules. A simulation model is built by placing those modules on a special wall board. Each processor module automatically exchanges information with neighbouring modules.
The processor modules, together with the wall boards, are the fundamental elements of the MOSES system (see fig. 1). Input, output and storage of information are accomplished by the front-end system (a minicomputer), the TV screens, the CRT terminal, the digitizer and the plotter.
MOSES is modular at several levels:

a) As was mentioned above, the computing capacity is embedded in small modules, each processor being interconnected when placed on the board with neighbouring modules and carrying out its computations simultaneously with all other modules.
b) There is no restriction on the number of boards that can be used at the same time. The system is expanded simply by adding more wall boards.
c) Two MOSES simulators can be combined by af very simple wiring.

The fact that the modules work in parallel and the modularity of MOSES account for its very high simulation speed regardless of the size of the model to be investigated.

The idea on which MOSES is based is in one respect identical to that of the vector and array processors i.e. building powerful computers out of many identical elementary processors that have the capability of simultaneously producing many identical results. In MOSES this idea was pushed to

The existing MOSES II system consist of:

- a number of
  electronic boards,
- eleven types of
  processor modules,
  each being hybrids of
  analogue and digital IC's,
- a front-end
  communication system,
- two colour TV screens,
- a CTR terminal,
- a digitizer
  and a plotter.

Fig. 1

its extreme, since not only one part of the computation is done in parallel, but all of the computations of the whole system are performed in parallel. This makes MOSES faster than the usual computers (in simulation).

The wall boards are built like a telephone net, and every module is, apart from being connected to all its neighbouring modules, a subscriber in this telephone net. Four different modules can be connected simultaneously to the front-end system that plays the role of the operator of the net. In this way the user can see the output of each of the modules, and using the computation capacity of the front-end system, also all possible common outputs of the models under investigation.

4. Lattice theory on MOSES

In order to test ability of the MOSES computer system for the simulation of lattice theories, we performed a series of computer measurements (ref. 3) on the smallest possible lattice in 2 dimensions, a 2x2 lattice. The reason for not being able to use a bigger lattice arrangement is the lack at the moment of special modular elements that can handle trigonometric couplings, but new elements will shortly be available. One might wonder whether it is possible at all to obtain an equilibrium state on a 2x2 lattice, i.e. to obtain a well defined temperature and potential energy over a large number of iterations. However, the data we present here show that it is in fact possible to define an equilibrium state even on such a small lattice provided that the system has time to stabilize itself, which in our case means roughly 10000 iterations. The power of MOSES is particularly seen here. It can produce 10000 iterations in 10 ms, which is rougly 10^5 times faster than on ordinary computers.

We chose to study the X-Y-lattice model since it is the simplest model that is similar to gauge theories. We chose the nearest neighbour action to be the Wilson form:

$$S=\beta\sum_i[\cos(\varphi_i-\varphi_{i+x})+\cos(-\varphi_i+\varphi_{i-x})+\cos(\varphi_i-\varphi_{i+y})+\cos(-\varphi_i+\varphi_{i-y})] \quad (1)$$

where the field indices are arranged on the lattice with i labelling each cross and with x and y indicating the two directions. In the case of a 2x2 lattice our action becomes

$$S=\beta[\cos(\varphi_1-\varphi_2)+\cos(\varphi_1-\varphi_4)+\cos(\varphi_4-\varphi_3)+\cos(\varphi_2-\varphi_3)] \quad (2)$$

4.1. Data from the micro-canonical ensemble approach

The equations of motion for a 2x2 lattice in the micro-canonical approach and with the action from eq. (2) are

$$\frac{d^2}{dt^2}\varphi_1=-[\sin(\varphi_1-\varphi_2)+\sin(\varphi_1-\varphi_4)], \quad \frac{d^2}{dt^2}\varphi_2=[\sin(\varphi_1-\varphi_2)-\sin(\varphi_2-\varphi_3)]$$

$$\frac{d^2}{dt^2}\varphi_3=[\sin(\varphi_4-\varphi_3)+\sin(\varphi_2-\varphi_3)] \quad , \quad \frac{d^2}{dt^2}\varphi_4=[\sin(\varphi_1-\varphi_4)-\sin(\varphi_4-\varphi_3)] \quad (3)$$

The set up on the MOSES system according to eq. (3) is shown in fig. 2. In figs. 3 and 4 we have presented the data corresponding to computer simulations during roughly 10 s. Each 20 ms a picture on the display is formed of a potential energy curve and a kinetic energy curve (temperature curve) versus computer time. The curves are shown in fig. 3 (the dotted curve is the potential energy and the dashed curve is the kinetic energy) which tells us that one can obtain an equilibrium state even on a small lattice provided that we include a sufficient number of iterations (5000-10000). Each iteration is produced by changing the initial values of the field parameters φ_i and derivatives p_i, which can be done either manually by adjusting the potentiometers on the modular elements, or automatically by using the random noise generator attached to each element. Resorting to the latter possibility we speed up considerably

the process of collecting data since the computer can now automatically produce around 10000 iterations randomly in every interval of 10 ms. The two curves are taken at different times, which is the reason why they do not add up to a constant energy before they have stabilized. We have checked that the total energy remains constant over many thousand iterations.

Going back to fig. 3 we take the last stable point from each energy curve and plot one (the potential energy) versus the other (the kinetic energy) which corresponds to plotting the average action versus the temperature. The full curve in fig. 3 corresponds to approximately 1000 such points and therefore represents the desired average action versus temperature plot for a lattice study. This plotting procedure is automatically done on the computer display in approximately 10 s during which the 1000 points have been accumulated.

In fig. 4 the specific heat C is shown versus the temperature

$$C \equiv -\beta^2 \frac{d}{d\beta}\langle S \rangle = \frac{d\langle S \rangle}{dT}$$

It has a peak around high values of T.

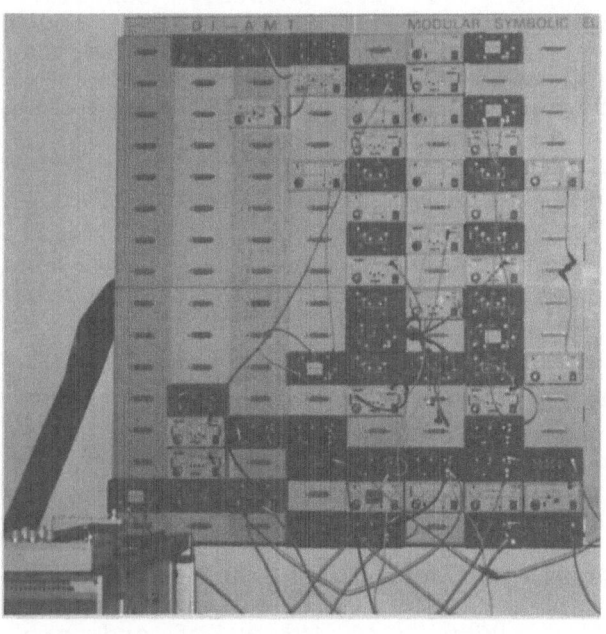

Fig. 2.   Arrangement of 2x2 X-Y model on MOSES in the micro-canocial approach.

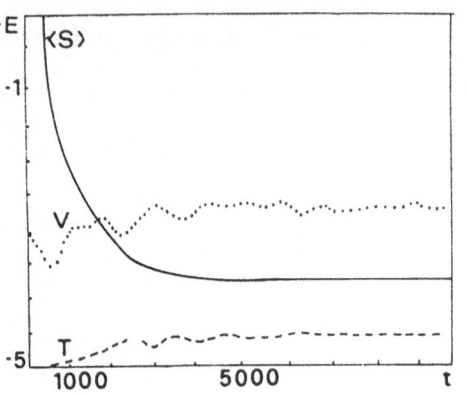

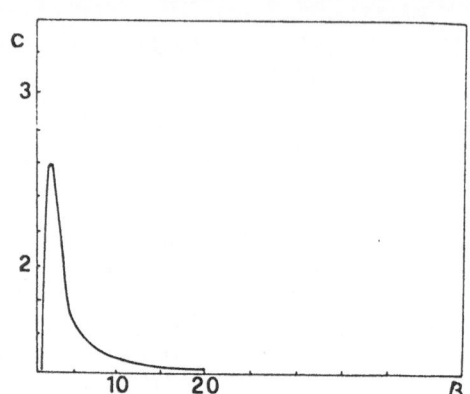

Fig. 3. *X-Y*-model in the micro-canonical ensemble. The full curve represents the average action versus (β
(1000 points, where each point corresponds to 10000 iterations). The
dotted line represents the potential
energy versus τ and the lower dashed
line represents the kinetic energy
versus τ(0⩽τ⩽*10000 iterations*).

Fig. 4. Curve of the specific heat for
the *X-Y*-model in the micro-canonical
ensemble (1000 points).

anonical ensemble (1000 points).

5. Stability of SU(2) monopoles with chiral fermions analysed on MOSES

Next we shall study a more complicated and intricate
differential equation system (ref. 4) on MOSES. The problem
is to find out if it is possible to have stable SU(2) monopoles in the presence of chiral fermions. The problem which
is a good example of MOSES capacity involves 6 highly non-linear coupled differential equations that are impossible
to solve analytically. The question of stability requires
a specification of all the boundary conditions and here
is MOSES superior to other computers since one can interac-
tively find the boundary conditions by adjusting the voltage
potentiometres while simultaneously checking the output
on a monitor screen and thus find the desired stable solutions.

We study the coupled system of SU(2) gauge potentials (A_α
$= A_\alpha^a t^a/2$), a triplet of Higgs scalars ($\varphi = \varphi^a t^a/2$) and a
doublet of chiral fermions ($x = x_\alpha^a$) described by the Lagrangian density

$$L=-\tfrac{1}{4} \mathrm{Tr}(FF)+\tfrac{1}{2}(D_\alpha\varphi)(D^\alpha\varphi)-V(\varphi\varphi)+i\dot{x}g_\alpha D^\alpha x-\tfrac{1}{2}(\dot{x}g_\alpha\delta^\alpha x+\delta^\alpha\dot{x}g_\alpha x) \qquad (4)$$

where

$$F_{\alpha\beta}=\delta_\alpha A_\beta-\delta_\beta A_\alpha-ig[A_\alpha,A_\beta] \quad (5) \quad, \quad D_\alpha\varphi=\delta_\alpha\varphi-ig[A_\alpha,\varphi] \qquad (6)$$

$$D_\alpha x=(\delta_\alpha-i\tfrac{1}{2}gA_\alpha)x \qquad (7) \quad, \quad V(z)=\tfrac{\varepsilon}{4}(1-z)^2 \qquad (8)$$

The field equations are

$$D_\alpha F_{\alpha\beta}=\tfrac{1}{2}gxg_\beta\Upsilon x+\tfrac{i}{2}g[D_\beta\varphi,\varphi] \quad (9), \quad g^\alpha D_\alpha x=0 \quad (10), \quad D^\alpha(D_\alpha\varphi)=-2V'(\varphi\varphi)\varphi \quad (11)$$

Eqs (9-11) are reduced to a system of ordinary differential
equations under substitution of the following ansatz (which
is justified on the following page):

286

$A_0^a = 0$ (12), $gA_i^a = \varepsilon^{aij}\dfrac{x^j}{r^2}(1+h_1(r)) + \delta^{ai}\dfrac{1}{r}h_2(r) + \dfrac{x^a x^i}{r^3}h_3(r)$ (13)

$\varphi^i = x^i/r^2 k(r)$ (14) , $g\chi_\alpha^a = 1/r^{3/2}((f_1(r)/r)i(\sigma^m)_{a\alpha}x^m + f_2(r)\delta_{a\alpha})$ (15)

where $r = |\bar{x}|$ and h_1, h_2, h_3, k, f_1, f_2 are real functions of one variable (r) which satisfy the couple non-linear ordinary differential equations given below:

$r^2 h_1'' - rh_2'(3h_2+2h_3) - rh_3'h_2 + h_1 + h_2^2 + h_2 h_3$

$\quad -2f_1 f_2 - h_1(h_1^2 + 2h_2^2 + h_3^2 + 2h_2 h_3 + k^2) = 0$ (16)

$r^2 h_2'' + 2rh_1'(h_2+h_3) + r(h_2'+h_3')h_1 + h_2 - h_1(h_2+h_3)$

$\quad -f_2^2 + f_1^2 - h_2(h_1^2 + 2h_2^2 + h_3^2 + 2h_2 h_3 + k^2) = 0$ (17)

$3rh_1'h_2 - 2rh_2'h_1 - 2h_2 - 2h_1 h_2 - f_1^2 - f_2^2 - 2h_2^3$

$\quad -3h_2^2 h_3 - 2h_1^2 h_2 - 2h_1^2 h_3 = 0$ (18)

$r^2 k'' - 2k(h_1^2 + h_2^2) - \varepsilon(k^2 - r^2)k = 0$ (19)

$rf_1' - (\tfrac{1}{2} + h_1)f_1 - \tfrac{1}{2}(3h_2 + h_3)f_2 = 0$ (20)

$rf_2' - (\tfrac{1}{2} - h_1)f_2 - \tfrac{1}{2}(h_2 - h_3)f_1 = 0$ (21)

The energy of the system for the static field configuration described by the ansatz (12-15) is calculated to be

$H = 4\pi \int \dfrac{1}{r^2}[\tfrac{1}{2}(r^2 k'^2 - 2rkk' + k^2) + r^2(h_1'^2 + h_2'^2) - 2r(h_2+h_3)(h_1'h_2 - h_2'h_1)$

$\quad + (h_1^2 + h_2^2)(h_2+h_3)^2 + (h_1^2 + h_2^2)k^2 + \tfrac{1}{2}(h_1^2 + h_2^2 - 1)^2 + \tfrac{\varepsilon}{4}(k^2 - r^2)^2]dr$ (22)

Notice that if $h_2 = h_3 = 0$ then Eq. (18) would imply $f_1 = f_2 = 0$. Thus $(h_2, h_3) \neq (0,0)$ is a necessary condition for the existence of solutions with a non-trivial fermion condensate. In wiew of this necessary condition we are led to consider the more general ansatz for A given by Eq. (13). From the integral for the energy given by Eq. (22) it is seen that necessary conditions for finiteness of energy are $k(r) \to r$ and $h_1(r)$, $h_2(r) \to 0$ as $r \to \infty$. We have used these conditions as primary signals for emergence of finite energy solutions in our search for solutions of the system qualitatively different from the usual monopole solution.

The full system of monopole, Higgs and fermions is simulated on MOSES, Fig. 5. This figure is the best of many simulations of the full monopole system and seem to indicate a possibility of a stable solution for the field system, since all h_i functions are going to zero at $r \to \infty$ as they should for obtaining finite energy solutions. Also

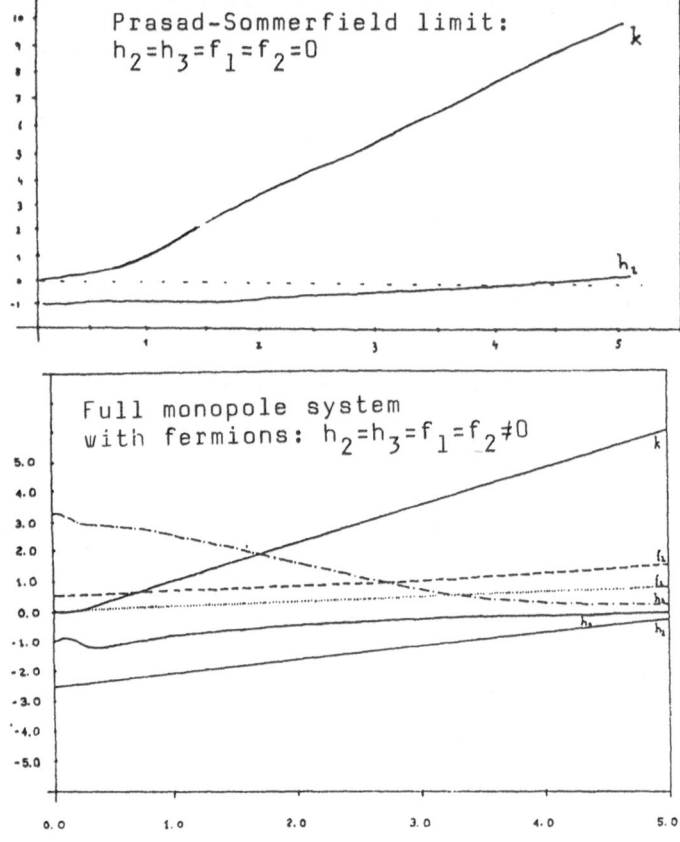

Fig.5 MOSES simulation of monopole system.

partial differential equations were studied on MOSES with an extra second order time derivative included (ref. 5) in the monopole equations in the Prasad Sommerfield limit. This was done by varying the initial conditions in time by a special sample module. The time dependent solutions were found to oscilate.

6. Digital parallel processors

Next we shall present a newly build parallel processor computer (ref. 1) that is fully digital. Since analogue computing lack the precision required in quantitative scientific analysis we thought it necessary to continue with digital parallel processors. We shall here report on such system, PALLAS (Parallel lattice Simulator), especially suited for lattice calculation. The parallel computer system PALLAS 0 (ref. 6) (see fig. 6.) is basically containing four elements: The central Processing Unit CPU that process the data, The front-end computer (an IBM PC), that reads in and reads out data, a colour monitor and a plotter that both display the results. The interesting part, the CPU unit, is build up of 16 equal processor cards arranged in a square lattice with nearest neighbour connections (see fig. 6). Furthermore the 16 processor cards are connected to the IBM PC equipped

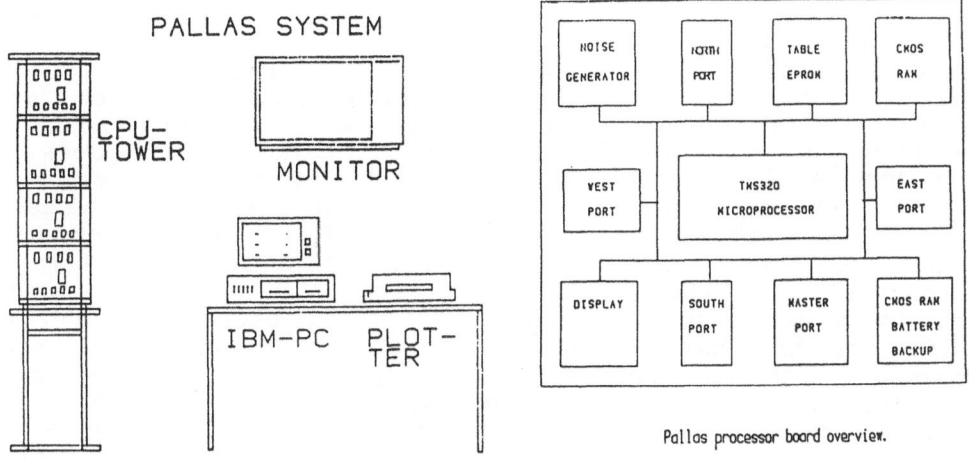

PALLAS SYSTEM

CPU-TOWER

MONITOR

IBM-PC PLOT-TER

Pallas processor board overview.

Fig.6

with a special interface card. Finally the system contains a clockgenerator which delivers synchronous clock signals of 20 Mhz to all the processor cards. Each processor card consist of a TMS 320 microprocessor from Texas Instrument, communication ports to the four nearest neighbours, a port to the main bus, a hardware noise generator, a display circuit, a tabel-eprom, memory (4K Byte distributed on user ram and system ram) and finally a battery back up. The memory is limited by the adressing capacity of the processors.

The communication network basically consist of nearest neighbour buses and a global bus that connects the front-end computer to all the processor card. The topology of the 2-dimensional processor network is a torus since the boundary processor cards on the hardwired lattice are connected to each other in the same row or column. The local communication is governed by communication ports (4 for each TMS 320) and the IC's contain a couple of 16-bit latches that store respectively the incoming and outgoing data, thus providing a full duplex communication path between processors. The status of the ports (empty/filled) is available to the software, but the ports cannot interrupt the processors. This restriction should be removed in future PALLAS editions.

The clock generator delivers synchronous signals to all processor cards so the communication flow becomes fully synchronised.

Concerning the hardware on each processor card as much CMOS technology as possible has been utilized since the LS-TTL logic that else is required would make the hardware much more complicated and with a higher power consumption. The new high speed CMOS logical circuits circumvent these problems.

7. The Performance of PALLAS 0

The aim for the PALLAS project has mainly been to handle simulation of dynamical systems in a reasonably fast and efficient way.

The types of problems, some which will be described later, that most easily can be handled i.e. divided into sub prob-

lems and processed in a parallel way are lattice problems
with neighbour interactions.

The last restriction ensures that the communication flow
is basically local and not producing inhomogeneities or
"bottlenecks" in communication between the processors. (Future editions of the PALLAS computer will be designed especially to handle these difficulties by means of special communication network similar to the so called (ref. 7) "Omega"
network). The lattice models can easily be programmed on
a machine, having the lattice structure of PALLAS, simply
by partitioning the physical lattice and assigning each
port to a processor. In this case each processor runs a
replica of the same program, and the coordination of the
processors activity reduces to the problem of sharing the
data related to the boundaries of the partitions of the
physical problem. We must stress that running the same programs on all the processors does not automatically imply
running the same instructions as in a vector processor because the execution can depend on the data. The main difference
between the two case studies we discuss below is exactly
related to this point.

In the next chapter the data from some lattice models
will be presented. One model was analysed by the Langevin
method which is a very tedious technique that requires many
million iterations in order to reach stable configurations.
The same program was run on a CRAY XMP computer after being
fully vectorized. The program that was executed on PALLAS
0 in 6.2 seconds CPU time took 3.9 seconds in CRAY XMP.
If we set the price of the PALLAS 0 system to be around
50.000 U.S. dollars, which is approximately the hardware
price times five, this gives a cost/performance ratio that
is 5 promille compared to that of CRAY for these types of
simulation problems.

8. Data from a syncronous parallel lattice set-up of the X - Y model

In this chapter we shall present the data (ref. 5) from
a typical synchronous lattice simulation that was executed
in a parallel way by PALLAS. By a synchronous program we
mean a program executed in such a way that all processors
are transfering the same amount of data between each other
in equal time intervals. Such a synchronous procedure is
easily controlled by a clock generator that sends out signals
synchronously to all processors and such program is obviously
suited for a parallel set-up. It utilizes predomenantly
the local comminication network and runs very quickly.

We choce the planar x-y spin model for this case and
shall briefly present the data as a test of the reliability
of the parallel set-up.

The x-y lattice model has the action of eq (1).

We have used the Langevin approach to analyse this model
which means that we had to solve the following first order
differential equation system

$$\frac{d\varphi_i}{dt} = -\frac{\delta S}{\delta\varphi_i} + \ni_i \qquad (23)$$

where the noise function $\ni$ involves fluctuations that lead
to a desired probability distribution for the fields

$$P(\varphi_i) = \exp(-S(\varphi_i)/\sigma) \qquad (24)$$

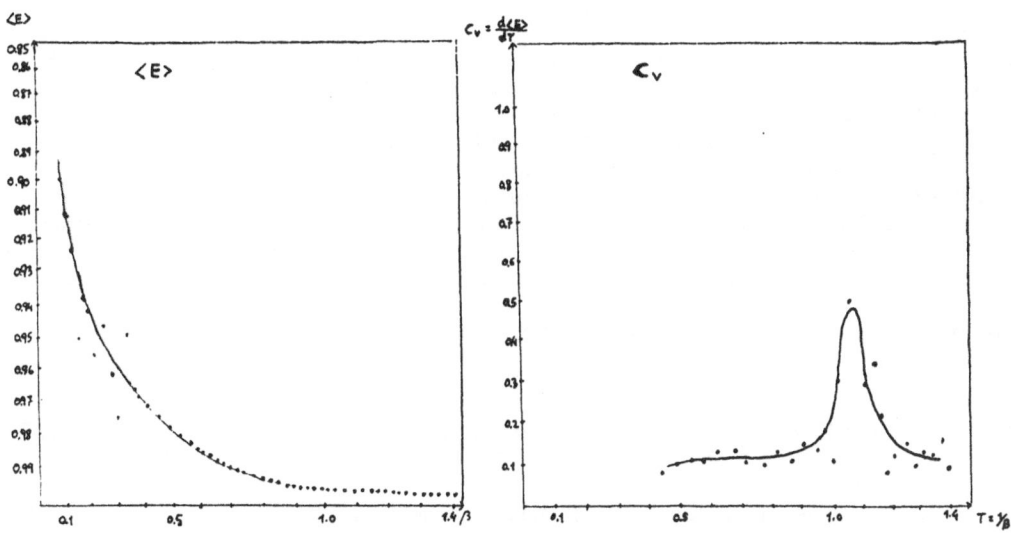

Fig.7 X-Y model results from PALLAS O.

where the noise functions $\ni_i$ are Gaussian random variables normalized to the width σ by

$$\overline{\ni_i(t)\ni_j(t')} = 2\sigma\delta_{ij}\delta(t-t') \qquad (25)$$

In our case the differential equations in eq(2) are solved numerically on PALLAS 0 by solving the following difference equation system iteratively

$$\varphi_{i_{new}} = \varphi_{i_{old}} + \sum_j \beta^{-1}\sin(\varphi_{i_{old}} -\varphi_{i+j_{old}})\Delta t + \ni_i \qquad (26)$$

The index j represent in this case the nearest neighbour spins to the i'th spin (4 to each i-site) and they could be handled altogether by a group of five nearest processors. In our case each processor took care of an aray of 7x7 spins so altogether the lattice size was 16x7x7. On fig. 7 we show the data for the average action per site as a function of the inverse temperature β in a cooling/heating cycle. The simulation required 100.000 iterations within the time interval of 8 minutes. The curve is in agreement with known results in the literature (ref. 8).

In fig. 7 we also present the curve of the specific heat as a function of β . The curve has a peak around $\beta = 1.04$

9. Data from an asynchronous parallel lattice set-up of a molecular gas

The PALLAS 0, computer system was originally designed for synchronous programs so it caused some problems to have asynchronous programs handled, i.e. programs that involve a different set of informations to be transfered between the processors in a given time interval according to the position of the processors. This means that if no precautions are taken data might be read on top of current data on one particular input port of a certain processor.

We managed these problems by a hardware handshake on each port that could delay new data in a "would be" bottleneck and together with software precautions could avoid any problems of overflow.

We chose a molecular lattice gas model (ref. 9) to represent such asynchronous programs that should work as test programs on the PALLAS 0 computer beside the wish to gain more insight in the model. Actually this lattice gas model was not the most general asynchronous program that could be thought of since the amount of data exchanged and their order are fixed and only the arrival time is undefined. The model is an Ising type theory with the following Bond-Hamiltonian.

$$H - \mu N = -\frac{J}{2} \sum_{(ij)} \prod_{k(i)} (1-c_k) c_i c_j \prod_{l(j)} (1-c_l) - \mu \sum_i c_i \qquad (27)$$

where c stands for an occupation number, $c \in [0\ 1]$ and a chemical potential μ controlling the average particle number $<N>$. Futhermore i, j are neigbouring sites of a lattice, $k(i)$ are all first neighbours of sites i which are different from j and similarly $l(j)$ are all first neighbours of j different from i. The two projectors $\prod_{k(i)} (1-c_k)$ $\prod_{l(i)} (1-c_l)$ ensure that a bond connecting two nieghbouring particles will pay energy - J only if each of the particles involved have no other bonds thus giving the model a next to the nearest neighbour interaction. The simulation is carried out using a standard Metropolis Monte Carlo algorithm. The model contains in principle 4 phases:
a molecular crystal phase MC, an atomic crystal phase AC, a molecular fluid phase MF, and an atomic fluid phase AF.

The purpose of the simulations was especially to analyse in detail (i.e. determine the phase transitions) the phase diagram that partly could be estimated (see fig. 8) by a mean field approximation. The results (see fig. 9) came out in resonable agreement with previous results (ref. 10) beside the higher statistics here.

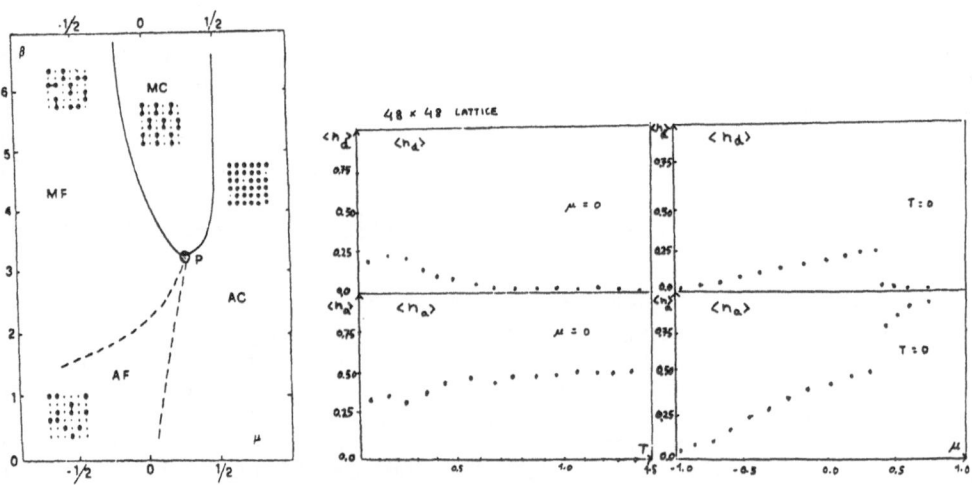

Fig.8,9 Lattice gas results from PALLAS 0.

REFERENCES

1) K. Jensen, "The MOSES concept", DTH preprint, July 1983. H. Bohr and K. Jensen, "PALLAS, a cellular computer project for fast simulation of dynamical systems", Internal report, SISSA preprint 26/84/EP (1984)

2) E.J. Lerner, High Technology/July 1985, 20.

3) H. Bohr et al., Com. Phys. Com. 30 (1983) 337.

4) H. Bohr an I. Khan, ICTP preprint, IC/85/34 (1985).

5) P. Forgacs, Privat communication.

6) H. Bohr et al., SISSA preprint, 59/85/EP (1985).

7) W. Fisher, Febr. 1985, IEEE.

8) J. Tobochnik and G.V. Chester, Phys. Rev. 20 B, 3761 (1979).

9) M. Parrinello and E. Tosatti, Phys, Rev. lett. 49, 1165 (1982).

10) G.M. Florio et al., SISSA preprint, 24/84/CM (1984).

ACKNOWLEDGEMENT

Jan Arenskov is acknowledged for editing the manuscript and the Institute for Inductrial Management for help and use of their equipment.

THE APE COMPUTER AND LATTICE GAUGE THEORIES

E. Marinari

Dipartimento di Fisica, Università di Roma 2 (Tor Vergata)
Via Orazio Raimondo, 00173 Roma , Italy

In this note I will say something about **building** a computer dedicated to Theoretical Physics. Ref.1 is a technical report about the Ape computer (Array Processor with Emulator), and the interested reader should refer to it for complete technical details. Here I will just try to give some feelings about the strategic decisions the Ape collaboration people did, and about the kind of physics people will be able to do with such a machine. I will mainly be concerned (both because of the nature of the Ape and of the nature of this meeting) with numerical simulations of Lattice Gauge Theories (LGT).

We want very **fast** computers. We did a lot of simulations, and we understand that LGT are a very powerful tool, but the numerical approach needs very much computer time. The simulations done in the last, let us say, year, took an amount of time of the order of 100 Cray-1 or Cyber 205 hours (up to 1000 hours) each, and they have given quite positive results, in the sense that the region of asymptotic scaling seems to be reached (for quenched QCD: the crucial discussion about the fermionic theory is more involved). We would like a factor 100 more of computer time, in order to get results that will be surely very precise (at the level of the percent statistical and systematic error) for the lower lying states of the hadronic spectrum of quenched QCD, and for determining the chiral properties of the theory. That is, order one year of Cray-1 for quenched QCD. Let me say that with a factor ten more in computer time we are sure we will get good hints about QCD with the internal quark loops. The point is that we will not get 10 years of a supercomputer for doing simulations, and that we have a possible better way to improve the situation. What I mean is that because of some peculiar features of our problem (locality, intrinsic parallelism, ...) a home made computer can be welcome, and will be able to solve a large class of problems in Theoretical Physics, with a very impressively low price / (floating point operations per second) ratio. Our Ape will be a **homogeneous differential equation solver** (stochastic or deterministic equations).

The second crucial evidence calling for a specialized machine is the **large memory** we need and we do not find on commercial supercomputers. The rush to big lattices has become more than one time the triumph of completely uneffective Fortran and Assembler codes, blocking beautiful vector machines in reading and writing from disks.

If we want to match the speed demand I discussed before (the commercial supercomputers we are talking about run at the speed of 10^2 Mflops, and in 1 year our QCD study should be done) and the memory request I was talking about (and that I will quantify in the following), we get a goal of

$$10^9 \text{ floating point operations per second} = 1 \text{ Gflops,}$$
and
$$10^9 \text{ byte of memory} = 1 \text{ Gbyte.}$$

These two requirements, together with the nature of the problems we will have to deal with, will decide the design of the machine.

On these basis, and on the premises of some very recent and crucial technical developments in the features of commercially available chips, a group of physicists supported by **INFN (Istituto Nazionale di Fisica Nucleare – Italy)**, decided to build **Ape**, a specialized computer, running at 1 Gflops, with a memory of 1 Gbyte (Ref. 2). A further requirement for such a machine is clearly that we have to be able to realize it in a reasonable amount of time, order of 2 years. Let me say now that the cost of Ape will be of the order of the .5 Giga Italian Lire. That is very little money, and I will leave the translation to dollars to the interested reader.

Ape is going to be a **parallel** processor, matching the features of differential equation solving, in general, and of LGT simulating. It is very natural, in these cases, to divide your lattice in S sublattices, on wich you will have to perform sequentially the same operations. S will indicate in the following the level of parallelism of the machine.

So Ape is going to be a Single Instruction Multiple Data machine (**SIMD**). All the S processors will be performing the same instruction at the same moment, processing (hopefully) different data. This feature assures a large deal of simplicity, and fits exactly with our typical requirements (S sublattices spanned in parallel by means of the same algorithm).

The basic chips that constitute our Floating Point Units (**Fpu**), are floating point multipliers and adders. They deal with 32 bit numbers, they have a basic cycle of 120 ns, and they are able to produce one result of a floating point operation at each clock cycle. This last feature is the new and remarkable one, the **pipelining**. On a scalar machine you spend the most part of your time in computing addresses, and searching for the operands of the operation. Here you can fetch two operands at each cycle, getting the first result some cycles after the first fetch (latency time, or startup time), and the other results sequentially, one per cycle. All that gives a speed, for the Weitek 1032/1033 chips we are going to use, of 8 Mflops. For reaching our goal we need **128** of such chips (adders and multipliers, in a suitable proportion: a computer with 128 adders and no multipliers would not do it). If we have S parallel Fpu we will have s chips on each Fpu, with the relation

$$s \cdot S = 128,$$

giving a theoretical speed of 1 Gflops. For a too large s (too many chips on each card) the theoretical speed will be very difficult to attain, since one will not find work to be done for s arithmetic elements together. The processor would be, for larger s, less and less flexible. For a too large S, on the contrary, the level of parallelism would be too high, and the hardware complexity of the machine would grow. One would need a more sophistcated switching network, more connectors, crates, chips and similar things. One has to find a good compromise, also accounting for the size of the memory cards.

I should underscore that in designing such an architecture one has to foresee a big **software effort**. Speed obtained through pipelining can only be supported from a clever optimizer. The pipe has to be filled with arithmetic operations. A crucial difference between parallelism and pipelining is that the user will have to take care about the parallelism of his code, but a compiler will have to optimize the pipelining.

Our choice has been to implement on each Fpu the **complex operation**

$$z = x \cdot y + w ,$$

that is $\underline{s = 8}$ (4 adders and 4 multipliers per Fpu). With $\underline{S = 16}$, that is a level of parallelism of 16, we will get our 1 Gflops machine. On the machine there will also be some specialized hardware such that the analogous real computation will be performed at .5 Gflops (one just looses a factor 2 on the theoretical speed for the complex floating point arithmetics). This division matches very well the memory request (1 Gbyte), that will be satisfied with **16 memory cards** done with 1 Mbit chips. Each card will be of 64 Mbyte in the final version of the machine. The first protoytpe (done by 4 Fpu and 4 memories) is on the process of being built with 16 Mbyte memory cards, based on 256 kbit chips (for economic reasons: a substantial decrease of the price of the 1 MBit chips is expected in the next months).

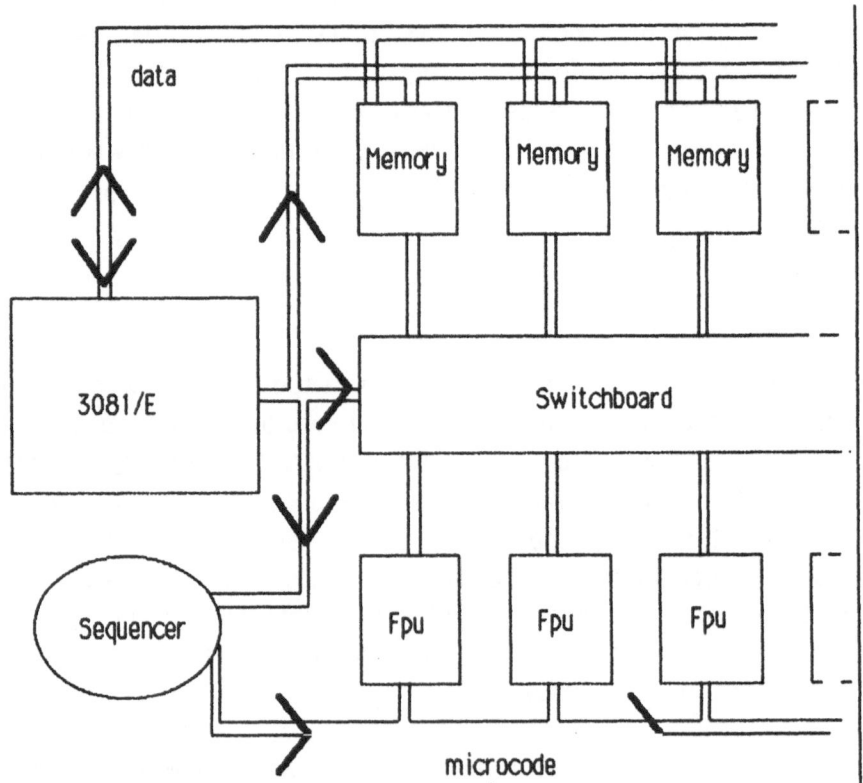

Fig. 1: The architecture of Ape.

With such an architecture a linear chain between the Fpu and the memory cards suffices, and one does not need a two dimensional or more complicated network. What I mean is that one can assume he will pick up a **single** dimension of his problem (the time dimension, for example), that will be discretized by $n \cdot 16$ sites.

In fig. 1 I show the general architecture of the machine, just including the most important features.

The **controller** of the machine is the 3081/E, a sinchronous machine built in CERN and SLAC (Ref. 3). The basic cycle of the 3081/E beautifully matches the on of Ape (120 ns). The 3081/E has to compute the memory addresses for all the memories, that all receive the same address. It plays the role of the clever controller of the 16 powerful Fpu. That is why we are building an Ape.

One should look at a program running on Ape as to 2 synchronous programs (one of which is executed 16 times): a program running on the controller, computing the addresses for memory references (location of the neighboring sites and links), and the true floating point code, stored in a memory on the **sequencer**. The sequencer takes care of sending the microcode strings to the 16 Fpu. The 2 programs, integer address computation and floating point number crunching, can be fairly independent. They just need to be synchronized at the moment in which an input or an output is needed. At the start of a run the sequencer memory is loaded with the floating point code, and the program memory of the controller is loaded with the code computing the addresses. The 16 memories are loaded with floating point data (the initial configuration). Eventually a Start is given to the sequencer. Now the Fpu will operate up to the moment in which a Stop is given by the sequencer. The status of the machine will have to be recorded from time to time. This is not a trivial operation, since one wants downloading 1 Gbyte of memory. A Vme interface will guarantee downloading on a mass storage.

The **memories** are dynamic memories. They have to be, because of problems of cost and of power dissipation. That means they forget, if not refreshed, their content. So we have refresh and **error correcting code**. The access time is 3 cycles in random access mode, and 1 cycle in sequential mode. The **switchboard** connects Fpu and memories. It is very simple, because of the reasons I explained before. Each Fpu can be connected, in the normal mode of operation, to 3 memories, the one in front of it, the one to the right or the one to the left. The internal design of a floating point unit is schematized in Fig. 2.

These basic connections are what one needs for performing a complex multiplication with an accumulation. The two 32 bit wide busses shown in the upper part of the figure allow comunication with the external world (the memory boards). The register files allow the storage of 64 complex words (64 real and 64 imaginary parts, each of 32

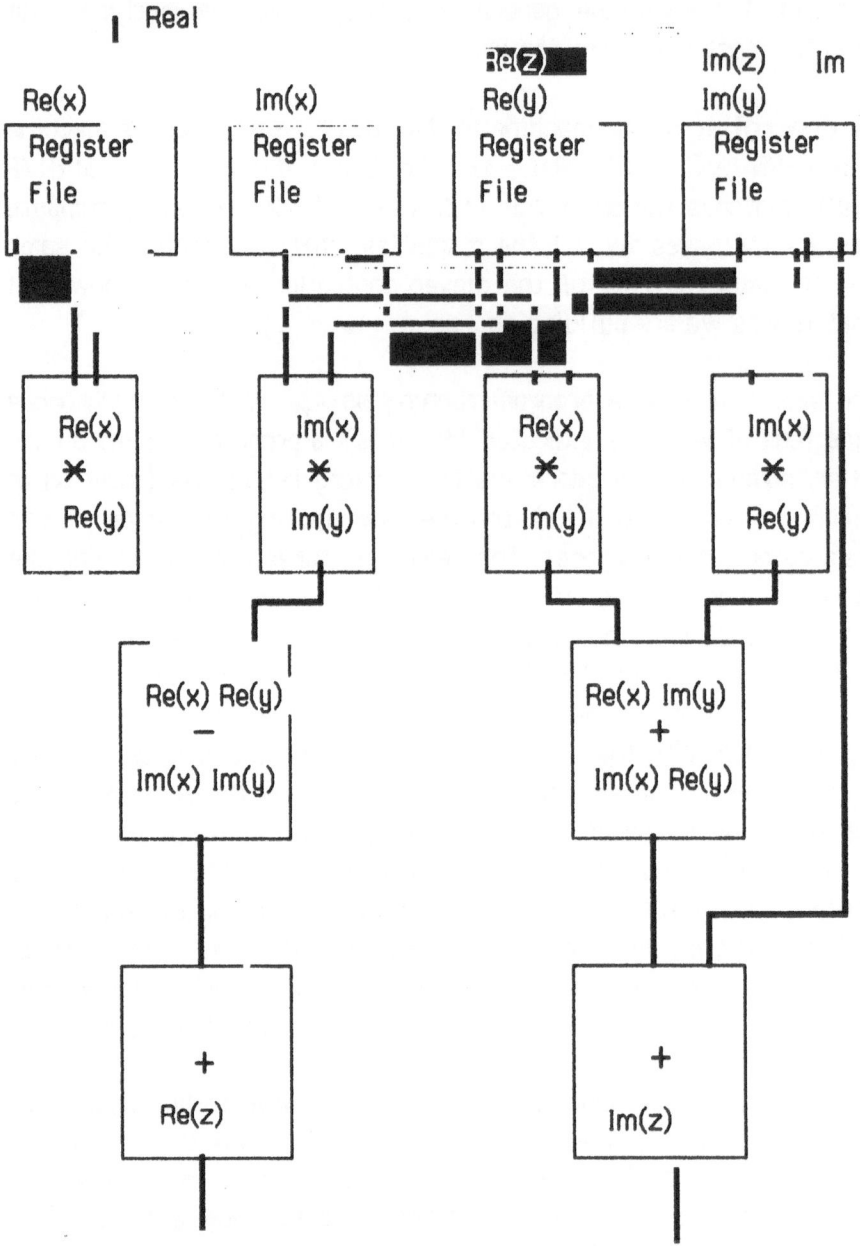

Fig. 2: The architecture of the Floating Point Unit. This is
just a scheme of the main features.

bits). They are very useful objects, matching the basic speed of the
adders and of the multipliers. They have six ports (3 inputs and 3
outputs), and all of them can be used at the same time. In the typical
pipelined operating mode of the machine at a given cycle an input or an
output is done from memory, an operand is fetched to the multipliers and

a result is read from the accumulators. In such a way a single floating point unit can reach the speed of **64 Mflops** in complex mode. A set of switches (not shown in the figure) allow the Fpu to run in real mode. In this case 2 parallel real multiplications with accumulation can be performed in parallel on each unit, giving a loose of a factor 2 on the speed of the complex case.

Register Files contain 2 look-up tables, that give the approximate result for $1/x$ and $1/\sqrt{x}$ (with full accuracy in the exponent and 7 bit of accuracy in the mantissa). With 3 iterations of a suitable iterative procedure the exact result is found. On the Fpu there is also some dedicated hardware, for computing exponential and logarithmic functions. An IF construction is also supported. It is a parallel IF (the two expressions are both executed, but just one of the two assigments is performed), that can be nested up to 4 levels.

Full software support to Ape is on the process of being built. An high level Ape language has been defined, supporting parallel constructions. The user code is interpreted and checked for consistency by a parser. An optimizer tries to fill the pipelining, producing a temporized code, that is eventually translated to loadable microcode. During this process one always deal with 2 parts of code. The one that will eventually run on the controller (address computation) and the one that will run on the 16 Floating point units (floating point part of the code). These two parts will have to be synchronized from the optimizer, and the presence of the address computation introduce new constraints on the optimization of the floating point code. If we have in our code, for example, the 2 lines

$$a = b * c$$
$$d = e * f$$

our compiler can, if he thinks that will help in producing a more effective code, interchange them without any problem. But if we write

$$a[i] = b[i] * c[i]$$
$$i = 3 * j$$
$$d[i] = e[i] * f[i]$$

the two vector multiplications **cannot** be interchanged (in the code we have given here the vector part will be executed on the floating point units, while $i=3*j$ is computed once, on the controller).

Let us discuss now what we can exactly do on Ape, as far as Lattice Gauge Theories are concerned. We start by discussing the size of the lattices we will be able to fit in Ape. For the pure gauge case we have to count N sites, 4 links/site, 12 real numbers/(gauge field) since we just store the 2 upper lines of the SU(3) matrix, 4 byte/real number. We give in Table 1 the memory needed for 8^4, 16^4, 32^4 lattices and for the bigger lattice one can fit in a 64 Mbyte machine, in a .5 Gbyte and in a 1 Gbyte machine.

Table 1: Amount of memory needed for a
SU(3) pure gauge Monte Carlo as a
function of the lattice size.

N	Memory
8^4	.8 Mbyte
16^4	12.6 Mbyte
$32 \cdot 21^3$	56.9 Mbyte
32^4	201.3 Mbyte
$48 \cdot 37^3$	466.8 Mbyte
$48 \cdot 47^3$	956.8 Mbyte

As far as the theory with the internal quark loops is considered, we give here the memory needs for the case of the **pseudofermions** (see ref. 4). Other stochastic methods for simulating the dynamical effect of fermions would give substantial analogous estimates. We have to store the pseudofermionic fields

$$\varphi^a(n) , \quad \xi^a(n) = (-D + m)\, \varphi^a(n) ,$$

and the currents (see ref.5)

$$G_\mu^{ab}(n) .$$

a and b are colour indices, n labels the site, μ the direction. D is here the Susskind Laplacian, and φ, ξ and G are complex fields. That gives the memory requirements shown in table 2. In this estimate we are using Susskind fermions since this is the most reliable way we see at the moment for including internal loops. Wilson fermions, that seem more reliable as far as the quantum number identification is concerned, would be likely dangerous in the internal loops, because of the explicit breaking of chiral symmetry.

Table 2: Amount of memory needed for a SU(3) pseudofermionic Monte Carlo as a function of the lattice size.

N	Memory
8^4	2.2 Mbyte
16^4	34.6 Mbyte
$32 \cdot 14^3$	46.4 Mbyte
$32 \cdot 30^3$	456.2 Mbyte
32^4	553.6 Mbyte
$48 \cdot 34^3$	996.1 Mbyte

The phase of the computation of the hadronic mass spectrum has, for Wilson fermions, the memory requirements shown in table 3. The memory needed for Susskind masses (external loops) is by far smaller than in the Wilson case.

We can summarize by saying that, as far as the memory size is concerned, we will be able to study lattices up to $48 \cdot 40^3$ for quenched QCD, and up to $48 \cdot 34^3$ for the complete theory.

As far as timing is concerned we can rely, at the moment, on preliminary runs on the first protoype, and on runs on a software simulator of the machine. A reasonable estimate is that a **link update** should take order **7 μs** . The precise timing will depend from the exact algorithm one will use. Assuming 7 μs per link update, we give in table 4 the number of configurations we will be able to update per day.

Table 3: Amount of memory needed for the mass spectrum computation using Wilson fermions.

N	Memory
8^4	1.2 Mbyte
16^4	18.9 Mbyte
$32 \cdot 19^3$	63.2 Mbyte
32^4	302.0 Mbyte
$48 \cdot 33^3$	496.8 Mbyte
$48 \cdot 41^3$	952.8 Mbyte

Table 4: Number of pure gauge configurations Ape can update per day, as a function of the lattice size.

N	#confs/day
8^4	$.8 \ 10^6$
16^4	$.5 \ 10^5$
$32 \cdot 21^3$	$.1 \ 10^5$
32^4	$.3 \ 10^4$
$48 \cdot 37^3$	$.1 \ 10^4$
$48 \cdot 47^3$	$.6 \ 10^3$

References

1) P.Bacilieri, S.Cabasino, F.Marzano, P.Paolucci, S.Petrarca, G.Salina, N.Cabibbo, C.Giovannella, E.Marinari, G.Parisi, F.Costantini, G.Fiorentini, S.Galeotti, D.Passuello, R.Tripiccione, A.Fucci, R.Petronzio, F.Rapuano, D.Pascoli, P.Rossi, E.Remiddi and R.Rusack, Amsterdam Conference on "Computing in High Energy Physics", 1985.
2) P.Bacilieri et al., preprint ROM2F/85/6, march 1985.
3) P.F.Kunz et al., Slac 3069, CERN-DD-83-03 (1983).
4) F.Fucito,E.Marinari, G.Parisi and C.Rebbi, Nucl. Phys. B180(1981)369.
5) H.Hamber, E.Marinari, G.Parisi and C.Rebbi, Nucl. Phys. B225(1983)475.

CRAY AND QCD

Ph. de Forcrand

Cray Research, Inc.
Chippewa Falls, Wisconsin 54729

    Supercomputers now stand at the forefront of scientific research.
Engineers have always had a need for scientific calculators. But
computational methods have also become increasingly relevant to basic
research. Supercomputers now come as complementary tools to the analytic
approach of complex problems, allowing for numerical results over a full
range of parameters when the analytic solution can only be obtained for
a few limiting cases. In some particularly difficult non-linear problems,
computers are even used qualitatively to provide hints at the salient features of the solution. Computer modeling and simulations have become a
widely cultivated art, steadily invading every field of physics, chemistry,
biomedical and even social sciences. Universities are awakening to the
high inflation index on computer needs, and some are starting to plan fast
hardware obsolescence and successive machine upgrades. Since CRAY machines
represent roughly 2/3 of the worldwide supercomputer installations, and
since I happen to work there, I have been asked to present an overview of
the Cray line to all of you potential customers. I will try to tempt you
in Part I. I will show some recent results from my own research in lattice
gauge theory in Part II. QCD simulations represent an exciting showcase
for these computers and I would like to stress the quality of the results
which can be achieved through proper use with machines already available.
A caveat: this is an informal presentation and my viewpoint in Part I
may be somewhat different from the official one.

I. The supercomputers from Chippewa Falls

    My definition of a supercomputer is: a state-of-the-art calculator
optimized for floating point operations. Now Cray Research emphasizes the
paradoxical feature that their machines are general purpose supercomputers.
In practice it means that their machines perform well "in spite of the
user", i.e., regardless of the consideration of the programmer for the
particular architecture of the machine. Three major characteristics are
necessary to meet this goal:

    1) Excellent scalar performance (arguably the best selling
    point for Cray, at least in the industry),

    2) Automatic vectorization. No special vector statements
    are necessary. The compiler, imbued with some level of

smartness, recognizes vectorizable do-loops in a standard
FORTRAN program. This strategy has now been generally
adopted by all manufacturers and Cray is fighting to keep
its compiler as sophisticated as those of its competitors
from Asia (albeit with a German name).

3) Good performance on short vectors. If VL is the
vector length (i.e., index range) on a vectorizable
do-loop, and ST is the start-up time in clock cycles
necessary before the first result is obtained, the
computing rate on this loop obeys Amdahl's Law:

$$\text{MFLOPS (VL)} = \frac{\text{constant}}{1 + \text{ST/VL}}$$

The performance degradation is 50% for VL=ST and Cray
designers strive to reduce ST with the result ST $\sim$
0(6-10) in general.

Another distinguishing feature of the machine is the I/O management.
There is no such thing as paging on a Cray. Cray's are real memory
machines where central memory space and I/O operations are managed
explicitly by the programmer. This strategy seems opposed to 2) above.
It is justified, however, on the grounds that unconstrained programming
with virtual memory yields considerable I/O demands and eventually
unacceptable performance degradation.

The operating system supported by Cray is called COS (Cray Operating
System), with a recent shift toward Unix standardization (the homebrewed
version available so far is called Unicos). Simultaneously, the typical
working environment is shifting from batch (where you "talk" to a frontend
machine which transmits your jobs to the big supercomputer) to interactive
(where you "talk" directly to the big machine).

Finally, a unique feature presently offered by Cray machines is the
possibility of multitasking. The simulation by the physicist of local
interactions lends itself obviously to parallelism over distant, non-
interacting regions. Since in the vast majority of cases the equations
of motion to simulate are translation-invariant, it would seem sufficient
to operate all processors in lock-step with a multitasking of the SIMD type
(Single Instruction Multiple Data). However, this simpler approach taken
successfully for dedicated machines is incompatible with a multiuser (and
general purpose) environment. Therefore, Cray offers true multitasking
of MIMD type (Multiple Instruction Multiple Data), with all processors
operating asynchronously. The synchronization constructs are basically of
two kinds: fork ;and join, which mark respectively the beginning and the
end of parallel blocks of work split among many tasks. Two sets of constructs, multitasking and microtasking, are available depending on the
typical time interval between synchronization points (the "task granularity"). The speedup factor, defined up to some subtleties as the ration (wall
clock time on 1 CPU/wall clock time on N CPU's) is very often close to N. From
a wide range of applications that my group tested in multitasked mode
on two and four processor Cray-XMP's, I want to extract three examples:

1) The global weather model was among the first applications to
be multitasked because of the obvious concern that the weather
forecast for tomorrow should take less than a day to calculate.
A speedup of $\sim$1.9 was achieved on a two-processor machine by
splitting the North and South hemispheres over the two CPU's.

2) My Monte-Carlo program for pure gauge QCD shows a
speedup of 3.77 on a four processor XMP. Each processor
updates one quarter of the independent degrees of freedom
and the granularity is quite large.

3) The prestigious chess program Cray Blitz is multitasked and yields speedups of typically O(3-3.5) on four
processors. In some cases, the speedup may exceed four
since the exploration of the move tree proceeds in a
different order than on one CPU.

The research community has little concern for real-time performance of
their applications since anyway the "big" jobs submitted at a supercomputer
center may wait, say up to a week in the input queue. However, these jobs
are often big not only because of the time they take to complete, but also
because of the memory size they require. If a large memory job does not
leave enough free central memory space to accommodate smaller jobs from
the input queue (or if all jobs there are large), all processors but one
will remain idle in a waste of computer power, disastrous for the user
community. For that reason, the operating system—and the staff at the
computer center—place a high penalty on jobs which use more than their
democratic share of central memory. But then a multitasked job which
makes use of Q processors will normally be entitled to almost Q shares.
From that angle, parallelism becomes a necessity for big applications,
regardless of real-time constraints.

The history of Cray Research is still brief since its creation in
1972 by Seymour Cray, former head designer of CDC 6600 and 7600 fast
scalar computers. A rough genealogy is traced in Table 1.

Table 1. Rough genealogy of Cray S.C.'s (Super Computer) with
their CDC ancestors. The left branch is labeled S.C. after
Seymour Cray. The right branch S.C. after Steve Chen.

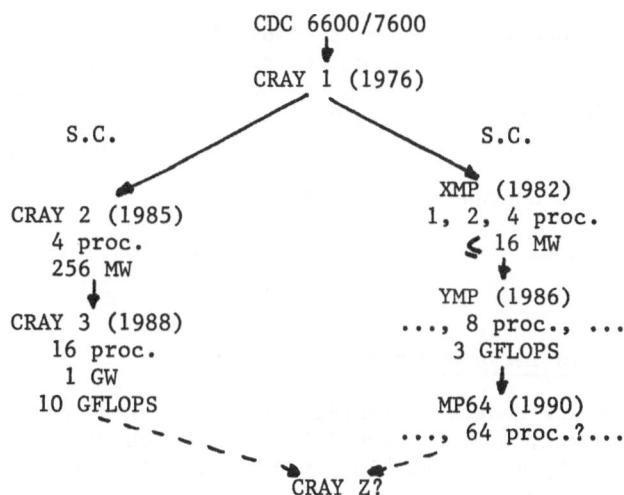

The performances of the the two best Cray machines, the XMP-4 and the
Cray 2, are compared in Table 2 on the following page.

Table 2.   Comparison between the XMP-4 and Cray 2 computers.

	XMP-4	vs.	Cray 2
Memory	"small", fast		"big", slow
Size	$\leqslant$ 16 MW		256 MW
Cycle time	4 CP		$\sim$ 57 CP
Clock Period	9.5 nsec.		4.1 nsec.
Upper bound on peak vector speed	$\dfrac{2(+,x) \times 4}{9.5 \times 10} = 840$ MFLOPS		$\dfrac{2(+,x) \times 4}{4.1 \times 10} = 1950$ MFLOPS
Practical peak vector speed	780 MFLOPS		1600 MFLOPS (so far)
Other type of memory	SSD		cache
Size	$\leqslant$ 128 MW (so far)		16 kw/proc.
Transfer rate	2.5 W/CP		1 W/CP scalar
Price tag	O(15-17 M$)		

I would like here to make two points, applicable to vector computers in general, not just to Cray machines.

1)   The peak vector performance is much greater than the scalar performance, by an order of magnitude (on the Cray's) or more. Hence the optimal algorithm is not necessarily the one which minimizes the operation count. In most cases the physicist can exploit the locality and the homogeneity of the problem to bring out vectorization, just like for multitasking. But it is necessary to think parallel from the start.

2)   Even though compilers get more sophisticated and vectorize an increasing fraction of existing codes, the programmer starting from scratch should strive to keep his code simple. A complicated loop, even vectorized, does not in general execute at peak speed. Three principal sources of compiler headaches are :

- indirect addressing, eg. $B(I) = A(L(I))$
- branches :   IF...
- recursion,   eg. $A(I) = FUNCTION(A(I-1))$

II.   Recent results in large QCD simulations

The quark-antiquark central potential has been extracted at 2 values of the coupling constant, β = 6.0 and 6.3, on lattices of sizes 16^4 and $24^3 \times 48$ respectively, by measuring elongated rectangular Wilson loops. A combination of large loop sizes (up to 24 x 12) and high statistics helps improve on previous measurements of the same kind. Since the size of the larger lattice is exceptional, I should explain how I accomodated the simulation within my not-so-exceptional computer resources.

The experiment was run at Cray Research on an XMP-48, with 4 processors and 8 MegaWords of central memory, during regular batch hours. An SSD (Solid State Storage Device) of 128 MegaWords attached to the computer appears to the user as a regular disk, with a data transfer rate improved by O(100). The data for the whole lattice was stored on the SSD as 48 time-slices of about 1 MegaWord each. The central memory space was organized as a circular buffer containing 4 time-slices. One of them, say time-slice i, is updated by Monte Carlo using information from its 2 neighbours (i-1) and (i+1). Meanwhile the fourth space is emptied of time-slice (i-2), updated two steps ago and now written to SSD, and then filled with time-slice (i+2) read from SSD. All I/O operations are therefore asynchronous. The data transfer rate to SSD is so fast (up to 2.5 words per clock period) that there is no danger of any I/O bottleneck.

The program is multitasked. All 4 processors, when available, team on the update then the loop measurement on time-slice i. The 3-dimensional time-slice is divided into 4 slabs of 24^3 x 6 sites. Each slab is assigned to a task. Links are updated plane by plane in each slab, with due care to interferences with neighbouring tasks at the boundaries. Essentially all the work is done in parallel. On a dedicated 4-processor machine, the speedup factor is 3.77 and the sustained computational rate 490 MFlops.

The algorithm used for the update is of the pseudo-heatbath type[1]. The minimum update time is less than 6 μsec/link on 4 processors (22 μsec on 1 CPU). Rectangular loops in one plane orientation, say (x,y), are measured every sweep (16^4) or every other sweep (24^3 x 48) in the following fashion. Links along the y-direction are replaced by their mean-field averages[2], computed by Gaussian quadrature[3]. Then all x-links are gauged to the identity, except for those in a narrow band $x = x_0$. Y-segments of length 1 to 24 are then assembled. And finally Wilson loops are computed from these segments, leaving out any loop crossing the band $x = x_0$. Thanks to the gauge transformation, the x-sides of all loops considered are 1 and do not need to be computed.

Measurements were taken over 20 Ksweeps (16^4) and 10 Ksweeps (24^3 x 48) after thermalization. I am deeply indebted to John Stack for their analysis and for providing the accompanying figures. Since loops of sizes up to $(N_S \times N_S/2)$ are measured on an $N_S \times N_T$ lattice, one might expect severe finite size effects on the longer loops which almost wrap around because of the periodic boundary conditions. This effect was studied by running yet another simulation at $\beta = 6.3$ on a 16^4 lattice for 20 K measurement sweeps. All (R x T) loops are consistent between the small and the large lattices until T $\gtrsim$ 15. Furthermore, on the large lattice, the quality of a linear fit of Ln W(R,T) versus T does not degrade when including larger and larger T's up to T $\gtrsim$ 23. This can be judged from Figure 1. Therefore such loops up to T = 22 were included in the fits to extract the potential V(R). The potential is plotted as a function of R in Figure 2, where the solid line is a Coulomb + linear fit for R = 3 to 8, with coefficients respectively

$$\alpha \sim -.34 \qquad \sigma a^2 \sim .0173$$

A similar analysis at $\beta = 6.0$, with a fit over R = 2 to 6, yields

$$\alpha \sim -.335 \qquad \sigma a^2 \sim .046$$

It is comforting that the 2 values of α be very close, although surprisingly far from the asymptotic $-\pi/12 \sim -.259..$ Assuming asymptotic scaling one gets

$$\frac{\sqrt{\sigma}}{\Lambda_L} \sim \quad 92 \qquad \text{at } \beta = 6.0$$
$$\qquad\qquad\qquad 79 \qquad \text{at } \beta = 6.3$$

A careful error analysis is necessary, but preliminary loop ratio tests also indicate a similar violation of asymptotic scaling between the 2 values of β.

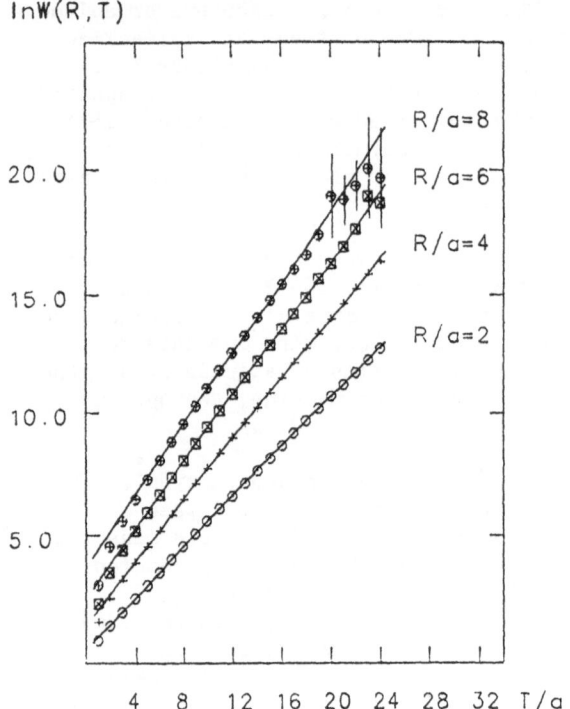

Fig. 1   Logarithms of the measured Wilson loops $W(R,T)$ versus T, on a 24^3 x 48 lattice. The slope for large T measures the potential $V(R)$.

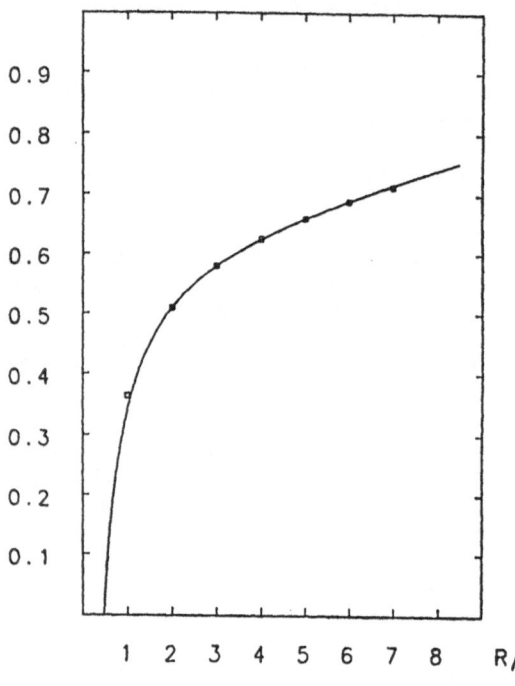

Fig. 2   Potential $V(R)$ extracted from the fitting procedure of Fig. 1.   The solid line is a (Coulomb + linear) fit to the points R = 3 to 8.

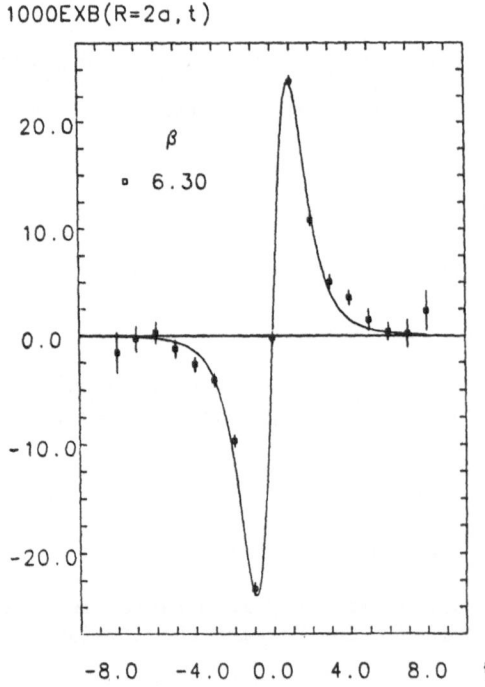

Fig. 3   E x B versus t at R = 2a, measured on a 24^3 x 48 lattice at β = 6.3.   The solid line represents perturbation theory.

310

Many more observables can be measured on the configurations saved on tape. One exciting possibility is to measure spin-dependent potentials, in the same manner as in Ref.4 (see also Ref.5 in these proceedings). Figure 3 shows the average value of a 2 x 12 loop with a chromo-electric and a chromo-magnetic insertions located T lattice units apart on opposite long sides. The solid curve is the appropriately rescaled perturbative prediction, and the agreement is remarkably good. It should therefore be expected that the integral of the quantity plotted, namely the spin-orbit potential

$$\hat{R}_k \frac{dV_1}{dR} = \lim_{T \to \infty} \frac{1}{T} \iint_{-T/2}^{T/2} dt_1\, dt_2\; \frac{t_1 \cdot t_2}{2}\; \varepsilon_{ijk}\; \frac{\langle E_i(\vec{o}, t_1)\, B_j(\vec{R}, t_2)\rangle}{\langle w(R, T)\rangle}$$

also agrees well with 1-gluon exchange. Hadronic observables can also be studied. Two groups, here in Wuppertal and in Los Alamos, have been presenting results for the quenched hadron spectrum obtained on configurations blocked from mine[6,7]. A major interest is to assess the quality of the blocking scheme by comparing hadron propagators on the blocked and on the original lattices.

Conclusion

I have the impression that still very few groups in the field of QCD simulations make full use of the computer resources they have access to. I hope to have demonstrated that very ambitious projects could be carried on today's supercomputers. The CPU time necessary for the 24^3 x 48 lattice simulation was on the order of 300 hours, which is now well within reach of many researchers. I hope similar size projects will come soon, which will probe deeper into the asymptotic scaling region.

Acknowledgements

I am very grateful to K. H. Mütter and K. Schilling for inviting me to participate in this very well organized and exciting conference.

References

1. N. Cabibbo and E. Marinari, Phys. Lett. 119B (1982) 387
2. G. Parisi, R. Petronzio and F. Rapuano, Phys. Lett. 128B (1983) 418
3. Ph. de Forcrand and C. Roiesnel, Phys. Lett. 151B (1985) 77
4. Ph. de Forcrand and J. Stack, Phys. Rev. Lett. 55 (1985) 1254
5. C. Michael, these proceedings.
6. K. H. Mütter, these proceedings.
7. G. Kilcup, these proceedings.

MEMORY AND LEARNING IN A CLASS OF NEURAL NETWORK MODELS

D.J. Wallace

Physics Department,
The University of Edinburgh
Edinburgh EH9 3JZ, U.K.

1. INTRODUCTION

Neural networks are massively parallel computational models which attempt to capture the "intelligent" processing faculties of the nervous system. They have been studied extensively for more than thirty years [1]. Apart from the longer term goal of understanding the nervous system, the current upsurge of interest in such models is driven by at least three factors. First, seminal papers by Hopfield [2] and by Hinton, Rumelhardt, Sejnowski and collaborators [3] exposed many salient properties of the models and extended their richness and potential in a significant way. Second, the developments in the theory of spin-glasses [4] and the discovery of replica symmetry breaking [5] in the long-range Sherrington-Kirkpatrick model [6] have led to an understanding in some depth of the Hopfield model [7]. Finally, there is now the expectation that the implementation of neural network models using VLSI technology may lead to significant computational hardware for a number of image and signal processing applications and for optimisation problems.

In this talk we discuss memory and learning properties of the model now identified with Hopfield's work [2]. In section 2 we describe the model, how it attempts to abstract some key features of the nervous system, and the sense in which learning (perhaps more appropriately, "training") and memory are identified in the model. Section 3 reports briefly the important role of phase transitions in the model [7] and their implications for memory capacity. The remaining sections are concerned mainly with the results of numerical simulations obtained using the ICL Distributed Array Processors at Edinburgh, in part in collaboration with Alastair Bruce and Elizabeth Gardner. Section 4 summarises work on how the fraction of "images" which are perfectly stored, depends on the number of nodes and the number of "nominal" images which one attempts to store using the prescription in Hopfield's paper. In section 5, I describe our results on the second phase transition in the model, which corresponds to almost total loss of storage capacity as the number of nominal images is increased; large finite size effects motivated us to simulate models with up to 4096 nodes - a very large scale simulation problem involving in principle 8 million words of computer storage. Finally, in section 6, I report results on the performance of a new iterative algorithm for exact storage of up to N images in an N node model.

2. THE HOPFIELD MODEL

The details of the behaviour of the nervous system are of course myriad and only incompletely understood. The kinds of models which concern us here are certainly extreme abstractions aiming to incorporate the following key features. The nervous system contains neurons which fire at a rate determined by their potential, as shown schematically in Fig.1.

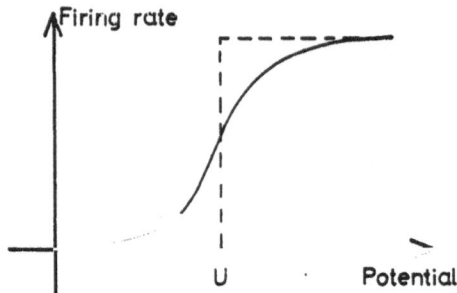

Fig.1 Schematic representation of the firing response of a neuron to its potential. The two state approximation involves a sharp threshold at potential U.

These neurons are connected in pairs by synaptic connections; if one neuron fires then it changes the potential of other neurons to which it is connected. The net potential of a neuron is determined by the sum of the contributions received from the (10^2 - 10^4) synaptic connections to that neuron.

This simple abstraction therefore represents a dynamical system of evolving firing patterns. The simplest interpretation of such a dynamical system is that the "facts" or "pictures" stored in memory correspond to the stable firing patterns of the system. Implicit in this is the idea that we are unaware of the transient evolution of patterns towards these stable fixed points of the dynamical system. One should also stress that much more complicated and interesting dynamical behaviour, such as limit cycles of length greater than 1, may be important in the neurophysiological context [8]; the focus on fixed point behaviour is only the simplest possibility.

The kinds of models we consider in this lecture have their origins in work done more than thirty years ago [1]. The particular model we consider is now associated with recent work of Hopfield [2]. In its basic form the Hopfield model makes a further simplifying assumption within the above general framework. Specifically, the neuron is represented as a two state system, which can be either firing or not firing. The rapid rise in response in Fig. 1 is therefore approximated by a step function at a threshold U, say. There are two implementations of this two state approximation, which denote the state of the ith neuron by say V_i (= 1 or 0) or S_i (= 1 or -1). An entire configuration of a firing pattern in these two cases is represented therefore by a vector of N bits where N is the total number of neurons; we shall also refer to these vectors as (black and white) images, or pictures. The influence which the firing of neuron j has on neuron i is represented by the value of the strength of the synaptic connection which we denote by T_{ij}. In the two implementations therefore the total potential experienced by neuron i is given respectively by

$$\phi_i = \sum_{i=1}^{N} T_{ij} V_j \quad : \quad \phi_i = \sum_{i=1}^{N} T_{ij} S_j \quad . \tag{2.1}$$

Finally, the new state of neuron i is determined by whether this potential is greater than or less than its threshold U_i, according to

$$V_i = \begin{matrix}1\\0\end{matrix} \quad \text{or} \quad S_i = \begin{matrix}1\\-1\end{matrix} \quad \text{for} \quad \phi_i \begin{matrix}>U_i\\<U_i\end{matrix} \quad . \tag{2.2}$$

We remark here that there is a 1 to 1 correspondence between the "V" and "S" implementations by appropriate tuning of the U's , in exact analogy with the equivalence of the Ising spin model and the lattice gas model. Henceforth we shall consider for simplicity the case where all thresholds are set to 0; we will then be dealing with two distinct models.

We have not yet specified the dynamics in full detail. There are various more or less natural possibilities. One may consider "random serial" updating, in which a single neuron is selected at random , and its new state determined before a new neuron is selected, with the updated firing pattern. One could also envisage a "lockstep parallel" update in which all of the neurons are updated simultaneously from the same firing pattern. Finally, an "asynchronous parallel" update is also conceivable. It is amusing that these three possibilities correspond loosely to the crude classification of computer architectures according to single instruction-stream single data-stream (SISD) , single instruction-stream multiple data-stream (SIMD) and multiple instruction-stream multiple data-stream (MIMD). Of course there are also many other possible updating schemes.

As a further simplification, Hopfield also presumes

(i) the diagonal elements are zero : $T_{ii} = 0$, i.e. the firing of a neuron at one instant has no influence on whether it continues in the firing state at the next update,

(ii) the synaptic connection strength is symmetric: $T_{ij} = T_{ji}$.

These assumptions are certainly not appropriate for the nervous system, but they have the great advantage that one can then identify an energy function

$$E = -\tfrac{1}{2} \sum_{i,j} V_i T_{ij} V_j \quad \text{or} \quad E = -\tfrac{1}{2} \sum_{i,j} S_i T_{ij} S_j \quad , \tag{2.3}$$

with the property that for the "random serial" update scheme the energy is guaranteed to be monotonic decreasing, i.e. $\Delta E \leqslant 0$ at each update. In this case therefore, the only possible asymptotic dynamics corresponds to fixed point behaviour at vectors which are extrema of E. In particular, the attractive stable vectors correspond to local minima of E.

In this context it is now easy to appreciate the significance of "learning" (or more precisely perhaps,"training") and of "memory". Learning means simply adjusting the energy surface i.e. choosing T_{ij} such that each of the <u>nominal</u> pictures which one wishes to store is indeed a (local) minimum of E. The memory aspect refers to the fact that if sufficient partial information about a stored picture is given to ensure that the initial configuration is in the basin of attraction of that stored picture, then the dynamics above will automatically iterate that initial configuration to the required minimum i.e. the complete picture will be exactly "recalled". It should be appreciated that this memory is of course very different from the standard memory on silicon which involves an address to a specific area of silicon which contains the variable of interest. In the neural model, memory is distributed in the values of the T_{ij}s and is recalled by specification of (part of) the <u>content</u> - content addressable memory. A very attractive feature <u>is</u> the robustness of such memory to damage or failure ; the stability of an image is unlikely to be significantly affected if some small fraction of the synaptic connections is broken for example .

Finally in this section we turn to the particular storage prescription adopted by Hopfield. Imagine that we wish to store a total of p nominal vectors $\{V^{(r)}\}$, r = 1,2, ... p. Then T_{ij} is defined in terms of an outer product as follows

$$
T_{ij} = \begin{array}{l} \sum_{r=1}^{p} (2V_i^{(r)}- 1)(2V_j^{(r)}- 1) \qquad i \neq j \\ \\ 0 \qquad\qquad\qquad\qquad\qquad\qquad i = j \ . \end{array} \qquad (2.4)
$$

This prescription adds (subtracts) 1 to the ij matrix element of T if the i and j bits of a nominal vector are the same (different). It is clear that this storage prescription should work provided that the number of nominal vectors p is much less than the number of nodes or pixels N. For, consider iterating a particular nominal vector $V^{(t)}$, in the "V" implementation . In T_{ij} the contribution from that nominal vector in the sum in Eq.(2.4) produces a signal which strongly reinforces the stability of that nominal vector. If the contributions from the other vectors in the sum in Eq.(2.4) are interpreted as additions of random and uncorrelated ± 1's, then they produce an "interference" term which competes with the signal. For large p and N this assumption implies:

$$
\sum_{j=1}^{N} T_{ij} V_j^{(t)} \simeq \tfrac{1}{2}N(2V_i^{(t)}- 1) \pm (pN/2)^{\tfrac{1}{2}} \quad . \qquad (2.5)
$$

Clearly, for p << N, the signal term wins and we should have good storage; correspondingly the interference terms will dominate for p ⪾ N. This simple analysis cannot of course provide an exact description of the storage properties because it neglects the correlations in the T_{ij}s.

3. PHASE TRANSITIONS

The loss of memory capacity as the number of nominal vectors is increased is associated with phase transitions in the usual sense of

statistical mechanics i.e. with changes in the minimum energy or ground
state structure . To appreciate this aspect of the model we first
consider the very close analogy between the neural models of this paper
and spin glass models in statistical mechanics. That analogy rests on
the identification of the neural variables S_i in terms of Ising spin
variables and the synaptic connections T_{ij} as random exchange constants.
The fact that we have a fully connected network (all T_{ij} potentially non-
zero) corresponds to a model with long range interactions. In fact the
only difference between the energy function in Eq.(2.3) and that of the
Sherrington-Kirkpatrick model [6] is that in the SK model the exchange
constants are chosen at random from some distribution whereas in the
neural interpretation the T_{ij}s are constructed to store specific
pictures; even if the nominal pictures are indeed random, in general this
does not imply that the T_{ij}s are random of course.

The SK model exhibits several well known features which are relevant
to appreciating the properties of the neural model however. First, the
fact that there are competing interactions gives rise to frustration [9] -
it is impossible to ensure in any configuration of spins that each term
in the energy has its minimum value. Second, the number of metastable
minima in the model increases exponentially as exp(0.1992 N) where N is
the number of nodes or sites, see e.g. Moore [4]. Third, even in the
long range model considered by SK, where mean field theory should be
adequate, the ground state is not elementary in the low temperature spin
glass phase. The method of replicas [10] which is used to deal with the
quenched nature of the random T_{ij} (in any given sample they are chosen
from the distribution and then held <u>fixed</u> while only the spin variables

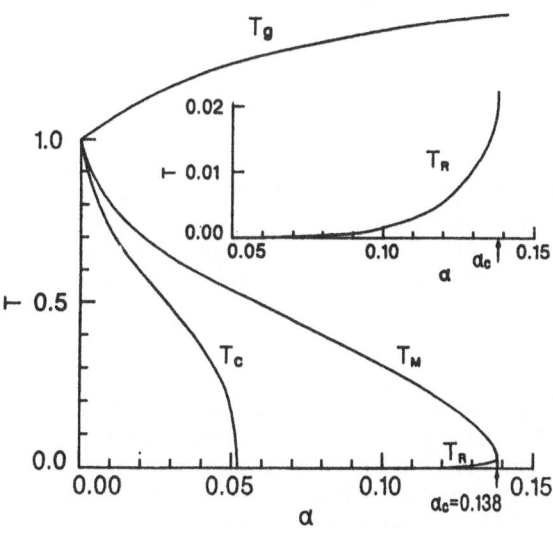

Fig. 2   Phase transition lines in the $\alpha = p/N$ and temperature T plane,
for the Hopfield model, as calculated by Amit et al [7]. The T_g
line marks the spin glass transition. The minima associated
with nominal vectors exist below the T_M line and are global
minima below the T_C line. The low temperature replica symmetry
breaking regime is expanded in the inset.

317

fluctuate) reveals the importance of sample to sample fluctuations at low temperatures where the solution breaks the symmetry amongst the replicas.

Finally, the exponentially large numbers of minima are organised into an ultrametric structure [11] which classifies the minima in a hierarchical way.

Returning to the neural model with T_{ij} defined by Eq.(2.4) with random nominal vectors, the existence of phase transitions in this model has been explored by Amit, Gutfreund and Sompolinsky [7]. They considered specifically the "S" implementation and studied the phase diagram in the plane of $\alpha = p/N$ and temperature T, corresponding to the addition of stochastic noise in the dynamics (2.2). We reproduce their calculation of the phase diagram Fig. 2. For the purposes of this talk we summarise their T = 0 results as follows. For $\alpha \ll (4 \ln N)^{-1}$, all the nominal vectors are stored essentially perfectly. As α increases through this value then the first thing that happens is that errors appear in some of the pixels of some of the vectors. As α increases to a finite number, spurious minima begin to appear. The first phase transition occurs when these spurious minima become of lower energy than the minima associated with the nominal vectors we are trying to store. The spurious minima and this phase transition clearly influence the content addressable memory properties of the network. As α is increased further there is a second phase transition at which the minima strongly correlated with the nominal vectors actually disappear. The Hamming distance (number of different bits) between an initial nominal vector and the final stable vector to which it iterates is predicted to jump discontinuously. This phase transition corresponds to almost complete loss of memory capacity (some residual correlation may remain in the T = 0 dynamics, but its practicality is obscure).

It is clear that the values of α at which these phase transitions occur are crucial to the performance of the network. Within the approximation of no replica symmetry breaking, Amit et al showed that at zero temperature these two phase transitions occurred at

$$\alpha_1^c \simeq 0.051 \quad ; \quad \alpha_2^c \simeq 0.138 \qquad\qquad (3.1)$$

Gardner [12] has extended these calculations to the "V" implementation showing that the critical values are half those above. However all of these calculations at low temperature are unstable to replica symmetry breaking, although the effect is expected to be numerically small. For further details the reader is referred to [7].

4. PERFECT MEMORY FRACTION

We now turn to summarising the results of theoretical calculations, and numerical simulations obtained using the ICL Distributed Array Processor (DAP). In this section we consider how the number of perfectly stored vectors depends upon p and N. This work, and that of section 5, was done in collaboration with Alastair Bruce and Elizabeth Gardner; for further details see [12]. Some preliminary results are presented in [13]. In this and the next section, we consider only the "V" model.

The DAP consist of 4096 bit-serial processing elements, hard-wired in a 64 × 64 square array, hosted on an ICL mainframe. In the two machines at Edinburgh, each processing element has 4K bits of local memory and data can be communicated amongst the processing elements by global North, South, East and West shifts. The array is one example of SIMD architecture; 4096 identical operations are performed simultaneously on the local data. An additional powerful feature is an activity register with which any subset of processing elements can be masked out. For 32-bit floating point arithmetic the machine performs at roughly 20 M flops. For short word or bit-manipulation problems it is extremely powerful: a 3-d Ising model runs at 200 million update attempts per second [14]. The language used is a convenient parallel extension of Fortran called DAP Fortran. The Edinburgh machines have been used to study a range of problems in condensed matter and particle physics; for a review and other references see Bowler and Pawley [15]. For the neural network simulations, the parallelism of the machine is fairly efficiently exploited by the parallel masking and sum involved in each iteration (2.1), (2.2) and by running several independent simulations simultaneously to increase statistics.

We start by showing in Fig.3 the raw data for the fraction of nominal vectors which are perfectly stored i.e. for which all nodes are stable under the iteration (2.1), (2.2) for the "V" model. It is plotted as a function of p/N , the ratio of nominal vectors to nodes, for 64, 128, 256 and 512 nodes. Our 64-node results are in qualitative agreement

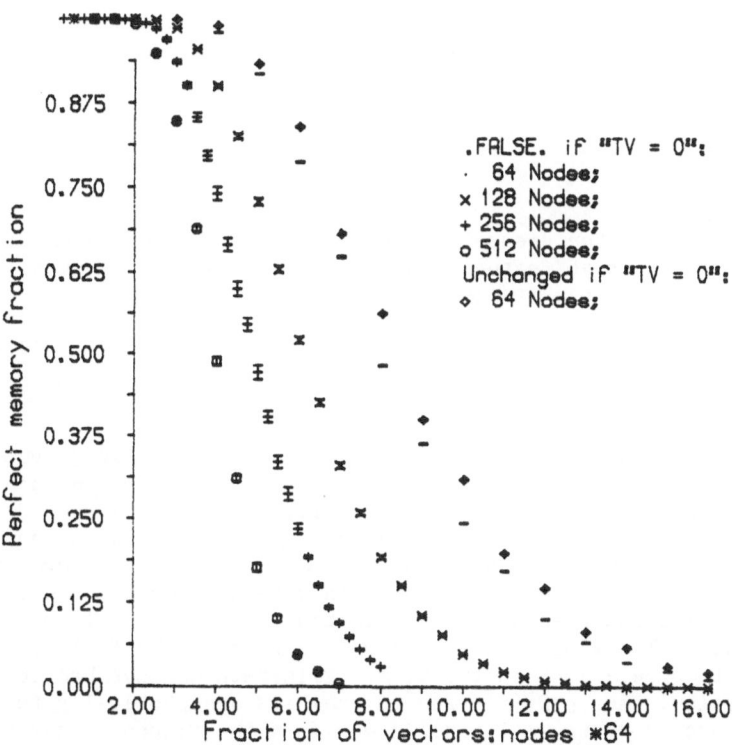

Fig. 3. Perfect storage fraction as a function of the number of nominal vectors (p) and nodes (N); the error bars on some data points are smaller than the resolution of the graph.

with an interpolation of Hopfield's 30 and 100 node data [2], but with much higher statistics. In order to begin exploring finite size effects, we show the results of two 64-node runs which differ according to the prescription adopted for the new state in the marginal case where $\sum_j T_{ij}V_j$ is zero. In one, the new state is unchanged, (i.e. the neuron continues in its previous state), in the other the new state is 0, i.e. the neuron fires only if it is strictly above threshold. The effect is significant with the "unchanged" data higher, as anticipated. Note that the discrepancy changes by a factor of roughly two in going from odd to even p; for odd p, zero can be obtained only if the number of 1-bits in the nominal vector is even. We have verified that the magnitude of the discrepancy decreases as the number of nodes increases.

In order to understand these data, let us consider first the "signal plus interference" analysis in Eq. (2.5). With this simplification one estimates the probability P_f that any node in any nominal vector will flip in terms of the probability that $(p-1)N/2$ numbers each randomly ± 1 will add up to a number greater than the signal $N/2$. For large p and N the result is

$$P_f = \frac{1}{\sqrt{2\pi}\sigma} \int_{N/2}^{\infty} \exp(-x^2/2\sigma^2)dx \quad : \quad \sigma^2 = pN/2$$

$$= \tfrac{1}{2} \, \mathrm{erfc}(\tfrac{1}{2}(N/p)^{\frac{1}{2}}) \quad . \tag{4.1}$$

Notice that this function scales i.e. depends only on the ratio p/N. If none of the nodes flips in a nominal vector of length N we obtain an approximation for the perfect storage fraction F :

$$F = (1-P_f)^N \quad . \tag{4.2}$$

This "signal plus interference" analysis is certainly wrong but it has some useful qualitative features which help to analyse the data. In particular it suggests a scaling form

$$F = [f(p/N)]^N \quad . \tag{4.3}$$

To test this form we show in Fig.4 the same data as in Fig.3, but taking the square root of the 128 data, the 4th root for 256 nodes and the 8th root for 512. If Eq.(4.3) is valid, all the data should, up to finite size effects, fall on a single scaling curve. The evidence from Fig.4 that this indeed happens is fairly convincing. However, on the same plot we also show the prediction obtained for this function from the "signal plus interference" approximation , which clearly disagrees with the data. The discrepancy is due of course to the failure to take into account correlations in the T_{ij}s. Elizabeth Gardner has done the exact calculation [12]; it indeed exhibits asymptotically the scaling form (4.3) and is shown as the solid curve in the figure. This result is in good agreement with the numerical data, even up to the finite size effects which are calculable and large for small p [12].

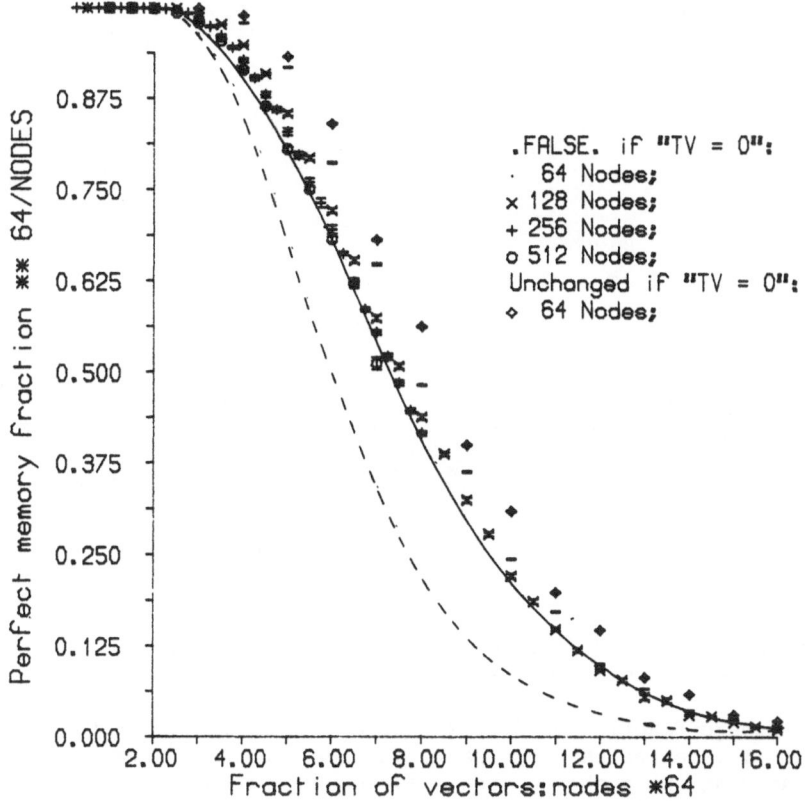

Fig. 4 The data of Fig. 3 raised to the power of 64/N, to expose the
scaling property (4.3). The dashed line is the "signal plus
interference" result (4.1), (4.2). The solid line is the
exact calculation for infinite N [12].

5. THE SECOND PHASE TRANSITION

All of the results in section 4 are concerned with the stability
property of nominal vectors and can be obtained by a single updating
sweep through the network. We turn now to the probability distribution
for the Hamming distance between an initial nominal vector and the stable
vector to which it iterates. Further details of this work can again be
found in Ref [12].

In contrast to the results of section 4, in principle the details of
the dynamics may now be important. We have explored this aspect by
studying both "random serial" updating and a variant of "lockstep
parallel" updating in which approximately half of the nodes, randomly
chosen, are updated simultaneously. Within the error bars quoted in
subsequent figures the two results are in agreement with each other; this
may signal a surprising and potentially useful universality. Another
interesting feature is the increase in the number of sweeps to stability
in the transition regime - up to 100 or more for 4096 nodes.

In Fig. 5 we show the 64 node data in a two dimensional plot for
Hamming distances 0 - 64 with 1 - 16 nominal vectors in T_{ij}; the

square root of the distribution is shown in order to enhance the
interesting features. The section corresponding to Hamming distance 0 is

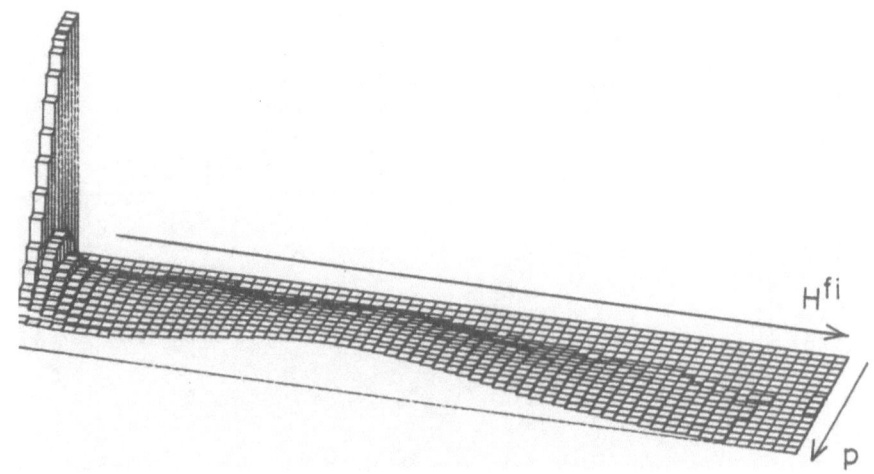

Fig.5. Histogram of Hamming distance H^{fi} between initial nominal and
final vector for 64 nodes, for p = 1 to 16 nominal vectors.

the perfect storage fraction of Fig. 3 (state is 0 if $\sum T_{ij}V_j$ = 0).
The histogram illustrates rather clearly the"leakage" away from zero
Hamming distance as the number of nominal vectors increases. There is
also just visible in this plot a non-zero value for Hamming distance 64.
This effect, in which a nominal vector iterates to its complement occurs
only in an intermediate range of p. The reason is as follows. It is
easy to see on the basis of the signal plus interference analysis that
the complement of a nominal vector is also a likely memory state. For
small p however the nominal vector is likely to be stable , and if it is
unstable will still be typically highly correlated with the final vector.
For large p, both nominal vectors and their complements are unlikely to
be stable. Only in the intermediate regime is there a finite probability
that a nominal vector will be unstable, and iterate to its stable
complement. (Note that "complement" is not an exact symmetry of the (1,0)
model). This is a classic finite size effect which one might associate
with a phase transition, as reviewed briefly in section 3.

   In Fig. 6 we show how the probability distribution for the Hamming
distance sharpens into a two-peaked distribution when the number of nodes N
is increased to 1024. Such a bimodal distribution does not of course
guarantee a first order phase transition; what is also required is a
jump in the area under each distribution as the value of α changes. In
order to estimate the critical value of α we proceed as follows. From
the Hamming distance distributions, we can calculate the fraction of
nominal vectors which iterate a Hamming distance less than N/4 i.e. the
area under the first peak of the distribution out to Hamming fraction
1/4. The choice of N/4 is of course arbitrary; it clearly effects the
values obtained for small N but becomes insignificant as N increases.

   In Fig. 7 we plot these fractions as functions of p/N for values of
N from 64 to 4096. Perhaps the most striking feature of these results is
the very strong dependence upon N. Qualitatively the steepening of the
approximately linear regions as N increases is evidence for a first
order phase transition. We remark also that in this regime there are
O(1) sample to sample fluctuation effects in the sense that for all 300

322

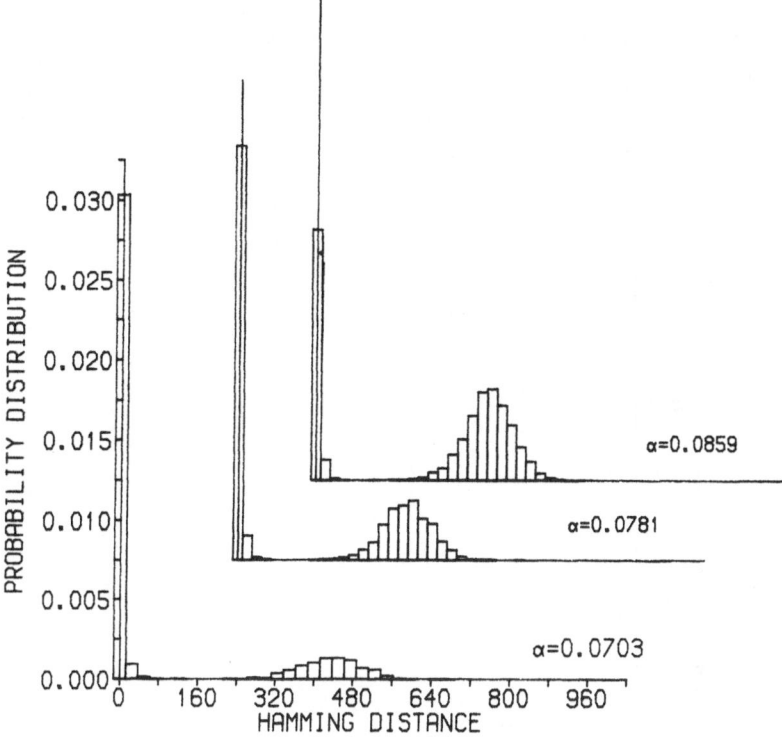

Fig. 6 Hamming distance distributions for 1024 nodes.

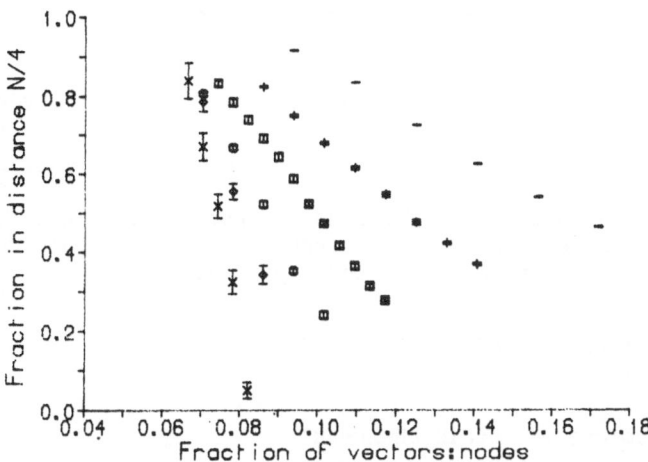

Fig. 7   Fractions of nominal vectors iterating a Hamming distance
less than N/4, as a function of p/N, for various N values:
•, 64 nodes; +, 128; □ , 256; o,512; ◇ , 1024; ×, 4096.
Critical values of p/N for a given N are estimated from the
intercepts with a horizontal line at 0.5.

or so random nominal vectors in a <u>given</u> sample for 4096 nodes, the
average of this fraction has a range exceeding 0.2 to 0.8; it is only the
averages over different samples which reduce the error bars to the values
shown in the data. In order to estimate a critical value of α we must
extrapolate this data in some way for N → ∞. Our prescription
is to interpolate the data for a given value of N to find an estimate for
α where 0.5 of the distribution is within each peak i.e. where the
approximately straight lines sections intercept 0.5 in Fig. 7. The
results are plotted in Fig. 8 versus 1/N. For a first order phase

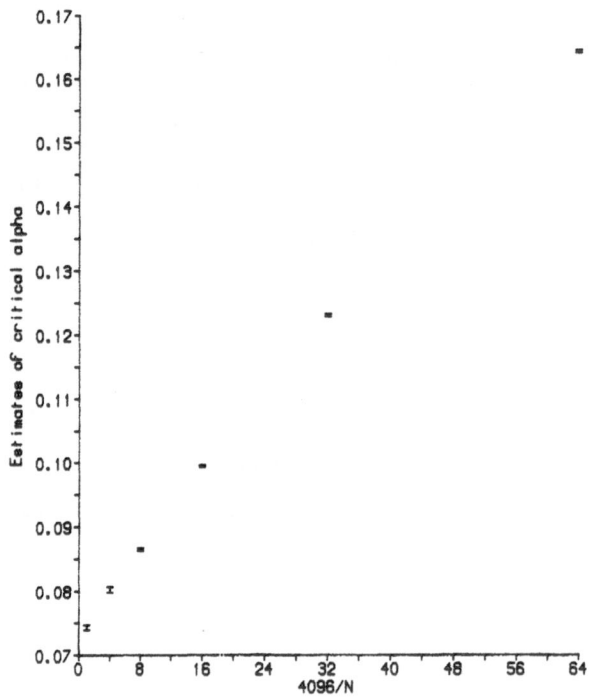

Fig. 8   Plot of the estimates of critical α from Fig. 7, versus 1/N.

transition, one expects a straight line asymptotically [16]. The data is
sufficiently accurate over a large enough range to show some deviation
from this behaviour. If we take a three parameter fit

$$\alpha(N) = \alpha_2^c + b/N + c/N^2 \qquad\qquad (5.1)$$

of the 6 data points we obtain the result

$$\alpha_2^c = 0.0731(5) \qquad\qquad (5.2)$$

with an acceptable X^2 of 1.1. We have also looked at fits of the form
$\alpha + b\,N^{-c}$. The optimal value of c is 0.85 and the estimate for α
is 0.0710(8); however the X^2 is 2.95 and the fit shows a systematic
discrepancy so we favour the fit (5.1) and result (5.2). Nevertheless
the moral in the uncertainty induced by the extrapolation should be
noted.

324

With this qualification, we note that (5.2) shows a small but significant discrepancy from the replica symmetric prediction [12] $\alpha_2 \simeq 0.069$. Since the presentation of these results, we have received a preprint from Amit et al [17] which reports similar numerical results for the (1,-1) model, with the estimate 0.145(1) for the critical point. It is interesting that this result and (5.2) are again in the ratio 2 within error.

6.   AN ITERATIVE LEARNING ALGORITHM

As the two previous sections indicate, the storage prescription (2.4) rapidly becomes inexact and in fact totally overloaded at rather small values of p/N. The model as presented is therefore very far from realising the exponential capacity which one might anticipate from the number of minima, In fact, for random vectors, the maximum capacity is $p \simeq N$ [18]; exponential capacity can be achieved presumably only for ultrametrically correlated nominal vectors. In this section we describe a simple iterative training algorithm for the Hopfield model which is guaranteed to store exactly any p pictures, in a finite number of steps, provided only that it is known that a solution is possible. We explore, again on the DAP, the performance of the algorithm in learning up to 512 random pictures on 512 nodels, and give some learning curves for random and recognisable pictures. Details will be published elsewhere [19].

As the first step in the procedure it is sensible (but not necessary) to take the storage prescription (2.4) as the starting approximation for T_{ij}. Then all of the nominal vectors are tested by iterating once the dynamics (2.1), (2.2). This enables one to cal-culate a mask ε^r for each nominal vector S^r:

$$\varepsilon_i^r = \begin{matrix} 1 \\ 0 \end{matrix} \quad \text{if} \quad \begin{matrix} S_i^r \text{ flips} \\ S_i^r \text{ is stable} \end{matrix} \quad . \tag{6.1}$$

The storage prescription is then reinforced for those pixels which are wrongly stored i.e. T_{ij} receives the additional (symmetric) contribution

$$\Delta T_{ij} = \begin{matrix} \sum_{r=1}^{N} S_i^r S_j^r [\varepsilon_i^r + \varepsilon_j^r] & \quad i \neq j \\ \\ 0 & \quad i = j \end{matrix} \quad . \tag{6.2}$$

All the nominal vectors are tested again, with the new T_{ij}, and T_{ij} modified until convergence has been achieved.

One may readily arrive at this kind of algorithm by the notion that if we are failing to retain information, we are better to realise it and to focus revision on those parts we have failed to retain. The single iteration to determine the ε^r is also clearly aimed at enforcing a local minimum of the energy at the rth nominal picture. In fact, the algorithm is a natural generalisation of the perceptron learning techniques [20]. Using these techniques one can readily establish the above claim that the procedure (6.1), (6.2) is guaranteed to converge provided that a solution for T_{ij} is known to exist. Note also the existence of other training algorithms [21].

In Fig. 9, we show the number of training cycles required to store p random pictures each of N pixels, and hence on N nodes, for N = 128, 256 and 512. Specifically, this is for the "S" (1,-1) model, and for a slightly modified training reinforcement:

$$\Delta T_{ij} = \sum_{r=1}^{N} s_i^r \, s_j^i \, [\epsilon_i^r + \epsilon_j^r - \epsilon_i^r \, \epsilon_j^r] \qquad i \neq j \qquad (6.3)$$

$$0 \qquad \qquad i = j \ .$$

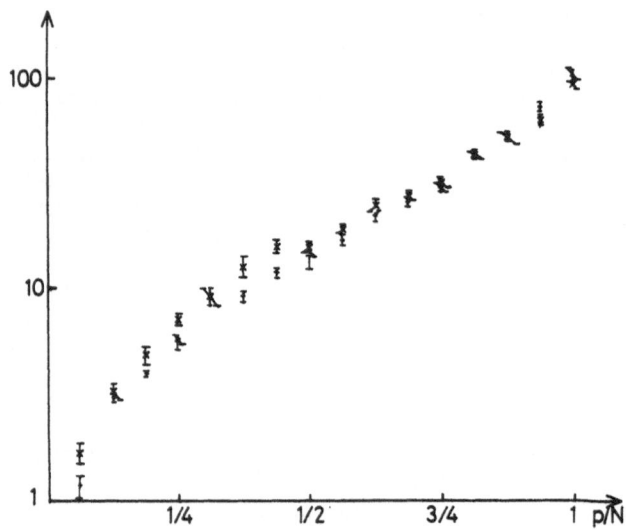

Fig. 9. Number of training cycles to store perfectly p pictures on N nodes, versus p/N, using the algorithm (6.3): •, 128 nodes; +, 256; ×, 512.

It is not clear how significant the scaling behaviour of Fig. 9 is, but it is at least useful empirically. It is not a feature of the algorithm (6.2), which needs fewer training cycles but took as much CPU time on the DAP.

In Fig. 10, we show the learning curves for N = 256, and for 16, 32, 48, ..., 256 nominal vectors. For p large every vector starts with some errors and is gradually corrected. The bit error fraction (fraction of pixels wrong) is shown with a normalisation of 1 initially. Since randomly half the pixels would be correct for zero storage capacity, it is clear that the storage prescription is a good starting point for the training routine, even for large p.

Finally, in Fig. 11 we show the five successive training cycles required to store exactly 256-pixel representations of the numbers 0 to 9, and the resulting content addressable memory property when the nominal pictures are presented with statistically 25% noise. To emphasise the very significant improvements achieved by the training algorithm over the storage prescription (which gives the first columns of pictures after one iteration), we have used there the "V" (1,0) model; in the "S" model the storage prescription almost does the job. The elimination of the confusion and inability to discriminate is illuminating. Of course these are only "toy" tests of the algorithm and the reader should note earlier impressive results using orthogonal projection methods [22].

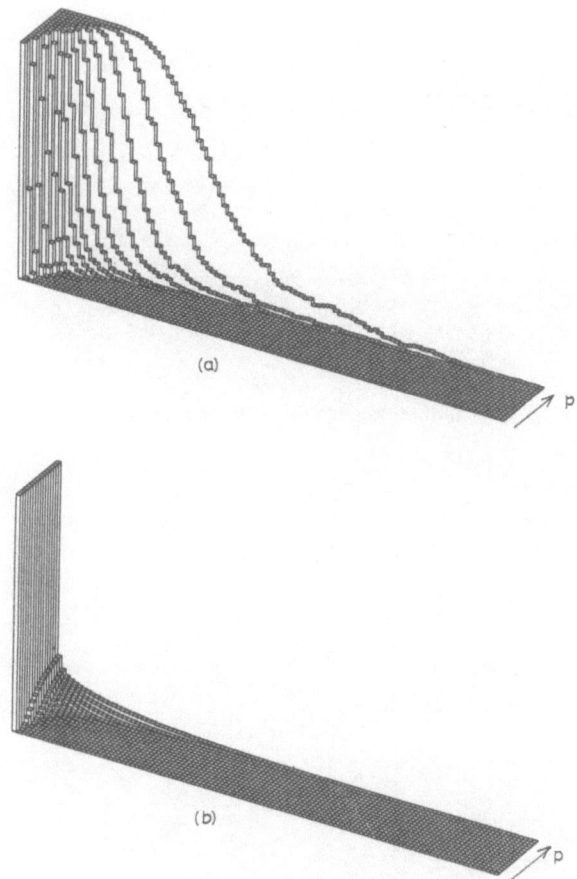

(a)

(b)

Fig. 10.   Learning curves for 100 cycles showing (a) the fraction of
vectors which are not perfectly stored and (b) the fraction
of bit (pixel) errors, for 256 nodes, for p = 16, 32, 48,
..., 256 pictures.

7.   CONCLUDING REMARKS

     In this paper we have reviewed briefly some aspects of the Hopfield
neural network model. The most important point to stress here is that
many questions remain unanswered about it, and that it is only one of a
host of models which have potentially even more interesting properties.

     The content addressable memory properties and the first phase
transition in the Hopfield model remain to be explored in detail, and
replica symmetry breaking calculations remain to be done. The
exponential storage potential of the model might be unlocked if the
suggestion of Parga and Virasoro [23] to preprocess nominal images into
an ultrametric structure can be implemented. Exponential storage is
also the aim of the hierarchical model of Dotsenko [24]. The iterative
training algorithm in section 6 effectively removes the second phase
transition but the properties of spurious states and optimisation of the
algorithm remain to be explored.

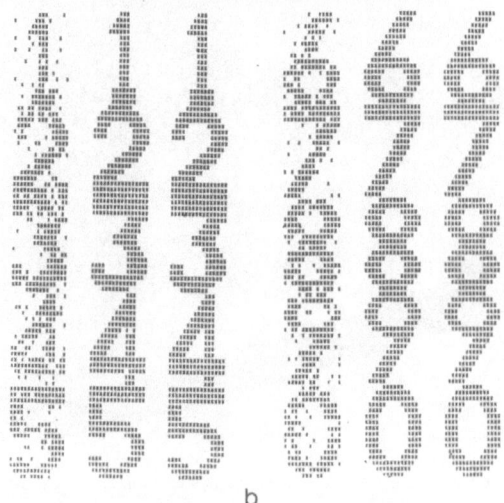

CONTENT-ADDRESSABLE MEMORY, 25% NOISE

b

Fig. 11. (a) Learning process for 256 pixel representations of the numbers 0 – 9 ("V" model) and (b) content addressable memory property with 25% noise.

An obvious limitation of the Hopfield model is that with zero diagonal entries in T_{ij} it cannot discriminate two vectors which differ only in a single pixel. With diagonal entries this limitation can be overcome; indeed $T_{ij} \propto \delta_{ij}$ stores any picture, but with zero content addressable capacity. A much more interesting generalisation of the model, to allow <u>hidden</u> units which are not clamped during any training algorithm, can be motivated just by noting that such a model can effect this discrimination by means of different configurations of these hidden units. The hidden units can also be viewed as generating effective multi-neuron interactions amongst the visible units, thus bypassing some of the criticisms of earlier perceptron models [20]. Such models (see particularly Rumelhardt et al [3]) have been shown to possess very interesting abilities to learn rules by example, to capture structure and to generalise from examples.

This brief review is a far from complete account of current interests; this is a fascinating area for future work, with many potential application areas.

ACKNOWLEDGEMENTS

I thank David Bounds, Alastair Bruce and Elizabeth Gardner for many fruitful discussions and comments.

REFERENCES

1. W.S. McCulloch and W.A. Pitts, Bull. Math. Biophys. $\underline{5}$ (1943) 115; D.O. Webb, The organisation of behaviour (Wiley, New York, 1949). Useful review articles are contained in G.E. Hinton and J.A. Anderson, eds., Parallel models of associative memory, (Lawrence Erlbaum, Hillsdale, New Jersey 1981).

2. J.J. Hopfield, Proc. Nat. Acad. Sc. USA $\underline{79}$ (1982) 2554, $\underline{81}$ (1984) 3088. A closely related model is described by W.A. Little, Math. Biosc. $\underline{19}$ (1974) 101 and W.A. Little and G.L. Shaw, Math. Biosc. $\underline{39}$ (1978) 281.

3. See for example D.H. Ackley, G.E. Hinton and T.J. Sejnowski, Cog. Sc. $\underline{9}$ (1985) 147 and D.E. Rumelhardt, G.E. Hinton and R.J. Williams, to appear in Parallel distributed processing: explorations in the microstructure of cognition, Vol. 1, eds. D.E. Rumelhardt and J.L. McClelland (Bradford Books/MIT Press, Cambridge MA).

4. For recent reviews see M.A. Moore in Statistical and particle physics: common problems and techniques, eds. K.C. Bowler and A.J. McKane, Proc. 26th Scottish Universities Summer School in Physics (SUSSP Publications, 1984) and C. De Dominicis in Applications of field theory to statistical mechanics, ed. L. Garrido, Proc. Sitges Conf., (Springer Verlag, 1985).

5. G. Parisi, J. Phys. $\underline{A13}$ (1980) L115; G. Parisi, Phys. Rev. Lett. $\underline{50}$ (1983) 1946; A. Houghton, S. Jain and A.P. Young, J. Phys. $\underline{C16}$ (1983) L375. For reviews and further references, see [4].

6. D. Sherrington and S. Kirkpatrick, Phys. Rev. Lett. $\underline{35}$ (1975) 1792; S. Kirkpatrick and D. Sherrington, Phys. Rev. $\underline{B17}$ (1978) 4384.

7. D.J. Amit, H. Gutfreund and H. Sompolinsky, Phys. Rev. Lett. $\underline{55}$ (1985) 1530; see also D.J. Amit, H. Gutfreund and H. Sompolinsky, Phys. Rev. $\underline{A32}$ (1985) 1007.

8. Limit cycle behaviour is explored and reviewed in J.W. Clark, J. Rafelski and J.V. Winston, Phys. Rep. $\underline{123}$ (1985) 215.

9. G. Toulouse, Commun. Phys. $\underline{2}$ (1977) 115.

10. S.F. Edwards and P.W. Anderson, J. Phys. $\underline{F5}$ (1975) 965.

11. M. Mezard, G. Parisi, N. Sourlas, G. Toulouse and M. Virasoro, J. Physique $\underline{45}$ (1984) 843 and Phys. Rev. Lett. $\underline{52}$ (1984) 1156.

12. E. Gardner, D.J. Wallace and A.D. Bruce, Memory and phase transition properties of the Hopfield model, Edinburgh preprint (1986).

13. D.J. Wallace, in Proc. Conf. Advances in Lattice Gauge Theory, eds. D.W. Duke and J.F. Owens (World Scientific, 1985).

14. S.F. Reddaway, D.M. Scott and K. Smith, Proc. VAPP II Conf., Comp. Phys. Comm. 37 (1985) 239 and 351.

15. K.C. Bowler and G.S. Pawley, Proc. IEEE 72 (1984) 42.

16. A detailed discussion and further references are given in K. Binder and D.P. Landau, Phys. Rev. B30 (1984) 1477; E. Brezin and J. Zinn-Justin, Nucl. Phys. B257 [FS14] (1985) 867.

17. D.J. Amit, H. Gutfreund and H. Sompolinsky, Statistical mechanics of neural networks near saturation, Hebrew University preprint (1986).

18. Y.S. Abu-Mostafa and J-M St. Jacques, IEEE Trans. IT-31 (1985) 461.

19. D.J. Wallace, Performance of an iterative learning algorithm for the Hopfield model, Edinburgh preprint (1986).

20. M.L. Minsky and S. Papert, Perceptrons (MIT Press, Cambridge MA, 1969).

21. W. Kinzel, IFK Julich preprint (1985).

22. See e.g.T. Kohonen, in Parallel models of associative memory, eds. G.E. Hinton and J.A. Anderson, Ref. [1].

23. N. Parga and M. Virasoro, The ultrametric organisation of memories in a neural network, Trieste preprint (1986).

24. Vik.S. Dotsenko, J. Phys. C10 (1985) L1017.

PARTICIPANTS

D. Arnaudon, University of Paris, France
N. Attig, University of Bielefeld, West-Germany
R. Badke, University of Bonn, West-Germany
R. Baier, University of Bielefeld, West-Germany
I.M. Barbour, University of Glasgow, Great Britain
C. Bernard, University of California, Los Angeles, USA
W. Bernreuther, University of Heidelberg, West-Germany
G. Bhanot, SCRI, SFU, Tallahassee, USA
J.M. Blairon, University Libre de Bruxelles, Belgium
H. Bohr, Niels Bohr Institute, Copenhagen, Denmark
B. Bunk, University of Wuppertal, West-Germany
A.N. Burkitt, University of Liverpool, Great Britain
R.H. Dalitz, University of Oxford, Great Britain
P. van den Doel, University of Amsterdam, Netherlands
J. Engels, University of Bielefeld, West-Germany
H.G. Evertz, University of Aachen, West-Germany
M. Faber, University of Graz, Austria
W. Feilmair, University of Graz, Austria
K. Fabricius, University of Wuppertal, West-Germany
Ph. de Forcrand, CRAY Research, Wisconsin, USA
H. Gausterer, University of Graz, Austria
G. v. Gehlen, University of Bonn, West-Germany
P. Gibbs, University of Glasgow, Great Britain
H. Gietl, Siemens AG, München, West-Germany
M. Göckeler, University of Heidelberg, West-Germany
M. Goltermann, University of Amsterdam, Netherlands
M. Grady, University of Chicago, Argonne, Illinois, USA
V. Grösch, University of Aachen, West-Germany
R. Gupta, Los Alamos National Laboratory, Los Alamos, New Mexico, USA
F. Gutbrod, DESY, Hamburg, West-Germany
O. Haan, University of Wuppertal, West-Germany
P. Hasenfratz, University of Bern, Switzerland
H.C. Hege, University of Berlin, West-Germany
D.W. Heys, University of Liverpool, Great Britain
D. Horn, University of Tel-Aviv, Israel
A. Horowitz, University of Bonn, West-Germany
A.C. Irving, University of Liverpool, Great Britain
K. Jansen, University of Aachen, West-Germany
J. Jersák, University of Aachen, West-Germany
F. Karsch, University of Illinois, Urbana, USA
K. Kanaya, University of Aachen, West-Germany
H.A. Kastrup, University of Aachen, West-Germany
E. Katznelson, University of Bonn, West-Germany
R. Kenway, University of Edinburgh, Great Britain
G. Kilcup, Harvard University, USA
A. König, Prakla-Seismos, Hannover, West-Germany

C. Korthals-Altes, University of Marseille, France
M. Kremer, Johannes-Gutenberg-University, Mainz, West-Germany
A.S. Kronfeld, DESY, Hamburg, West-Germany
W. Langguth, University of Karlsruhe, West-Germany
F. Langhammer, University of Aachen, West-Germany
M. Laursen, NORDITA, Copenhagen, Denmark
P.G. Lauers, University of Bonn, West-Germany
C. Linhares, University of Heidelberg, West-Germany
G. Mack, University of Hamburg, West-Germany
P. Mackenzie, Princeton, USA
M. Marcu, University of Freiburg, West-Germany
E. Marinari, University of Roma, Italy
A. Markum, University of Graz, Austria
M. Meinhart, University of Graz, Austria
S. Meyer, University of Kaiserslautern, West-Germany
H. Meyer-Ortmanns, MPI für Physik, München, West-Germany
C. Michael, University of Illinois, USA
I. Montvay, DESY, Hamburg, West-Germany
A. Morel, CEN, Saclay, Gif-sur-Yvette, France
K.H. Mütter, University of Wuppertal, West-Germany
A. Nakamura, FU Berlin, West-Germany
O. Napoly, CEN, Saclay, Gif-sur-Yvette, France
Th. Neuhaus, University of Aachen, West-Germany
D. Petcher, NIKHEF, Amsterdam, Netherlands
K. Peters, University of Wuppertal, West-Germany
B. Petersson, University of Bielefeld, West-Germany
R. Petronzio, University of Roma, Italy
H. Piel, University of Wuppertal, West-Germany
M. Rafiq, University of Glasgow, Great Britain
C. Rebbi, BNL, Upton, Long Island, USA
H. Rollnik, University of Bonn, West-Germany
H.J. Rothe, University of Heidelberg, West-Germany
K. Rothe, University of Heidelberg, West-Germany
W. Rühl, University of Kaiserslautern, West-Germany
H. Satz, University of Bielefeld, West-Germany
G. Schierholz, DESY, Hamburg, West-Germany
I. Schmitt, University of Wuppertal, West-Germany
W. Schoenmaker, University of Kaiserslautern, West-Germany
G. Seegmüller, University of München, West-Germany
J. Smit, University of Amsterdam, Netherlands
P. Speier, Ministry of Research NRW, Düsseldorf, West-Germany
I.O. Stamatescu, FU Berlin, West-Germany
M. Testa, Lab. Naz. dell'INFN, Frascati, Italy
D. Wallace, University of Edinburgh, Great Britain
J. Vink, University of Amsterdam, Netherlands
D. Weingarten, IBM, Yorktown Heights, USA
U.J. Wiese, University of Hannover, West-Germany
R. Wigge, Ministry of Research NRW, Düsseldorf, West-Germany
M. Wolff, University of Bielefeld, West-Germany
J. Wosiek, Jagellonian University, Krakow, Poland
S.K. Yang, Nils Bohr Institut, Copenhagen, Denmark
J. Zinn-Justin, CEN, Saclay, Gif-sur-Yvette, France

INDEX